T0138134

Canadian Mathematical Society
Société mathématique du Canada

Editors-in-Chief
Rédacteurs-en-chef
K. Dilcher
K. Taylor

Advisory Board
Comité consultatif
G. Bluman
P. Borwein

For further volumes:
http://www.springer.com/series/4318

Colin C. Graham • Kathryn E. Hare

Interpolation and Sidon Sets for Compact Groups

Colin C. Graham
Department of Mathematics
University of British Columbia
Vancouver, British Columbia
Canada

Kathryn E. Hare
Department of Pure Mathematics
University of Waterloo
Waterloo, Ontario
Canada

ISSN 1613-5237
ISBN 978-1-4899-7360-3   ISBN 978-1-4614-5392-5 (eBook)
DOI 10.1007/978-1-4614-5392-5
Springer New York Heidelberg Dordrecht London

Mathematics Subject Classification (2010): 42A55, 43A46, 42A61, 42A38, 43A25, 43A05,
42A75, 43A60, 42A16

Printed on acid-free paper

Springer is part of Springer Science+Business Media (www.springer.com)

Preface

About ten years ago we started an investigation into some classes of "thin" subsets of discrete abelian groups. These classes were subclasses of "interpolation sets", which are themselves a subclass of "Sidon sets". While we did not always articulate them, we were motivated by several unsolved problems related to these sets. The most important are:

- Is every Sidon set a finite union of interpolation sets (or sets of some related class) [**P 1**]?
- Can a Sidon set be dense in the Bohr group [**P 2**]?

This book is our attempt to present what is known about interpolation sets in the context of those (and related) problems. We give the necessary background on Sidon sets and results related to both problems. Neither problem has been solved for $\mathbb{Z}$ (though the answer to the first is yes for the duals of certain groups of bounded order and no to the second for a slightly smaller set of dual groups).

A surprise for us in writing this book was to see that what is known about Sidon sets is mostly algebraic (this relates to the first problem) and what is known about interpolation sets is mostly topological (the second problem). A theme of the book is thus the interplay of the algebraic and topological.

There are many questions to which we do not know the answer. The ones we think are the most important are flagged with a [**P nn**]. An index of open problems is at the end of the book.

We hope that this book will stimulate work on these questions.

Haines Junction, YT, Canada   Colin C. Graham
Waterloo, ON, Canada   Kathryn E. Hare

Contents

Introduction

A Brief Summary

Let G be a compact abelian group and $\boldsymbol{\Gamma}$ its discrete dual group, the group of continuous characters on G. An important example is the circle group, $G = \mathbb{T}$, with dual group, $\boldsymbol{\Gamma} = \mathbb{Z}$. By a *measure* on G we mean a finite, regular, Borel measure with total variation norm $\|\mu\|$. The discrete measures are those of the form $\mu = \sum_{j=1}^{\infty} a_j \delta_{x_j}$ where $\|\mu\| = \sum |a_j| < \infty$. The set of measures concentrated on $U \subseteq G$ is denoted by $M(U)$ and the discrete measures by $M_d(U)$. A superscript r or $+$ on a space of measures denotes the real (respectively, non-negative) measures within that class, for example, $M^+(U)$.

A subset $\mathbf{E}$ of $\boldsymbol{\Gamma}$ is an *interpolation set* (or I_0 *set*) if every bounded function on $\mathbf{E}$ is the restriction of the Fourier–Stieltjes transform of a discrete measure on G. The set $\mathbf{E}$ is a *Sidon set* if every bounded function on $\mathbf{E}$ is the restriction of the Fourier–Stieltjes transform of a measure on G, discrete or not. Every I_0 set is Sidon, but not conversely.

Here are three examples of I_0 sets:

- *Hadamard* sets in $\mathbb{Z}$, that is, sets $\mathbf{E} = \{n_j\} \subset \mathbb{N}$ such that $\inf n_{j+1}/n_j > 1$
- ε-*Kronecker* sets with $\varepsilon < \sqrt{2}$, that is, $\mathbf{E} \subset \boldsymbol{\Gamma}$ with the property that for every $\{t_\gamma\} \subset \mathbb{T}$ there exists $x \in G$ such that $|\gamma(x) - t_\gamma| < \varepsilon$ for all $\gamma \in \mathbf{E}$
- *Independent* sets, that is, $\mathbf{E} \subset \boldsymbol{\Gamma} \smallsetminus \{1\}$ with the property that $\prod \gamma_j^{m_j} = 1$ if and only if all $\gamma_j^{m_j} = 1$, whenever $\{\gamma_j\}$ is a finite subset of $\mathbf{E}$.

Special sets of integers, such as Hadamard, ε-Kronecker and independent sets, have long been of interest to mathematicians and continue to be of current interest. Many properties of Hadamard (ε-Kronecker or independent) sets, and properties of classes of functions with frequencies supported thereon, hold more generally when "Hadamard" (or "ε-Kronecker" or "independent") is replaced with "Sidon" or "I_0" and, as we illustrate, some classical results can be proved more easily when viewed in this abstract setting.

We gather and unify results about these classes of sets, both recent and early. Of particular interest to this book are structural problems, such as characterizations of Sidon and I_0 sets in terms of arithmetic properties, and the question of decomposing the special set into a finite union of simpler sets. We are motivated in part by two open problems that have been outstanding for more than 40 years:

- Is every Sidon set a finite union of I_0 sets [**P 1**]?
- Can a Sidon set be dense in the Bohr group [**P 2**]?

These problems are related since a yes to the first implies a no to the second. They have led to other problems, both solved and unsolved, also discussed here.

A main contribution of this book is to give the first complete (all in one place) proof in English of Pisier's characterization of Sidon sets as proportional quasi-independent.[1] We also detail the proofs that Sidon sets are proportional I_0 (in a strong sense) and proportional ε-Kronecker, in the latter case for subsets of $\mathbb{Z}$ and other similar groups.

We know of no book covering I_0 sets, and the only book on Sidon sets, [123], was published more than 35 years ago. There have been many new results about I_0 and Sidon sets since [123], and it is the purpose of this book to

- Present the new developments
- Provide a thorough treatment of I_0 sets
- Describe what is known about the relationship between I_0 and Sidon sets
- State with context the major open problems concerning that relationship.

I_0 sets are perhaps easier than Sidon sets, and they are covered, along with Hadamard and ε-Kronecker sets, in the first part of the book, Chaps. 1–5. Sidon sets are addressed in the second part, Chaps. 6–8; these latter chapters are essentially independent of Chaps. 1–5. Chaps. 6–8 include that proof of Pisier's proportional quasi-independent characterization of Sidon sets, as well as other characterizations.

Chap. 9 discusses the relationship between Sidon and I_0 sets, including characterizations as proportional I_0 and proportional ε-Kronecker (when Γ does not have too many elements of order two). Chap. 10 presents a different way to approach the two outstanding open problems mentioned above.

The first appendix discusses two related issues: Sidon and I_0 subsets of *non-discrete* abelian groups and Sidon and I_0 subsets in the *non-abelian* compact group setting.

The remaining appendices provide some of the tools we need from combinatorics, harmonic analysis and probability.

This book is addressed to researchers, both active and prospective. Each chapter includes exercises and extensive references to the literature. Many

---

[1] The only other complete proof of which we are aware at this time is in the excellent French book of Li and Queffélec [119].

of the exercises can be found in [51–55, 57–59]. These are not individually attributed.

We only assume the reader is familiar with graduate level real and functional analysis and has a basic understanding of Fourier analysis on compact abelian groups, such as can be found in [167, Ch. 1], although, as has become common in harmonic analysis, techniques from combinatorics, probability and topology are used in some proofs.

In the remainder of this chapter we give a more detailed description of the chapters, remarks on our notation, and acknowledgements.

Chapter Summaries

We begin, in Chap. 1, with a discussion of Hadamard sets. Some of the more important classical results in complex and Fourier analysis, involving Hadamard sets, are reviewed. We show that Hadamard sets are I_0 and establish properties that Hadamard sets have in common with other classes of I_0 sets, but which are more easily proved in the Hadamard case.

ε-Kronecker sets are studied in Chap. 2. This class includes many Hadamard sets and can be characterized by good interpolation of arbitrary choices of signs. Their interesting analytic, arithmetic and topological properties are explored. Large ε-Kronecker sets exist inside most subsets of $\boldsymbol{\Gamma}$ and they are shown to be I_0 sets with very strong properties.

Analytic, function algebra and topological characterizations of I_0 sets are established in Chap. 3. Analytically, a set $\mathbf{E}$ is I_0 if every bounded function defined on $\mathbf{E}$ is "almost" a trigonometric polynomial whose degree depends only on $\mathbf{E}$ and the error specified. The set $\mathbf{E}$ is I_0 if certain pairs of function algebras on $\mathbf{E}$ agree. Topologically, a set is I_0 if disjoint subsets have disjoint Bohr closures. We also show that an I_0 set does not cluster, in the Bohr topology, at any element of $\boldsymbol{\Gamma}$.

The subclasses of I_0 sets where the interpolating measure can be taken to be real, positive and/or concentrated on $U \subseteq G$ (the RI_0, FZI_0 or $I_0(U)$ sets) are studied in Chap. 4. These subclasses are shown to be distinct and criteria are given for a set in a larger class to belong to a smaller. Every infinite group $\boldsymbol{\Gamma}$ is shown to contain a subset that is $FZI_0(U)$ with "bounded length" (a uniformity condition, defined in Chap. 3) and of the same cardinality as the group. Every infinite $\mathbf{E} \subseteq \boldsymbol{\Gamma}$ is shown to contain a Kronecker-like subset that is $I_0(U)$ with bounded length and has the same cardinality as $\mathbf{E}$.

In Chap. 5, it is shown that if U is a non-empty, open subset of a connected group G, then every I_0 set $\mathbf{E} \subset \boldsymbol{\Gamma}$ is $I_0(U)$. Without the assumption of connectedness, it is still true that every I_0 set is a finite union of $I_0(U)$ sets with bounded length. I_0 sets fail to satisfy the finite union property, and in this chapter conditions are given for the union of two $I_0(U)$ sets to be $I_0(U)$. Necessary and sufficient conditions for an I_0 set to be a finite union of $RI_0(U)$ (or $FZI_0(U)$) sets are also established.

Chapter 6 begins with a survey of characterizations, examples and well-known properties of Sidon sets. Quasi-independent sets and Rider sets are defined and shown to be Sidon. Sidon sets are $\Lambda(p)$ for all $1 \leq p < \infty$. Properties of $\Lambda(p)$ sets are also surveyed. Every Sidon set is shown to be a finite union of k-independent sets and also a finite union of sets that are Sidon(U) for all non-empty, open U.

A complete proof is given of Pisier's arithmetic characterization of Sidon sets as those sets that are proportional quasi-independent in Chap. 7. As part of the proof, we also prove Pisier's result that Sidon sets can be characterized as those $\Lambda(p)$ sets with the minimal growth in their $\Lambda(p)$ constants. Yet another part of the proof is a refined version of proportional quasi-independence due to Bourgain. The proof of this theorem takes up most of the chapter and involves probabilistic, combinatorial and analytic arguments. An immediate corollary is the union theorem for Sidon sets. Sidon sets are also shown to satisfy a separation property, known as the Pisier ε-net condition, which will be proven to be another characterization in Chap. 9.

Chapter 8 addresses the question, "Can a Sidon set be dense in the Bohr group?", in two ways. In the first section we give Ramsey's result that if there is a Sidon set in $\mathbb{Z}$ which clusters at one continuous character, then there is another that is dense in $\overline{\mathbb{Z}}$. In the second section, statistical evidence, due to Kahane and Katznelson, is given for non-density.

The nature of the relationship between Sidon and I_0 sets is explored in more detail in Chap. 9. We prove that Sidon sets are proportional I_0 and, using that fact, show that satisfying a Pisier ε-net condition is equivalent to Sidonicity, adding another circle of equivalences to that of Pisier's characterization theorem. If $\boldsymbol{\Gamma}$ has only a finite number of elements of order two, every Sidon subset is also proportional ε-Kronecker, with $\varepsilon < \sqrt{2}$ depending only on the Sidon constant. The Ramsey–Wells–Bourgain $B_d(\mathbf{E}) = B(\mathbf{E})$ characterization of I_0 is also proved in this chapter, completing the set of function space characterizations of I_0 sets given in Chap. 3.

Chapter 10 deals with a generalization of I_0: A subset $\mathbf{E}$ of the dual of a connected group is said to have the "zero discrete harmonic density property (zdhd)" if every element of $B_d(\mathbf{E})$ can be interpolated by discrete measures whose support is concentrated arbitrarily near the identity. We show that finite unions of I_0 sets have zdhd when G is connected, though not all zdhd sets are finite unions of I_0 sets. If it could be determined whether all Sidon sets have zdhd, we could either prove that Sidon sets cannot be dense in $\overline{\boldsymbol{\Gamma}}$ (if the answer were yes) or prove that not every Sidon set is a finite union of I_0 sets (if the answer were no). The zdhd property is also used to prove the Hadamard gap theorem.

Appendix A looks at interpolation and Sidon sets for groups that are abelian, but non-compact, and for groups that are compact, but not abelian. In its first two sections, we show that in the abelian, non-compact but σ-compact case, I_0 and Sidon sets may be perturbed and remain I_0 (resp., Sidon). In the duals of non-σ-compact abelian groups, the situation will be seen to be quite different. The second section of Appendix A presents a guide to the literature on I_0, Sidon and $\Lambda(p)$ sets in the dual object of a compact, non-abelian group.

Appendix B gives combinatorial results needed for the proportional characterizations of Sidonicity of Chaps. 7 and 9. The first section of Appendix C gives an overview of harmonic analysis on abelian groups, and the second reviews basic probability.

Remarks on Notation

The circle group $\mathbb{T}$ is sometimes represented as the multiplicative group, $\{z \in \mathbb{C} : |z| = 1\}$, and other times as the additive group $[0, 1)$ (or $[0, 2\pi)$) with addition mod 1 (or mod 2π), depending on which is most convenient at the time.

We follow the usual custom of letting $\mathbb{N}$ denote the strictly positive integers, $\mathbb{Z}$ the integers, $\mathbb{R}$ the real numbers with the usual topology, $\mathbb{Q}$ the rational numbers and $\mathbb{C}$ the complex numbers. For these groups and their finite product groups, the group operation will be $+$. For all other groups the group operation will be multiplication.

We shall frequently use the product group $\mathbb{D} := \mathbb{Z}_2^{\mathbb{N}}$ and its direct sum dual, $\widehat{\mathbb{D}}$. In that context $\pi_n \in \widehat{\mathbb{D}}$ will be the element given by projection on the nth factor of $\mathbb{D}$. We will call these the *Rademacher functions*[2] and the set of them the *Rademacher set*.

The identity of G will be denoted e and that of Γ by $\mathbf{1}$, unless $G = \mathbb{T}$, $\mathbb{Z}$, or $\mathbb{R}$, in which case the identity is (respectively) 1, 0, 0. We use boldface to distinguish the trivial character $\mathbf{1}$ from the integer 1. We also put subsets of Γ in bold face. Normalized Haar measure on G is denoted m_G, while m_Γ denotes counting measure on Γ. Integration on G is often written, "$\int f(x)\mathrm{d}m$" or "$\int f(x)\mathrm{d}x$", rather than "$\int f(x)\mathrm{d}m_G(x)$".

Many spaces of functions or measures are used throughout the book. Some important ones are listed below. Others will be introduced as they arise.

$A(\Gamma) = \{\widehat{f} : f \in L^1(G)\}$—the Fourier algebra—the Fourier transforms of integrable functions on G

$B(\Gamma)$ (or $B_d(\Gamma)$)—the Banach algebra of Fourier Stieltjes transforms of (discrete) measures on G

$C(X)$ (or $C_0(X)$)—the Banach space of bounded continuous functions (vanishing at infinity) on a topological space X

$\ell^\infty(\mathbf{E})$—the Banach space of bounded functions on the set $\mathbf{E}$

$\mathrm{Trig}(G)$—the space of trigonometric polynomials on G

---

[2] This is a slight abuse. Let $r(x)$ equal 1 on $(0, 1/2]$ and -1 on $(1/2, 1]$ and be periodic with period 1 on all of $\mathbb{R}$. The classical Rademacher functions are then $r_n(x) = r(2^{n-1}x)$ for $x \in \mathbb{R}$ and $n \geq 1$. Dyadic expansion of $x \in [0, 1)$ gives a correspondence between the r_n and π_n.

Often we will be interested in functions with Fourier transforms supported on a specified subset $\mathbf{E} \subseteq \mathbf{\Gamma}$. The subspace of these functions in a function space X is denoted $X_{\mathbf{E}}$. For example, $\mathrm{Trig}_{\mathbf{E}}(G)$ is the space of trigonometric polynomials with transforms supported on $\mathbf{E}$. We refer to elements of $\mathrm{Trig}_{\mathbf{E}}(G)$ as $\mathbf{E}$-*polynomials*.

Finally, $\mathrm{Ball}(X)$ is the closed unit ball of the Banach space X.

Throughout the book we will also be restricting functions or measures to subsets and taking the norms of those restrictions in various algebras. To avoid long, hard-to-read expressions, we will omit the usual subscript that denotes the restriction when the subset is clear from the norm expression. For example, in place of $\||\varphi|_{\mathbf{E}} - \widehat{\mu}|_{\mathbf{E}}\|_{\ell^\infty(\mathbf{E})}$ we will write $\|\varphi - \widehat{\mu}\|_{\ell^\infty(\mathbf{E})}$.

Acknowledgements

This book could not have been written without the support of many people and particularly of our families, friends and colleagues. To them our heartfelt thanks.

Several students at the University of Waterloo read portions of the typescript and gave us useful comments. They are: C. Bruggeman, T. Johansson, C. Naymie, S. Tan, M. Wiersma and Yang Xu. E. Crevier kindly allowed us to use his solution to a homework problem, which appears as the proof of Proposition 6.2.14.

We also thank the authors and publishers who have allowed us to use their copyrighted material here:

R.C. Blei and the American Mathematical Society for the use of material [13] on sup-norm partitions in our Sect. 9.4.1

J. Bourgain and la Societé Mathématique de France for permission to include the statement and proof of [17, Proposition 3.2]; J. Bourgain and l'Annales de l'Institut de Fourier for permission to include from [19] what is now part of the proof of Theorem 7.2.1 and much of the material here from Theorem 6.4.1 to Corollary 6.4.7

J.-P. Kahane, Y. Katznelson and the Journal d'Analyse Mathématique [104] for what appears in Sect. 8.3

T.W. Körner and the Mathematical Proceedings of the Cambridge Philosophical Society for the use of material from [61]

D. Li, H. Queffélec and la Societé Mathématique de France for permission to include a translation of most of their proof from [119, pp. 482–498] of Theorem 7.2.1

J.-F. Méla and la Societé Mathématique de France for the proofs [130] of Theorems 1.3.9 and 5.3.1

L.T. Ramsey and *Colloquium Mathematicum* for the use of the proof [157] of Theorem 8.2.1 and for the proof of Theorem 9.2.1 in [158], which material is also important in the proof of Theorem 9.3.2. Also to L. T. Ramsey and *Acta*

Scientiarum Mathematicarum (Szeged) for the use of material from [62, 63] which appears in Chap. 5

Colloquium Mathematicum for the use of the statement of [107, Theorem 2], which appears in several guises in Chap. 3 The *Mathematical Proceedings of the Cambridge Philosophical Society, Colloquium Mathematicum, Studia Mathematica*, the *Journal of the Australian Mathematical Society* and *the Rocky Mountain Journal of Mathematics* for permission to use material previously published in [51–55, 57–59].

Chapter 1
Hadamard Sets

Hadamard sets before 1960. Interpolation on Hadamard sets. Sums and differences of a Hadamard set with itself. Bohr cluster points of those sum and differences.

1.1 Introduction

Mathematicians have long been fascinated by Hadamard sets.

One reason for their interest is that power series or trigonometric series which have their frequencies supported on a Hadamard set have unusual behaviour. Hadamard sets were the first examples of I_0 sets and many of the properties that Hadamard sets possess are held by general I_0 sets.

In this chapter we recall some of the notable classical results about Hadamard sets and their associated power series or trigonometric series. We show that Hadamard sets are I_0 in Sect. 1.3 and investigate the combinatorial and topological "size" of Hadamard sets in Sect. 1.4 and 1.5. In particular, we show that a sum (or difference) of a Hadamard set with itself does not cluster in the Bohr topology at any integer (other than 0).

Later in the book we will present proofs of generalizations of some of these classical results and other properties of Hadamard sets.

Definition 1.1.1. A subset $\mathbf{E} = \{n_j\}_{j=1}^{\infty}$ of $\mathbb{N}$, with $n_1 < n_2 < n_3 < \ldots$, is said to be a *Hadamard set* if there exists some $q > 1$ (called a *Hadamard ratio*) such that $n_{j+1}/n_j \geq q$ for all $1 \leq j < \infty$.

Hadamard sets are also called *lacunary* [199].[1] As indicated, the elements of a Hadamard set are written in increasing order.

---

[1] "Lacunary" is also used for increasing sequences $\{n_j\} \subset \mathbb{N}$ with $n_{j+1} - n_j \to \infty$. Also, some authors, for example, [101], say a symmetric $\mathbf{E} \subset \mathbb{Z}$ is "Hadamard" if $\mathbf{E} \cap \mathbb{N}$ is Hadamard in our sense.

C.C. Graham and K.E. Hare, *Interpolation and Sidon Sets for Compact Groups*,
CMS Books in Mathematics, DOI 10.1007/978-1-4614-5392-5_1,
© Springer Science+Business Media New York 2013

1.2 Classical Results Related to Hadamard Sets

An early and famous use of Hadamard sequences is in Weierstrass's 1872 example of a nowhere differentiable function.

Theorem 1.2.1 (Weierstrass). *Let $0 < b < 1$ and a be an integer such that $ab > 1 + \frac{3\pi}{2}$. Then $f(x) = \sum_{n=1}^{\infty} b^n \cos(a^n x)$ is continuous but nowhere differentiable.*

In 1892 Hadamard published his "gap" Theorem 1.2.2 from which we get the name, "Hadamard set", though "Weierstrass set" might be equally appropriate. In Chap. 10 we give a harmonic analysis proof of a generalization of Hadamard's gap theorem.

Theorem 1.2.2 (Classical Hadamard gap theorem). *Let $\{n_j\} \subseteq \mathbb{N}$ be a Hadamard set with ratio $q > 1$. Suppose the power series $f(z) = \sum_{j=1}^{\infty} c_j z^{n_j}$ has radius of convergence equal to 1. Then f cannot be analytically continued across any portion of the arc $|z| = 1$.*

Another interesting complex analysis theorem is a Picard-type result.

Theorem 1.2.3. *Suppose that $f(z) = \sum_{j=1}^{\infty} a_j z^{n_j}$ is analytic in the open unit disc, $\mathbf{E} = \{n_j\}$ is a Hadamard set and that $\limsup_{j \to \infty} |a_j| > 0$. Then $f(z)$ assumes every complex value infinitely often in the open unit disc.*

In 1918 Riesz gave the first example of a continuous measure whose Fourier–Stieltjes coefficients did not vanish at infinity. His construction was the following.

Theorem 1.2.4 (Riesz product). *The products $\prod_{n=1}^{N}(1 + \cos(4^n x))$ are bounded in $L^1(\mathbb{T})$-norm and converge weak* in $M(\mathbb{T})$ to a continuous probability measure μ such that*

$$\widehat{\mu}(n) = \begin{cases} 1 & \text{if } n = 0, \\ 2^{-K} & \text{for } n = \pm 4^{n_1} \pm \cdots \pm 4^{n_K}, n_1 < \cdots < n_K \text{ and} \\ 0 & \text{otherwise.} \end{cases}$$

The Hadamard property of $\{4^n\}$ is used to justify that the partial products converge in the weak* topology to a probability measure, called a *Riesz product*, whose Fourier-Stieltjes transform is as described. An application of Wiener's Lemma C.1.10 proves that the measure is continuous, meaning the measure of every singleton is zero. The details of this construction and generalizations to other Hadamard sequences can be found in Exercise 1.7.9.

Riesz products have been generalized to all compact abelian groups and play an important role in the study of Sidon sets. These generalizations will be described in detail in Sect. 6.2.2 and will be extensively used throughout the latter part of the book.

Classical harmonic analysts have long been interested in questions about almost everywhere convergence and whether a trigonometric series is that of a function in some L^p space. For series involving only Hadamard frequencies the answers are often simple, as the following three results, due to Kolmogorov, Sidon and Zygmund, respectively, demonstrate.

Theorem 1.2.5. *Suppose $\{n_k\}$ is a Hadamard set and the coefficients a_k are real. The trigonometric series $\sum_k a_k \cos(n_k x)$ converges (diverges) a.e. if and only if $\sum a_k^2$ converges (diverges).*

Theorem 1.2.6. *Suppose $\{n_k\}$ is a Hadamard set, the coefficients a_k, b_k are real and $\sum_k a_k \cos(n_k x) + i b_k \sin(n_k x)$ is the Fourier series of a continuous function. Then*

$$\sum_k |a_k| + |b_k| < \infty.$$

This theorem implies that Hadamard sets are Sidon. Exercise 1.7.9 provides an alternate proof using Riesz products.

Theorem 1.2.7. *Suppose $\{n_k\}$ is a Hadamard set and the coefficients a_k, b_k are real. The trigonometric series $\sum_k a_k \cos(n_k x) + i b_k \sin(n_k x)$ is the Fourier series of an integrable function f if and only if $\sum a_k^2 + b_k^2$ converges, in which case $f \in L^p(\mathbb{T})$ for all $1 \leq p < \infty$.*

This result is known to be true whenever the set $\{n_k\}$ is Sidon; see Theorem 6.3.9.

There are many other results that illustrate the unexpected behaviour of functions associated with a Hadamard sequence. We have only given a sample here and refer the reader to the section Remarks and Credits at the end of this chapter for further discussion and references.

1.3 Interpolation Properties

In this section it will be shown that every bounded function on a Hadamard set is the restriction of a Fourier–Stieltjes transform of a discrete, positive measure, and that even the support of these measures can be controlled. This means Hadamard sets are I_0 sets.

Lemma 1.3.1. *Suppose $\mathbf{E} = \{n_j\}_1^\infty$ is a Hadamard set with ratio $q \geq 2$. Then for each $\varphi : \mathbf{E} \to \mathbb{T}$ there exists*

$$\theta \in \left[-\frac{2\pi}{n_1}, \frac{2\pi}{n_1} \right] \quad \text{with} \quad |\varphi(n_j) - e^{i n_j \theta}| < |1 - e^{i\pi/(q-1)}| \qquad (1.3.1)$$

for all j.

Proof. Let $\varphi : \mathbf{E} \to \mathbb{T}$ be given. Since $e^{in_j x}$ takes on all values in $\mathbb{T}$ on any subinterval of length at least $2\pi/n_j$, we can pick $\theta_1 \in [-\pi/n_1, \pi/n_1]$ such that $e^{in_1 \theta_1} = \varphi(n_1)$ and then inductively choose

$$\theta_j \in [\theta_{j-1} - \pi/n_j, \theta_{j-1} + \pi/n_j] \text{ with } e^{in_j \theta_j} = \varphi(n_j) \text{ for all } j > 1. \quad (1.3.2)$$

By (1.3.2),

$$|\theta_j - \theta_k| \leq \sum_{\ell=j+1}^{k} \frac{\pi}{n_\ell} \leq \sum_{\ell=1}^{\infty} \frac{\pi}{n_j q^\ell} = \frac{\pi}{n_j(q-1)} \text{ for } 1 \leq j < k < \infty. \quad (1.3.3)$$

Since $n_j \to \infty$, we see that θ_j is a Cauchy sequence. Let θ be its limit. Two more geometric series calculations show that $|\theta| \leq \pi q/(n_1(q-1)) \leq 2\pi/n_1$ and $|\theta - \theta_j| < \pi/(n_j(q-1))$ for all j. Now (1.3.1) follows. $\square$

   The next result is an example of what we call the *standard iteration*. It will appear often, and we give several variations on the result, the second variation being the application to single point masses that will be used in this chapter.
   We define

$$\Delta = \{z \in \mathbb{C} : |z| \leq 1\}.$$

A subset X of a group is *symmetric* if $X = X^{-1}$. It is *asymmetric* if all the elements of $X \cap X^{-1}$ (if any) have order 2. In particular, $1 \notin X$.

Proposition 1.3.2 (Basic standard iteration). *Let $\mathbf{E} \subseteq \Gamma$ and U be a Borel subset of G. Assume $\varepsilon < 1$ and C is a constant.*

1. *[Complex version]. Assume that for each $\varphi : \mathbf{E} \to \mathbb{T}$ there exist $\mu \in M_d(U)$ (resp., $M_d^+(U)$) such that $\|\mu\| \leq C$ and*

$$|\varphi(\gamma) - \widehat{\mu}(\gamma)| < \varepsilon \text{ for all } \gamma \in \mathbf{E}.$$

   Then for each $\varphi : \mathbf{E} \to \Delta$ there exists $\mu \in M_d(U)$ (resp., $\mu \in M_d^+(U)$) such that $\widehat{\mu} = \varphi$ on $\mathbf{E}$ and $\|\mu\| \leq C/(1 - \varepsilon)$.
2. *[Real version]. Assume that U is symmetric and that for each $\varphi : \mathbf{E} \to [-1, 1]$ there exists $\mu \in M_d(U)$ (resp., $M_d^+(U)$) such that $\|\mu\| \leq C$ and*

$$|\varphi(\gamma) - \widehat{\mu}(\gamma)| < \varepsilon \text{ for all } \gamma \in \mathbf{E}.$$

   Then for each $\varphi : \mathbf{E} \to [-1, 1]$ there exists $\mu \in M_d(U)$ (resp., $\mu \in M_d^+(U)$) with $\widehat{\mu}$ real-valued, such that $\widehat{\mu} = \varphi$ on $\mathbf{E}$ and $\|\mu\| \leq C/(1 - \varepsilon)$.

Proof. The proofs for $M_d(U)$ and $M_d^+(U)$ are identical. We give the proof for the latter. The interpolating measure will be constructed through an iterative procedure.

(1) We begin by observing that every $\varphi : \mathbf{E} \to \Delta$ is the average of two functions mapping $\mathbf{E} \to \mathbb{T}$. Thus, we may assume that every $\varphi : \mathbf{E} \to \Delta$ may be approximated by $\mu \in M_d^+(U)$ with $\|\mu\| \leq C$.

Fix $\varphi : \mathbf{E} \to \Delta$. Choose $\mu_1 \in M_d^+(U)$ such that $\|\mu_1\| \leq C$ and $|\varphi - \widehat{\mu_1}| < \varepsilon$ on $\mathbf{E}$. Now suppose $\mu_1, \ldots, \mu_J \in M_d^+(U)$ have been found such that, for $1 \leq j \leq J$, we have $\|\mu_j\| \leq C$ and

$$\left| \varphi - \sum_{j=1}^{J} \varepsilon^{j-1} \widehat{\mu_j} \right| < \varepsilon^J \text{ on } \mathbf{E}.$$

Then $\varepsilon^{-J}(\varphi - \sum_j \varepsilon^{j-1} \widehat{\mu_j}|_{\mathbf{E}})$ maps $\mathbf{E}$ to Δ, so another application of the hypothesis gives a measure $\mu_{J+1} \in M_d^+(U)$ with $\|\mu_{J+1}\| \leq C$ and such that

$$\left| \varepsilon^{-J} \left(\varphi - \sum_{j=1}^{J} \varepsilon^{j-1} \widehat{\mu_j} \right) - \widehat{\mu_{J+1}} \right| < \varepsilon \text{ on } \mathbf{E}.$$

That completes the iterative construction.

Let $\mu = \sum_{j=1}^{\infty} \varepsilon^{j-1} \mu_j$. By construction, $\mu \in M_d^+(U)$, $\|\mu\| \leq C/(1-\varepsilon)$ and $\varphi = \widehat{\mu}$ on $\mathbf{E}$.

(2) The proof for this part is almost identical to the proof for (1). The key difference is that at each stage, the measure, μ_j, that is obtained from the hypotheses should be replaced by $\nu_j = \frac{1}{2}(\mu_j + \widetilde{\mu_j})$. (Here, $\widetilde{\mu_j}$ is the measure whose mass on each Borel set X is $\widetilde{\mu_j}(X) = \overline{\mu(X^{-1})}$.) This is where the symmetry of U is used. One should also note that whenever φ is real-valued, $|\varphi - \mathfrak{Re}\widehat{\mu_j}| \leq |\varphi - \widehat{\mu_j}|$, where $\mathfrak{Re}\widehat{\mu_j}$ denotes the real part of the Fourier–Stieltjes transform.

Since the Fourier–Stieltjes transform of $\widetilde{\mu}$ is $\overline{\widehat{\mu}}$, those observations allow us to assume that the approximate interpolation can be done at each stage with a measure which has real-valued Fourier transform. That permits the iterative construction to proceed. We omit the remaining details. $\qquad\square$

A computational trick enables us to improve (2), weakening the assumption only to require the interpolation of $\{-1, 1\}$-valued functions on $\mathbf{E}$ using the real parts of Fourier–Stieltjes transforms.

Corollary 1.3.3 (Improved standard iteration). *Let $\mathbf{E} \subseteq \Gamma$ and U be a symmetric subset of G. Assume $\varepsilon < 1$ and C is a constant. If for each $\varphi : \mathbf{E} \to \{-1, 1\}$ there exists $\mu \in M_d(U)$ (resp., $\mu \in M_d^+(U)$) such that $\|\mu\| \leq C$ and*

$$|\varphi(\gamma) - \mathfrak{Re}\widehat{\mu}(\gamma)| < \varepsilon \text{ for all } \gamma \in \mathbf{E},$$

then for each $\varphi : \mathbf{E} \to [-1, 1]$ there exists $\mu \in M_d(U)$ (resp., $\mu \in M_d^+(U)$), with $\widehat{\mu}$ real-valued, such that $\widehat{\mu} = \varphi$ on $\mathbf{E}$ and $\|\mu\| \leq C/(1-\varepsilon)$.

Proof. Let $\psi : \mathbf{E} \to [-1,1]$ be given. Define $\varphi : \mathbf{E} \to \{-1,1\}$ by

$$\varphi(\gamma) = \begin{cases} 1, & \text{if } \psi(\gamma) \in [0,1] \text{ and} \\ -1, & \text{otherwise.} \end{cases}$$

Then $|\psi - \frac{1}{2}\varphi| \le 1/2$ on $\mathbf{E}$ (that is the "trick"). Let $\mu \in M_d(U)$ (or $\mu \in M_d^+(U)$) be such that $\|\mu\| \le C$ and $|\varphi(\gamma) - \mathfrak{Re}\widehat{\mu}(\gamma)| < \varepsilon$ for all $\gamma \in \mathbf{E}$ and let $\nu = \frac{1}{2}(\mu + \widetilde{\mu})$. Then $|\varphi(\gamma) - \widehat{\nu}(\gamma)| < \varepsilon$ on $\mathbf{E}$ and $\widehat{\nu}$ is real. Also,

$$\left| \psi(\gamma) - \frac{1}{2}\widehat{\nu}(\gamma) \right| < \frac{1}{2} + \frac{\varepsilon}{2} \text{ for } \gamma \in \mathbf{E}. \tag{1.3.4}$$

We now apply the basic standard iteration using (1.3.4). $\qquad\square$

Corollary 1.3.4 (Standard iteration with point masses). *Let $\mathbf{E} \subseteq \Gamma$, $U \subseteq G$, $\varepsilon < 1$ and C be a constant.*

1. *[Complex version]. If for each $\varphi : \mathbf{E} \to \mathbb{T}$ there exists $x \in U$ such that $|\varphi(\gamma) - \gamma(x)| < \varepsilon$ for all $\gamma \in \mathbf{E}$, then for each $\varphi : \mathbf{E} \to \Delta$ there exists $\mu \in M_d^+(U)$ such that $\widehat{\mu} = \varphi$ on $\mathbf{E}$ and $\|\mu\| \le 1/(1-\varepsilon)$.*
2. *[Real version]. If U is symmetric and for each $\varphi : \mathbf{E} \to \{-1,1\}$ there exists $x \in U$ such that $|\varphi(\gamma) - \mathfrak{Re}\gamma(x)| < \varepsilon$ for all $\gamma \in \mathbf{E}$, then for each $\varphi : \mathbf{E} \to [-1,1]$, there exists $\mu \in M_d^+(U)$, with $\widehat{\mu}$ real-valued, such that $\widehat{\mu} = \varphi$ on $\mathbf{E}$ and $\|\mu\| \le 1/(1-\varepsilon)$.*

Theorem 2.3.1 gives a further improvement to Corollary 1.3.4: "$\varepsilon < 1$" may be weakened to "$\varepsilon < \sqrt{2}$".

Proof. (1) is immediate from Proposition 1.3.2 (1) since $\gamma(x) = \widehat{\delta_{-x}}(\gamma)$, where $\delta_x \in M_d(G)$ is the unit point mass measure at x.

(2) follows similarly from Corollary 1.3.3. $\qquad\square$

The following corollary is immediate from Lemma 1.3.1, just take $G = \mathbb{T}$, $\mathbf{E} \subset \mathbb{Z}$ the Hadamard set and $U = [-2\pi/n_1, 2\pi/n_1]$.

Corollary 1.3.5. *If $\mathbf{E}$ is a Hadamard set with ratio $q > 4$, then $\mathbf{E}$ is I_0 and the interpolation can be done using the Fourier–Stieltjes transforms of non-negative, discrete measures supported in $L = [-2\pi/n_1, 2\pi n_1]$. Furthermore, for each $\delta > 0$ there exists M (depending only on q and δ) such that for each $\varphi : \mathbf{E} \to \Delta$ there are $x_1, \ldots, x_M \in L$ and $c_1, \ldots, c_M \in [0,1]$ such that for all $n \in \mathbf{E}$,*

$$\left| \varphi(n) - \sum_{m=1}^{M} c_m e^{2\pi i x_m n} \right| < \delta.$$

To extend Corollary 1.3.5 to $q \le 4$, we will improve Lemma 1.3.1.

Lemma 1.3.6. *Suppose* $\mathbf{E} = \{n_j\}$ *is a Hadamard set with ratio* q *and satisfying* $\inf n_{j+1}/n_{j-1} > 6q/(q-1)$. *There is some* $0 < \varepsilon < 1/4$ *such that for all* $\psi : \mathbf{E} \to \mathbb{T}$ *and intervals* $I \subset \mathbb{R}$ *of length at least* $3\pi/(2n_1)$, *there exists* $\theta \in I$ *such that*

$$\left| \psi(n_j) - e^{2i\theta n_j} \right| \leq \left| 1 - e^{2\pi i\varepsilon} \right| \text{ for all } j. \tag{1.3.5}$$

Remark 1.3.7. $0 < \varepsilon < 1/4$ implies $e^{2\pi i\varepsilon}$ lies in the open right half of the complex plane. By using the real part of $e^{2i\theta n_j}$, the real version of the standard iteration will give us an approximation to $\mathfrak{Re}\psi$.

Proof (of Lemma 1.3.6). Fix a compact interval $I \subset \mathbb{R}$ of length at least $3\pi/2n_1$. Choose $\varepsilon > 0$ and close enough to $1/4$ such that

$$q > \frac{1 - 2\varepsilon}{2\varepsilon} \text{ and} \tag{1.3.6}$$

$$\inf \frac{n_{j+1}}{n_{j-1}} > \frac{6q}{q - 1 - (1 - 4\varepsilon)(q+1)}. \tag{1.3.7}$$

Let

$$L_j = \left\{ \theta \in \mathbb{R} : \left| e^{2i\theta n_j} - \psi(n_j) \right| \leq \left| 1 - e^{2\pi i\varepsilon} \right| \right\}.$$

Each set L_j is periodic with period π/n_j and consists of a union of disjoint intervals of lengths $2\varepsilon\pi/n_j$. The complement of each L_j is also a periodic set with period π/n_j and is a union of intervals of length $(1 - 2\varepsilon)\pi/n_j$.

Note that any interval of length at least $3\pi/(2n_j)$ must contain a (full) component of L_j and each interval of length more than $(1 - 2\varepsilon)\pi/n_j$ must intersect L_j non-trivially. In particular, I contains a component of L_1.

Because inequality (1.3.6) implies that

$$2\varepsilon/n_{j-1} > (1 - 2\varepsilon)/n_j,$$

each component of L_{j-1} will necessarily contain points of L_j. Indeed, such an interval will either contain a component of L_j or a subinterval L'_j of L_j with

$$\text{length}(L'_j) \geq \frac{\pi}{2} \left(\frac{2\varepsilon}{n_{j-1}} - \frac{1 - 2\varepsilon}{n_j} \right).$$

In the latter case, an easy computation using inequality (1.3.7) and the fact that $n_{j+1}/n_j \leq n_{j+1}/qn_{j-1}$ shows that $\text{length}(L'_j) \geq 3\pi/(2n_{j+1})$, and thus L'_j will contain a component of L_{j+1}. It follows that

$$I \cap \bigcap_{j=1}^{\infty} L_j \neq \emptyset,$$

and, of course, any $\theta \in I \cap \bigcap_{j=1}^{\infty} L_j$ will suffice. $\qquad\qquad\square$

Remark 1.3.8. If $\mathbf{E}$ is as in the lemma and I is any open interval in $\mathbb{R}$, then there is a finite subset $\mathbf{F} \subseteq \mathbf{E}$ such that given any $\psi : \mathbf{E} \to \mathbb{T}$, there is some $\theta \in I$ such that $\left|\psi(n_j) - \mathrm{e}^{2\mathrm{i}\theta n_j}\right| \leq \left|1 - \mathrm{e}^{2\pi \mathrm{i}\varepsilon}\right|$ for all $n_j \in \mathbf{E} \smallsetminus \mathbf{F}$. Indeed, just take $\mathbf{F} = \{n_j\}_{j=1}^{K-1}$ where $3\pi/2n_K \leq \mathrm{length}(I)$.

Here is the promised improvement of Corollary 1.3.5.

Theorem 1.3.9 (Weiss–Strzelecki–Méla). *Suppose* $\mathbf{E} = \{n_j\}$ *is a Hadamard set with ratio* q *and* $K \in \mathbb{N}$ *is chosen with* $q^K > 6q/(q-1)$.

1. *The set* $\mathbf{E}$ *is an* I_0 *set and the interpolation can be done with non-negative, discrete measures supported on*

$$L = \left[-\frac{6K\pi}{n_1}, \frac{6K\pi}{n_1} \right].$$

2. *The set* $\mathbf{E} \cup -\mathbf{E}$ *is an* I_0 *set and the interpolation can be done with discrete measures supported on*

$$L' = \left[-\frac{12K\pi}{n_1}, \frac{12K\pi}{n_1} \right].$$

3. *Furthermore, if* $\delta > 0$ *there exists* $M \geq 1$ *(depending only on* δ *and* q*) such that whenever* $\varphi : \mathbf{E} \to \Delta$ *(resp.,* $\varphi : \mathbf{E} \cup -\mathbf{E} \to \Delta$*), then there exist* $x_1, \ldots, x_M \in L$ *(resp.,* L'*) and* $c_1, \ldots, c_M \in [0,1]$ *(resp.,* Δ*) such that for all* $n \in \mathbf{E}$ *(resp.,* $n \in \mathbf{E} \cup -\mathbf{E}$*)*

$$\left| \varphi(n) - \sum_{m=1}^{M} c_m \mathrm{e}^{2\pi \mathrm{i}x_m n} \right| < \delta.$$

Proof. (1) For each $k = 1, \ldots, K$, let $\mathbf{E}_k = \{n_{k+jK} : j = 0, 1, \ldots\}$. Then $\mathbf{E}_k$ is Hadamard with ratio $q^K > 6$.

For each pair ℓ, k, with $\ell \neq k$, consider the Hadamard set $\mathbf{H} = \mathbf{E}_k \cup \mathbf{E}_\ell$ of ratio q and a bounded $\psi : \mathbf{H} \to \mathbb{C}$ that is equal to 1 on $\mathbf{E}_k$ and -1 on $\mathbf{E}_\ell$. If we suppose $\mathbf{H} = \{m_j\}$, with $m_{j+1} \geq m_j$, then since each pair m_{j-1}, m_{j+1} either belongs to $\mathbf{E}_k$ or to $\mathbf{E}_\ell$, one can see that $m_{j+1}/m_{j-1} \geq q^K > 6q/(q-1)$ for all j.

Thus, the previous lemma can be applied with $I = [-3\pi/4m_1, 3\pi/4m_1]$ to obtain $\varepsilon < 1/4$ and $\theta \in I$ satisfying

$$\left| \mathrm{e}^{2\mathrm{i}\theta m_j} - \psi(m_j) \right| \leq \left| 1 - \mathrm{e}^{2\pi \mathrm{i}\varepsilon} \right| \text{ for all } j.$$

Let $t = \cos 2\pi\varepsilon$. Then $t > 0$,

$$\mathfrak{Re}\,\mathrm{e}^{2\mathrm{i}n\theta} \geq \quad t \text{ for all } n \in \mathbf{E}_k \text{ and}$$

$$\mathfrak{Re}\,\mathrm{e}^{2\mathrm{i}n\theta} \leq -t \text{ for all } n \in \mathbf{E}_\ell.$$

Put $\nu = \frac{1}{2}(\delta_\theta + \delta_{-\theta})$. Then ν is a positive, discrete measure supported on $[-3\pi/4n_1, 3\pi/4n_1]$, with $\widehat{\nu}$ real, $\widehat{\nu} \geq t$ on $\mathbf{E}_k$ and $\widehat{\nu} \leq -t$ on $\mathbf{E}_\ell$. Since $\mathbf{E}_\ell$ is a Hadamard set with ratio at least 6, by Corollary 1.3.5, there is a $\tau_1 \in M_d^+([-2\pi/n_1, 2\pi/n_1])$ such that $\widehat{\tau_1} = \sqrt{-\widehat{\nu}}$ on $\mathbf{E}_\ell$. Let $\nu_0 = \nu + \tau_1 * \widetilde{\tau_1}$ so $\nu_0 \in M_d^+([-4\pi/n_1, 4\pi/n_1])$. Because the Fourier transform of $\tau_1 * \widetilde{\tau_1}$ is non-negative, ν_0 has transform 0 on $\mathbf{E}_\ell$ and at least t on $\mathbf{E}_k$.

But $\mathbf{E}_k$ is also a Hadamard set with ratio at least 6, and hence there is also a measure $\tau_2 \in M_d^+([-2\pi/n_1, 2\pi/n_1])$ such that $\widehat{\tau_2} = \widehat{\nu_0}^{-1}$ on $\mathbf{E}_k$. Putting

$$\nu_{k,l} = \tau_2 * \nu_0 \in M_d^+([-6\pi/n_1, 6\pi/n_1]),$$

we obtain a measure whose transform is 1 on $\mathbf{E}_k$ and 0 on $\mathbf{E}_\ell$. Finally, let ω_k be the convolution of the $K-1$ measures $\nu_{k,l}$ for $\ell \neq k$. This is a non-negative, discrete measure, supported on $[-6(K-1)\pi/n_1, 6(K-1)\pi/n_1]$, with $\widehat{\omega_k} = 1$ on $\mathbf{E}_k$ and equal to 0 otherwise on $\mathbf{E}$.

Now suppose $\varphi : \mathbf{E} \to \mathbb{C}$ is bounded. Once more use the fact that each $\mathbf{E}_k$ is a Hadamard set with ratio 6 to obtain $\sigma_k \in M_d^+([-2\pi/n_1, 2\pi/n_1])$ with $\widehat{\sigma_k} = \varphi$ on $\mathbf{E}_k$. Then

$$\mu = \sum_{k=1}^K \omega_k * \sigma_k \in M_d^+([-6K\pi/n_1, 6K\pi/n_1])$$

interpolates φ on $\mathbf{E}$, completing the proof of (1).

(2) For interpolation on $\mathbf{E} \cup -\mathbf{E}$, obtain $\omega \in M_d^+([-6K\pi/n_1, 6K\pi/n_1])$ such that $\widehat{\omega} = -i$ on $\mathbf{E}$. Because ω is a non-negative measure, $\widehat{\omega}(-n) = \overline{\widehat{\omega}(n)}$. Thus, $\frac{1}{2}(\delta_0 + i\omega)$ has transform 1 on $\mathbf{E}$ and 0 on $-\mathbf{E}$. It is easy to now find $\omega' \in M_d^+([-12K\pi/n_1, 12K\pi/n_1])$ with $\widehat{\omega'} = \varphi$ on $\mathbf{E} \cup -\mathbf{E}$.

(3) Follows from the proofs of (1)–(2). $\square$

Remark 1.3.10. It is easy to see that if interpolation can be done with measures on a set $L \subset \mathbb{T}$, then the interpolation can be done with measures concentrated on any translate of L; see Exercise 1.7.3.

1.4 Sums of Hadamard Sets

A common theme in the study of special sets is to quantify, in some fashion, the "smallness" or "thinness" of the set. This theme will be explored throughout the book. For Hadamard sets there are obvious ways to make this very explicit. For example, it is very easy to see that if $\mathbf{E}$ is a Hadamard set with ratio q, then

$$\sup_N |\mathbf{E} \cap [N, 2N)| < \infty,$$

with the supremum depending only on q. Here $|X|$ denotes the cardinality of the set X. Furthermore, there is a constant C_q such that $|\mathbf{E} \cap [-N, N]| \leq C_q \log N$ for all N. (We will see later (Corollary 6.3.13) that this density property is true, more generally, for Sidon sets.)

This fact generalizes to sums and differences of Hadamard sets.

Proposition 1.4.1. *Let* $\mathbf{E} = \{n_j\}$ *be Hadamard with integer ratio* $q \geq 2$ *and let* $\mathbf{F} = \{0\} \cup \mathbf{E} \cup -\mathbf{E}$. *Then there exists a constant* $c = c(q)$ *such that*

$$|(\overbrace{\mathbf{F} + \cdots + \mathbf{F}}^{q}) \cap [-N, N]| \leq c(\log N)^q \text{ for any } 2 \leq N < \infty.$$

Proof. Let $\mathcal{S} = \overbrace{\mathbf{F} + \cdots + \mathbf{F}}^{q}$. Consider a sum $s = \sum_{j=1}^{J+1} a_j n_j$ with $a_j \in \mathbb{Z}$ and $\sum_{j=1}^{J+1} |a_j| \leq q$. If $|a_{J+1}| = 1$, then

$$|s| \geq n_{J+1} - \sum_{1}^{J} |a_j| n_J \geq n_{J+1} - (q-1)n_J$$

$$\geq q n_J - (q-1)n_J = n_J.$$

If $|a_{J+1}| > 1$, we have $|s| > n_J$. It follows that $s \notin [-n_J+1, n_J-1]$ if $a_{J+1} \neq 0$. Furthermore, $s = \pm n_J$ can only be achieved if $a_{J+1} = \pm 1$ and $a_J = \mp(q-1)$. Because each $|a_j| \leq q$, there are at most $(2q+1)^q J^q$ sums of the form $\sum_{1}^{J} a_j n_j$ when $\sum_{1}^{J} |a_j| \leq q$, and therefore $|\mathcal{S} \cap [-n_J, n_J]| \leq 2 + (2q+1)^q J^q$, with the extra 2 accounting for the endpoints.

Let $n_{J-1} \leq N < n_J$. Then $\mathcal{S} \cap [-N, N] \subseteq \mathcal{S} \cap [-n_J, n_J]$, so $|\mathcal{S} \cap [-N, N]| \leq 2 + (2q+1)^q J^q$. Since $n_1 q^{J-2} \leq n_{J-1} \leq N$, $J^q \leq c(\log N)^q$ for a suitable constant c.   □

In particular, such a q-fold sum has density zero. If the terms n_j are not allowed to repeat, we can obtain this same sparseness with arbitrarily many summands.

Proposition 1.4.2. *Let* $\mathbf{E} = \{n_j\}$ *be Hadamard with ratio* q. *Let*

$$\mathbf{E}_k = \left\{ \sum_j a_j n_j : \sum_j |a_j| \leq k \text{ and } a_j = 0, \pm 1 \text{ for all } j \right\}. \qquad (1.4.1)$$

1. *If* $q \geq 2$, *then for each* $k \geq 1$ *there exists a constant* $c = c(q, k)$ *such that*

$$|\mathbf{E}_k \cap [-N, N]| \leq c(\log N)^k \text{ for } 1 \leq N < \infty. \qquad (1.4.2)$$

2. *If* $q \geq 3$, *then* $\mathbb{Z} \setminus \mathbf{E}_k$ *contains increasingly long sequences of consecutive integers, located between pairs* n_j *and* n_{j+1} *for all* j *sufficiently large.*

Proof. Both assertions are obvious if $k = 1$, so we may assume that $k \geq 2$.

(1) Find a positive integer $m = m(q, k)$ so large that

$$1 < \frac{q^m(q^2 - 2q + q^{-k+2})}{q - 1}.$$

A geometric series calculation shows that if $j + m \geq k$, then

$$n_{j+m+1} - n_{j+m} - \cdots - n_{j+m-k+2} \geq n_{j+m+1}\left(1 - \sum_{i=1}^{k-1} q^{-i}\right)$$

$$> n_j q^m \left(\frac{q^2 - 2q + q^{-k+2}}{q - 1}\right). \quad (1.4.3)$$

The inequality of (1.4.3) implies that all the elements of $\mathbf{E}_k \cap [-n_j, n_j]$ can be represented as sums of the form $\sum_{\ell=1}^{j+m} a_\ell n_\ell$, where $\sum |a_\ell| \leq k$ and $a_\ell = 0, \pm 1$ for all ℓ. Counting in a similar fashion to the proof of Proposition 1.4.1 gives $|\mathbf{E}_k \cap [-n_j, n_j]| \leq 3^k \binom{j+m}{k} \leq 3^k(j+m)^k \leq cj^k$ where the constant c depends only on k and q.

(2) Geometric series calculations show that

$$n_j + n_{j-1} + \cdots + n_{j-k+1} \leq n_j \frac{q - q^{-k+1}}{q - 1}. \quad (1.4.4)$$

Therefore, the gap between the smallest sum one can compute involving $+n_{j+1}$ and its predecessors, and the largest sum involving $+n_j$ and its predecessors is at least

$$n_{j+1} - 2(n_j + \cdots + n_{j-k+1}) \geq n_j \left(\frac{q^2 - 3q + 2q^{-k+1}}{q - 1}\right).$$

Since $q \geq 3$, this can be made arbitrarily large by choosing j sufficiently large. $\qquad\square$

However, this sparseness result depends crucially upon not allowing repetition. If we do allow repetition, then Proposition 1.4.1 is sharp. Indeed, there are Hadamard sets of integer ratio q with a $q+1$-fold sum or difference equal to all of $\mathbb{N}$, as the next result shows.

Proposition 1.4.3. *Let* $2 \leq q \in \mathbb{N}$. *There exists a Hadamard set* $\mathbf{E} = \{n_j\}$ *with ratio* q *such that*

$$\mathbf{E} - \overbrace{(\mathbf{E} + \cdots + \mathbf{E})}^{q} \supseteq \mathbb{N}. \quad (1.4.5)$$

Proof. The set $\mathbf{E}$ will be a union of doubletons, $\mathbf{E}_j = \{n_j, j + n_j q\}$, $1 \leq j < \infty$, where $n_1 = 1$ and $n_{j+1} = q(j + n_j q)$ for $1 \leq j < \infty$. That (1.4.5) holds is easy to see: just represent j as the difference $j + n_j q - (n_j + \cdots + n_j)$ (q copies of n_j) to obtain the positive integer j. Each $\mathbf{E}_j$ is Hadamard with ratio at least q and the union will be Hadamard with ratio q because of the choice $n_{j+1} = q(j + n_j q)$. $\square$

1.5 Bohr Cluster Points of Hadamard Sets and Their Sums

Another way to quantify the size of a set is to consider its closure in the Bohr topology. In this section we will prove that a sum (or difference) of a Hadamard set with itself has no continuous characters (other than 0) as Bohr cluster points. This fact will later be generalized to ε-Kronecker sets; see Proposition 2.7.4.

The Bohr topology on $\mathbb{Z}$ is the topology of pointwise convergence on $\mathbb{T}$. This topology makes $\mathbb{Z}$ a dense subgroup of the Bohr group, denoted $\overline{\mathbb{Z}}$, the compact group consisting of all group homomorphisms of $\mathbb{T}$ to $\mathbb{T}$ (including discontinuous ones). We refer the reader to Sect. C.1.3 for more details on the Bohr group and topology.

Theorem 1.5.1 (Kunen–Rudin). *Let* $\mathbf{E} = \{n_j\} \subset \mathbb{N}$ *be a Hadamard set and let* $\mathbf{F}_0 \subset \mathbb{Z}$ *be finite. Then*

1. $\mathbf{E} \cup -\mathbf{E}$ *has no Bohr cluster points in* $\mathbb{Z}$.
2. $(\mathbf{E} \cup -\mathbf{E}) + \mathbf{F}_0$ *has no Bohr cluster points in* $\mathbb{Z}$.
3. $\mathbf{E} - \mathbf{E}$ *has only 0 as a Bohr cluster point in* $\mathbb{Z}$.
4. $\mathbf{E} + \mathbf{E}$ *has no Bohr cluster points in* $\mathbb{Z}$.

Proof. Let $q > 1$ be the ratio of $\mathbf{E}$ and fix $m \in \mathbb{Z}$.

(1) It will suffice to show that m is not a Bohr cluster point of $\mathbf{E}$, for if m were a Bohr cluster point of $-\mathbf{E}$, then $-m$ would be a cluster point of $\mathbf{E}$. Since any Hadamard set is a finite union of Hadamard sets with ratio at least 16, and a finite union of sets clusters at m if and only if at least one of the sets in the union clusters at m, there is no loss of generality in assuming $q \geq 16$.

Since we can find θ such that $|1 + e^{i\theta n_k}| < 1/4$ for all $n_k \in \mathbf{E}$, $m = 0$ cannot be a cluster point.

Now suppose $m \neq 0$. If $\mathbf{E}$ clusters at m, so does any cofinite subset of $\mathbf{E}$, so we can assume $n_1 > 8\pi|m|$. By Lemma 1.3.1 we may find $\theta \in [-2\pi/n_1, 2\pi/n_1]$ such that $|1 + e^{in_j\theta}| < |1 - e^{i\pi/15}| < 1/4$ for all j. But $|1 - e^{im\theta}| < 2\pi|m|/n_1 < 1/4$. Therefore, $\mathbf{E}$ does not cluster at m.

(2) It will suffice to show that $\mathbf{E} + N$ has no Bohr cluster points in $\mathbb{Z}$ for each $N \in \mathbb{Z}$. By Exercise 1.7.2, $(\mathbf{E} \setminus \mathbf{F}) + N$ is Hadamard for a suitable finite set $\mathbf{F}$, and thus part (1) above gives this conclusion. Alternatively, just note that $\mathbf{E} + N$ clusters at integer N_0 if and only if $\mathbf{E}$ clusters at $N_0 - N$.

(3) For each finite set, $\mathbf{F}$, the closure of $\mathbf{E} - \mathbf{E}$ is equal to the union of the closure of $(\mathbf{E} \setminus \mathbf{F}) - (\mathbf{E} \setminus \mathbf{F})$ together with the closure of $(\mathbf{E} - \mathbf{F}) \cup (\mathbf{F} - \mathbf{E})$. Hence, the second part of the proposition implies that if $k \in \mathbb{Z}$, $k \neq 0$ is a cluster point of $\mathbf{E} - \mathbf{E}$, then k belongs to the closure of $(\mathbf{E} \setminus \mathbf{F}) - (\mathbf{E} \setminus \mathbf{F})$. It follows that for each finite set $\mathbf{F}$,

$$(k + \overline{\mathbf{E} \setminus \mathbf{F}}) \cap \overline{\mathbf{E} \setminus \mathbf{F}} \neq \emptyset. \tag{1.5.1}$$

Since $k \neq 0$, we may choose $x \in \mathbb{T}$ such that $e^{2ikx} = -1$. Let $K \in \mathbb{N}$ satisfy $q^K > 6q/(q - 1)$ and suppose M is given by Theorem 1.3.9 (3) with $\delta = \pi/4$. Let $L = [x - \varepsilon, x + \varepsilon]$ be a neighbourhood of x so small that $|1 + e^{2iku}| < 1/(4M)$ for all $u \in L$. Choose j_0 so large that $6K\pi/n_{j_0} < \varepsilon$ and let $\mathbf{F} = \{n_1, \ldots, n_{j_0}\}$.

Theorem 1.3.9 (3) applied to $\mathbf{E} \setminus \mathbf{F}$ tells us that there exist $x_1, \ldots, x_m \in L$ and $c_1, \ldots, c_m \in [0, 1]$ such that $|1 - \sum_1^M c_m e^{2\pi i x_m n}| < 1/4$ for all $n \in \mathbf{E} \setminus \mathbf{F}$. Combining these inequalities shows that

$$\left| 1 + \sum_{m=1}^M c_m e^{2i x_m (k+n)} \right| \leq \left| 1 + e^{2i x k} \right| + \left| e^{2i x k} \left(\sum_{m=1}^M c_m e^{2i x_m n} - 1 \right) \right|$$

$$+ \left| \sum_{m=1}^M c_m e^{2i x_m n} (e^{2i x_m k} - e^{2i k x}) \right|$$

$$< 0 + \frac{1}{4} + M \frac{2}{4M} = \frac{3}{4}.$$

Therefore, the continuous (on $\overline{\Gamma}$) function $\sum_1^M c_m e^{2i x_m (\cdot)}$ separates $\overline{\mathbf{E} \setminus \mathbf{F}}$ from $k + \overline{\mathbf{E} \setminus \mathbf{F}}$, contradicting (1.5.1).

(4) Similarly, it is not hard to see that $k \in \mathbb{Z}$ is a cluster point of $\mathbf{E} + \mathbf{E}$ if and only if for any finite set $\mathbf{F}$, $k - \overline{\mathbf{E} \setminus \mathbf{F}} \cap \overline{\mathbf{E} \setminus \mathbf{F}} \neq \emptyset$. We proceed as for (3), except that we choose an interval L around 0 and require that the c_m, x_m satisfy $|\sum_{m=1}^M c_m e^{2\pi i x_m n} - i| < 1/4$ for all $n \in \mathbf{E} \setminus \mathbf{F}$ and that $|e^{2\pi i k x} - 1| < 1/4M$ on L. This will give us a trigonometric polynomial on $\mathbb{Z}$ which is approximately i on $\mathbf{E} \setminus \mathbf{F}$ and approximately $-i$ on $k - \mathbf{E} \setminus \mathbf{F}$. That implies $\overline{\mathbf{E} \setminus \mathbf{F}}$ and $k - \overline{\mathbf{E} \setminus \mathbf{F}}$ are disjoint. We omit further details. $\qquad \square$

Example 1.5.2. Disjoint Hadamard sets with common Bohr cluster points: Let $\mathbf{E} = \{3^j : j = 1, 2, \ldots\}$ and $\mathbf{F} = \{3^j + j : j = 1, 2, \ldots\}$. Exercise C.4.16 says that 0 is a Bohr cluster point of $\mathbb{N}$. Take a net of positive integers $\{j_\alpha\}$

such that $j_\alpha \to 0$ in the Bohr topology. Because of compactness of the Bohr group, a subnet, $\{3^{j_\beta}\}$, converges in $\overline{\mathbb{Z}}$. Thus, $\mathbf{E}$ and $\mathbf{F}$ have common Bohr cluster points in $\overline{\mathbb{Z}}$.

We remark that this property implies that $\mathbf{E} \cup \mathbf{F}$ is not I_0. The function that is 1 on $\mathbf{E}$ and 0 on $\mathbf{F}$ cannot be the restriction of the Fourier–Stieltjes transform of a discrete measure on $\mathbb{T}$ since such transforms are continuous on $\overline{\mathbb{Z}}$. See also Proposition 3.4.1.

1.6 Remarks and Credits

Classical Results. For further classical results, see Zygmund's volumes [199, 200].

Theorem 1.2.1 appears in Weierstrass's [192]. According to Weierstrass, Riemann had proposed orally in 1861, or perhaps earlier, that $\sum(1/n^2) \sin n^2 x$ was continuous and nowhere differentiable. Weierstrass justified giving a new example on the basis that it appeared to him that a proof of the nowhere differentiability of Riemann's example would be "rather difficult". In fact, Riemann's sum *is* differentiable at all points of the form $c\pi/d$ where $c, d \in 2\mathbb{Z} + 1$ [44] but, at no other point, rational multiple of π [45] or not [72].

An English version of Weierstrass's proof of Theorem 1.2.1 is in the translation of Goursat's classic *Cours d'Analyse Mathématique* [48, pp. 423ff]. A twenty-first century proof of the non-differentiability of versions of Weierstrass's sum is given by Johnsen [97]; he also gives a good historical survey and writes that the first example of a nowhere differentiable function was given by Bolzano. Hardy [72] in 1916 improved Weierstrass's example and gave references to the intervening literature. Other ways of constructing nowhere differentiable functions have also been given, cf., [139] and its references.

Hadamard's gap theorem (Theorem 1.2.2) is in [70]. Hadamard points out that the gap theorem is easily proved when $n_j = a^j$ for a positive integer a. Mordell's lovely proof [135] can also be found in [92, pp. 88–89].

A much more general form of Theorem 1.2.2 is that of Fabry (which has, itself, been generalized in ways outside of harmonic analysis). The Fabry theorem states that if $\{\lambda_n\}_{n=1}^\infty \subset [0, \infty)$ is a real sequence with $\inf(\lambda_{n+1} - \lambda_n) > 0$ and $\inf_n \frac{n}{\lambda_n} = D$, then the series $f(z) = \sum_{n=1}^\infty c_n e^{-\lambda_n z}$ has at least one singularity in every interval of length exceeding $2\pi D$ on the abscissa of convergence; see [118, p. 89] for a proof of the Fabry theorem in this form and [92, p. 89ff] for a proof when the λ_n's are integers with $\lim_n n/\lambda_n = 0$. Interest in the relationship between analyticity and Hadamard sequences has never ceased; [144] is a recent example.

Theorem 1.2.3 is due to Fuchs [40], who was improving on Pommerenke [152], who had improved a result of G. and M. Weiss [193]. For related results, see [12, 47].

Riesz Products. Theorem 1.2.4 is due to Riesz [163]. Zygmund [199, p. 209] has a nice proof. The literature on Riesz products is extensive; [56, Chap. 7] has some of what was known 30 years ago, including most parts of Exercise 1.7.9. The terms $1 + a_j \cos(n_j x)$ can be replaced by trigonometric polynomials p_j if $p_j \geq 0$, $\int p_j \mathrm{d}x = 1$ and the frequencies of different p_j do not interfere with each other. These are called "generalized Riesz products" and more can be found on them in [94]. A recent use of Riesz products in Fourier analysis is [103] and another in dynamical systems is [154]. Further discussion of Riesz products can be found in Sect. 6.2.2.

Theorem 1.2.5 is in [115]. This result is, of course, an instance of the Carleson–Hunt a.e. convergence theorem [3, 22, 95]. Theorem 1.2.6 is due to Sidon [173, 174]. Theorem 1.2.7 comes from [197, 198]. In more modern language, the theorem says Hadamard sets are $\Lambda(p)$ sets; see [123, pp.54ff] for more on $\Lambda(p)$ sets and Theorem 6.3.9 for the proof that Sidon sets are $\Lambda(p)$ sets.

Interpolation properties. The idea of doing interpolation using measures supported in an interval of length proportional to $1/n_1$ seems to have first been used by Mary Weiss [194, Theorem 2]. Lemma 1.3.1 and Proposition 1.3.2 are standard. Lemma 1.3.6 is from [130, Proof of Theorem 1].

Theorem 1.3.9 was proved originally by Strzelecki [180]; other proofs can be found in [101, 116]. The proof we have given is from [130].

Sums of Hadamard Sets. The results of Sect. 1.4 are from [61, 116]. A version of inequality (1.4.3) appears in [199, pp. 208ff].

Proposition 1.4.1 is related to other results. The sum of M disjoint sets whose union is dissociate is p-Sidon [98] and hence has a logarithmic density as in Proposition 1.4.1 [35, Corollary 2.6]. See [35] and its references for earlier results of this type. Hadamard sets with sufficiently large ratio q are dissociate (Exercise 1.7.4), so Proposition 1.4.1 is an variation of the combination of [35, 98].

The sets $\mathbf{E}_k$ of Proposition 1.4.2 were also investigated by Déchamps and Selles [30, Theorem A] who proved that if $q \geq 3$, then the sets $\mathbf{E}_k$ have a very strong property (called "zhd" in Chap. 10): for each non-empty, open set $U \subset \mathbb{T}$, every restriction of a Fourier–Stieltjes transform to $\bigcup_{k=1}^{K} \mathbf{E}_k$ can be written as the restriction of a Fourier–Stieltjes transform of a measure concentrated on U. We use the same geometrical series calculation as [30]. See Chap. 10 for more on this property.

Bohr cluster points. Theorem 1.5.1 (1) is a special case of a result of Ryll-Nardzewski about I_0 sets in $\mathbb{R}$ [168] and holds more generally for I_0 sets; see Theorem 3.5.1. The other items of Theorem 1.5.1 are due to Kunen and Rudin [116]. (1)–(4) hold for ε-Kronecker sets (see Propositions 2.7.1, 2.7.2 and 2.7.4); for other classes of I_0 sets see Theorem 5.3.9. Example 1.5.2 appeared first in [129, p. 178].[2]

---

[2] Méla writes that "rapellons" there refers to an unpublished part of his thèse [131].

Periodic Extension. The subject of I_0 sets began with a question of Marczewski and Ryll-Nardzewski [137]: Does there exist an infinite subset $\{t_n\}$
of $\mathbb{R}$ with the property that for every choice $\epsilon_n = \pm 1$ there exists a periodic
function f such that $f(t_n) = \epsilon_n$ for all n? Mycielski [137] showed that every
Hadamard[3] subset $0 < t_1 < t_2 < \cdots$ with ratio $q > 3$ had that property.
Subsequently, Lipiński [121] showed that $q = 3$ was possible. If $q \leq 2$ then
there is a choice of $\epsilon_j = \pm 1$ for which there is no continuous periodic
$f : \mathbb{R} \to \mathbb{R}$ such that $f(n_j) = \epsilon_j$ for all j [168].

In [122] Lipiński constructed a union $\mathbf{E}$ of intervals tending to infinity such
that every uniformly continuous function on $\mathbf{E}$ has a periodic extension to all
of $\mathbb{R}$. He showed that the set of periods of the extensions of a given function
has cardinality $c = |\mathbb{R}|$ (it always has Lebesgue measure zero [79]). Versions
of these periodic extension results for $\mathbb{R}^n$ appear to be unknown [**P 18**].

It was soon apparent that studying *almost* periodic extensions (defined
in Chap. 3) would be more fruitful than periodic ones. (For example, the
restriction to $\mathbb{Z}$ of a periodic function on $\mathbb{R}$ is not always periodic.) This led
to the interest in I_0 sets, originally defined by Hartman and Ryll-Nardzewski
[81] as the sets $\mathbf{E}$ with the property that every bounded function on $\mathbf{E}$ was
the restriction of an almost-periodic function.

1.7 Exercises

Exercise 1.7.1. Suppose $\mathbf{E} \subset \mathbb{N}$ is Hadamard.

1. Show that $sup_N |\mathbf{E} \cap [N, 2N]| < \infty$.
2. Prove that if $\mathbf{F} \subset \mathbb{N}$ is finite, then $\mathbf{E} \cup \mathbf{F}$ is Hadamard.

Exercise 1.7.2. Suppose $\mathbf{E} \subset \mathbb{N}$ is Hadamard with ratio $q > 1$ and $q'' < q < q' < \infty$.

1. Prove that $\mathbf{E}$ is a finite union of Hadamard sets with ratio q'.
2. Show that for every $n \in \mathbb{N}$, there is a finite set $\mathbf{F} \subset \mathbf{E}$ such that $(\mathbf{E} \setminus \mathbf{F}) + n$
   is Hadamard with ratio q''.

Exercise 1.7.3. Let $\mathbf{E} \subset \mathbb{Z}$, $L \subset \mathbb{T}$ and $e^{ix} \in \mathbb{T}$. Suppose that every $\varphi :$
$\mathbf{E} \to \mathbb{T}$ can be interpolated by a measure in $M_d(L)$ (resp., $M_d^+(L)$). Show
that every $\varphi : \mathbf{E} \to \mathbb{T}$ can be interpolated by a measure in $M_d(L \cdot e^{ix})$ (resp.,
$M_d^+(L \cdot e^{ix})$).

Exercise 1.7.4. The set $\mathbf{E} \subset \mathbb{Z} \setminus \{0\}$ is called *quasi-independent* if whenever $m_1, \ldots, m_N$ are distinct elements of $\mathbf{E}$ and $e_1, \ldots, e_N \in \{-1, 0, 1\}$, then
$\sum e_n m_n = 0$ implies all $e_n m_n = 0$. $\mathbf{E}$ is *dissociate* if whenever $m_1, \ldots, m_N$ are
distinct elements of $\mathbf{E}$ and $e_1, \ldots, e_N \in \{-2, -1, 0, 1, 2\}$, then $\sum e_n m_n = 0$
implies all $e_n m_n = 0$.

---

[3] We define "Hadamard set" in $\mathbb{R}$ exactly as in $\mathbb{N}$.

Suppose **E** is a Hadamard set with ratio q. Show

1. **E** need not be either dissociate or quasi-independent.
2. **E** is quasi-independent if $q \geq 2$.
3. **E** is dissociate if $q \geq 3$.
4. Every Hadamard set is a finite union of dissociate sets.

Exercise 1.7.5.   1. Suppose ω is a real measure. Show that $\widehat{\omega}(\gamma) = \overline{\widehat{\omega}(\gamma^{-1})}$
for all $\gamma \in \Gamma$.

2. Show that for any measure μ, $\widehat{\widetilde{\mu}} = \overline{\widehat{\mu}}$.

3. Show that if **E** is an I_0 set and the interpolation can be done with a real (or non-negative), discrete measure supported on a symmetric set U, then the same is true for $\mathbf{E}^{-1}$.

Exercise 1.7.6. Let $\mathbf{E} \subseteq \Gamma$, U be a symmetric subset of G, $\varepsilon < 1$ and $N \in \mathbb{N}$. Assume that for each $\varphi : \mathbf{E} \to \mathbb{T}$ (resp., $\varphi : \mathbf{E} \to \{-1, 1\}$) there exists $x_1, \ldots, x_N \in U$ and $c_1, \ldots, c_N \in \Delta$ (resp., $\in [0, 1]$) with

$$\left| \varphi(\gamma) - \sum_{n=1}^{N} c_n \gamma(x_n) \right| < \varepsilon \text{ for all } \gamma \in \mathbf{E}.$$

Show that for every $\delta > 0$ there is a $M = M(\delta, \varepsilon, N)$ such that for each $\varphi : \mathbf{E} \to \Delta$ (resp., $[-1, 1]$) there exist $y_1, \ldots, y_M \in U$ and $d_1, \ldots, d_M \in \Delta$ (resp., $\in [0, 1]$) with

$$\left| \varphi(\gamma) - \sum_{m=1}^{M} d_m \gamma(y_m) \right| < \delta \text{ for all } \gamma \in \mathbf{E}.$$

Exercise 1.7.7. Let $\mathbf{E}, \mathbf{E}_k, q$ be as in Proposition 1.4.2.

1. Show that for each $k \geq 1$ there exist $1 < q_k < 2$ and constant $c = c(q, k)$ such that (1.4.2) holds for k and $q_k < q$.
2. Show that for each $k \geq 1$ there exists $1 < q_k < 3$ such that if $q_k < q$, then $\mathbb{Z} \smallsetminus \mathbf{E}_k$ contains increasingly large subsets of consecutive integers between successive elements of **E**.

Exercise 1.7.8. Prove that if the integers a_j of Proposition 1.4.2 are allowed to take on values in $[-L, L]$ for an integer $L \geq 2$, then there exist $q_{k,L} < L+1$ and constant c for which (1.4.2) holds when $q > q_{k,L}$.

Exercise 1.7.9. This exercise investigates Riesz products.

Suppose **E** is a Hadamard set with ratio $q \geq 3$. Suppose $-1 \leq a_k \leq 1$ for $k = 1, 2, \ldots$ and let $p_K(x) = \prod_{k=1}^{K}(1 + a_k \cos(n_k x))$.

1. Show $p_K \geq 0$, $\widehat{p_K}(0) = 1$ and $\|p_K\|_1 = 1$.

2. Prove the measures $p_K dm$ (where dm denotes normalized Lebesgue measure on $\mathbb{T}$) converge weak* to a probability measure μ on $\mathbb{T}$. Determine the Fourier transform of μ. (This measure is called a *Riesz product*.)

3. Show $\mu \in L^2(\mathbb{T})$ if and only if $\sum a_k^2 < \infty$.

4. Show μ is a continuous measure.

5. Prove that every bounded function on $\mathbf{E}$ is the restriction of the Fourier–Stieltjes transform of a non-negative measure. (This means that $\mathbf{E}$ is a Sidon set.)

6. Prove Theorem 1.2.6. Hint: For any trigonometric polynomial f, $\sum \overline{\hat{f}} \hat{\mu} = \int \overline{f} d\mu$.

Chapter 2
ε-Kronecker Sets

ε-Kronecker sets defined. Are I_0. Can be found in most infinite subsets of a discrete group. Defined by approximating ± 1. Arithmetical properties investigated. Have small sums.

2.1 Introduction

In this book we explore generalizations of Hadamard sets, both in $\mathbb{Z}$ and in discrete abelian groups other than $\mathbb{Z}$. One such generalization is the notion of an ε-Kronecker set, a set $\mathbf{E}$ with the property that for every $\varphi : \mathbf{E} \to \mathbb{T}$ there exists $x \in G$ such that $|\varphi(\gamma) - \gamma(x)| < \varepsilon$ for all $\gamma \in \mathbf{E}$.

Hadamard sets in $\mathbb{Z}$ with suitably large ratio are examples of ε-Kronecker sets (with ε depending on the ratio), but not all ε-Kronecker sets in $\mathbb{Z}$ are finite unions of Hadamard sets. Independent sets of characters (defined in the introduction) of sufficiently large order are another important class of ε-Kronecker sets. An ε-Kronecker set with $\varepsilon < \sqrt{2}$ is an I_0 set, and this is sharp. Thus, we have a particular interest in the case $\varepsilon < \sqrt{2}$. These facts and other basic properties are established in Sect. 2.2 and 2.3.

In $\mathbb{Z}$, infinite ε-Kronecker sets exist for each given $\varepsilon > 0$; just take a Hadamard set with large enough ratio. In contrast, $\boldsymbol{\Gamma} = \widehat{\mathbb{D}}$ contains no ε-Kronecker subsets if $\varepsilon \leq \sqrt{2}$. In Sect. 2.4 we will see that whenever $\mathbf{E}$ is an infinite set in a group that does not contain "too many" elements of order two, then $\mathbf{E}$ contains a $(1+\varepsilon)$-Kronecker subset of the same cardinality. If the torsion subgroup of $\boldsymbol{\Gamma}$ is finite, as with $\boldsymbol{\Gamma} = \mathbb{Z}$, then every infinite subset of $\boldsymbol{\Gamma}$ contains an ε-Kronecker subset of the same cardinality, for each $\varepsilon > 0$.

The related problem of interpolating arbitrary choices of signs, rather than all complex numbers of modulus 1, is investigated in Sect. 2.5. Conditions are given which ensure that this formally weaker interpolation implies the set is actually ε-Kronecker. Those ideas will be used in Sect. 9.3 to show that if $\boldsymbol{\Gamma}$

C.C. Graham and K.E. Hare, *Interpolation and Sidon Sets for Compact Groups*,   19
CMS Books in Mathematics, DOI 10.1007/978-1-4614-5392-5_2,
© Springer Science+Business Media New York 2013

does not contain too many elements of order 2, then every Sidon set in Γ is characterized by the property of being proportionally ε-Kronecker.

Provided $\varepsilon < 2$, ε-Kronecker sets have arithmetic properties similar to those possessed by Sidon sets (see Chap. 6). For instance, they do not contain long arithmetic progressions or large squares. When $\varepsilon < \sqrt{2}$, their step length must tend to infinity. Those properties are proven in Sect. 2.6. It is unknown if every ε-Kronecker set, for $\varepsilon < 2$, is Sidon [**P 4**].

ε-Kronecker sets have product properties similar to those of Hadamard sets, and these are discussed in Sect. 2.7. If $\mathbf{E}$ is ε-Kronecker, then $\mathbf{E} \cdot \mathbf{E}$ does not cluster at a continuous character, and the identity is the only continuous character at which $\mathbf{E} \cdot \mathbf{E}^{-1}$ clusters. When ε is small (depending on N) the closure of $(\mathbf{E} \cup \mathbf{E}^{-1})^N$ in the Bohr topology has zero $\overline{\Gamma}$-Haar measure.

2.2 Definition and Interpolation Properties

Definition 2.2.1. Let $U \subseteq G$ and $\varepsilon > 0$. A set $\mathbf{E} \subseteq \Gamma$ is ε-*Kronecker*(U) if for every $\varphi : \mathbf{E} \to \mathbb{T}$ there exists $x \in U$ such that

$$|\varphi(\gamma) - \gamma(x)| < \varepsilon \text{ for all } \gamma \in \mathbf{E}. \tag{2.2.1}$$

Weak ε-*Kronecker*(U) sets $\mathbf{E}$ are the same as ε-Kronecker(U) sets, except that the strict inequality in (2.2.1) is replaced with $\leq$. When $U = G$ we omit the writing of "(G)". The subset $\mathbf{E}$ is called *Kronecker* if it is ε-Kronecker for all $\varepsilon > 0$. By the *Kronecker constant* of $\mathbf{E}$ we mean

$$\varepsilon(\mathbf{E}) = \inf\{\varepsilon : \mathbf{E} \text{ is weak } \varepsilon\text{-Kronecker}\}.$$

A compactness argument shows that if $\mathbf{E}$ is ε-Kronecker(U) for all $\varepsilon > \varepsilon_0$, then $\mathbf{E}$ is weak ε_0-Kronecker$(\overline{U})$.

Remark 2.2.2. Every subset $\mathbf{E}$ of Γ is trivially weak 2-Kronecker. If $1 \in \mathbf{E}$ this cannot be improved, and thus our interest is in $\varepsilon < 2$. The case $\varepsilon < \sqrt{2}$ is particularly interesting, as will be seen in what follows.

It is often convenient to measure angular distances, that is distances in the metric space of the quotient group $\mathbb{R}/(2\pi\mathbb{Z})$, rather than the absolute value metric from $\mathbb{C}$ restricted to $\mathbb{T} = \{z \in \mathbb{C} : |z| = 1\}$. Thus, we denote by d the usual metric in the quotient space, $\mathbb{R}/(2\pi\mathbb{Z})$:

$$d(\theta, \phi) = \inf\{|\theta - \phi + 2\pi k| : k \in \mathbb{Z}\}, \text{ for } \theta, \phi \in \mathbb{R}.$$

Of course, $d(\theta, \phi) \in [0, \pi]$ for all $\theta, \phi \in \mathbb{R}$. Given $t \in \mathbb{T}$, we let $\arg(t)$ denote the element of $[0, 2\pi)$ satisfying $t = e^{i \arg(t)}$. This identifies $\mathbb{T}$ with $[0, 2\pi)$.

Definition 2.2.3. A set $\mathbf{E} \subseteq \Gamma$ is *angular* ε-Kronecker(U) if for every $\{\theta_\gamma\}_{\gamma \in \mathbf{E}} \in [0, 2\pi)^{\mathbf{E}}$ there exists $u \in U$ such that

$$d(\arg(\gamma(u)), \theta_\gamma) < \varepsilon \text{ for all } \gamma \in \mathbf{E}.$$

It is said to be *weak angular ε-Kronecker(U)* if the strict inequality is replaced with $\leq$. Lastly, denote by $\alpha(\mathbf{E})$ the *angular Kronecker constant*

$$\alpha(\mathbf{E}) = \inf\{\varepsilon : \mathbf{E} \text{ is weak angular } \varepsilon\text{-Kronecker}\}.$$

It is easy to see that $\mathbf{E}$ is angular ε-Kronecker if and only if $\mathbf{E}$ is $|1 - e^{i\varepsilon}|$-Kronecker. Consequently, angular ε-Kronecker for some $\varepsilon < \pi$ is equivalent to ε-Kronecker for some $\varepsilon < 2$ and angular ε-Kronecker for some $\varepsilon < \pi/2$ is equivalent to ε-Kronecker for some $\varepsilon < \sqrt{2}$.

Here are some easy facts whose proofs are left for the reader.

Lemma 2.2.4. *1. If $\mathbf{E}$ is (weak) ε-Kronecker, then so is $\mathbf{E}^{-1}$.*

2. If $\mathbf{E}$ is (weak) ε-Kronecker(U) and $x \in G$, then $\mathbf{E}$ is (weak) ε-Kronecker($x \cdot U$).

3. $\mathbf{E} \subseteq \mathbb{Z}$ is ε-Kronecker if and only if $n\mathbf{E}$ is ε-Kronecker for all $n \neq 0$.

Remarks 2.2.5. (i) The class of ε-Kronecker sets is not closed under translation since a set that contains the identity element is no better than weak 2-Kronecker. On the other hand, a Baire category theorem argument (see Exercise 2.9.1 (3)) shows that if Γ is countable and has no elements of finite order, then $\Gamma \smallsetminus \{\mathbf{1}\}$ is 2-Kronecker.

(ii) An ε-Kronecker set $\mathbf{E}$ with $\varepsilon \leq \sqrt{2}$ cannot contain a character of order two since that character can take on only real values. Such an ε-Kronecker set $\mathbf{E}$ must also satisfy $\mathbf{E} \cap \mathbf{E}^{-1} = \emptyset$.

2.2.1 Examples of ε-Kronecker Sets

It is immediate from Lemma 1.3.1 that Hadamard sets are ε-Kronecker for suitable ε. Indeed, with the terminology of this chapter, Lemma 1.3.1 may be restated as:

Proposition 2.2.6. *If $\mathbf{E} \subseteq \mathbb{Z}$ is a Hadamard set with ratio q, then $\mathbf{E}$ is weak angular ε-Kronecker with $\varepsilon = \pi/(q-1)$.*

Of course, this result is only of interest if $q > 2$ and only guarantees an ε-Kronecker set with $\varepsilon < \sqrt{2}$ when $q > 3$.

Since every Hadamard set is a finite union of Hadamard sets with large ratio, it follows that every Hadamard set is a finite union of ε-Kronecker sets, whatever the choice $\varepsilon > 0$. The converse is false.

Example 2.2.7. A set $\mathbf{E} \subseteq \mathbb{N}$ with the property that for each $\varepsilon > 0$ there is a finite subset $\mathbf{F} \subseteq \mathbf{E}$ such that $\mathbf{E} \smallsetminus \mathbf{F}$ is ε-Kronecker, but $\mathbf{E}$ is not a finite union of Hadamard sets: To construct such a set, inductively choose finite sets of

positive integers, $\mathbf{E}_j$, of increasing cardinality, with Hadamard ratios tending
to infinity, and an increasing sequence of positive integers m_j satisfying the
conditions

$$1 < \frac{\min \mathbf{E}_{j+1}}{m_j} \to \infty \text{ and } 1 < \frac{m_j}{\max \mathbf{E}_j} \to \infty.$$

Put $\mathbf{E} = \bigcup_j (m_j + \mathbf{E}_j)$. The sets $\mathbf{F}_j = m_j + \mathbf{E}_j$ are disjoint since each
element of $\mathbf{F}_{j+1}$ is greater than every element of $\mathbf{F}_j$. If $n_j = \max \mathbf{E}_j$ and
$l_j = \min \mathbf{E}_j$, then

$$\frac{n_j + m_j}{l_j + m_j} \leq \frac{n_j}{m_j} + 1 \to 1.$$

Thus, each Hadamard set will meet only a finite number of the sets $\mathbf{F}_j$ in more
than one point. That proves $\mathbf{E}$ cannot be a finite union of Hadamard sets.

But given any number $q > 1$, the assumptions on $\mathbf{E}_j$ and m_j certainly
ensure that there is an index J such that the set $\{m_j\}_{j=J}^\infty \cup (\bigcup_{j=J}^\infty \mathbf{E}_j)$ is
a Hadamard set with ratio at least q. By Proposition 2.2.6, for each $\varepsilon > 0$
there exists J such that $\{m_j\}_{j=J}^\infty \cup (\bigcup_{j=J}^\infty \mathbf{E}_j)$ is an $\varepsilon/2$-Kronecker set. That
$\bigcup_{j=J}^\infty \mathbf{F}_j$ is ε-Kronecker is now a straightforward matter of approximating 1
on the m_j's and φ on each $\mathbf{E}_j$.

A specific instance of this occurs with $\mathbf{E}_j = \{3^{2^{j^2}k} : k = 1, \ldots, j\}$ and
$m_j = 3^{(j+1)2^{j^2}}$.

Other examples of ε-Kronecker sets are the independent sets, defined on
p. 1. Before stating the precise Kronecker properties of independent sets, we
give a very useful lemma whose proof is asked for in Exercise 2.9.4.

Lemma 2.2.8. *Let $\mathbf{E} \subseteq \mathbf{\Gamma}$, $\varepsilon > 0$ and $\Lambda \subseteq \mathbf{\Gamma}$ be a subgroup.*

1. *Let $q : \mathbf{\Gamma} \to \mathbf{\Gamma}/\Lambda$ be the quotient homomorphism. If q is one-to-one on $\mathbf{E}$
   and $q(\mathbf{E})$ is (weak) ε-Kronecker, then $\mathbf{E}$ is (weak) ε-Kronecker.*
2. *If $\mathbf{E} \subseteq \Lambda$ then $\mathbf{E}$ is (weak) ε-Kronecker as a subset of $\mathbf{\Gamma}$ if and only if it
   is (weak) ε-Kronecker as a subset of Λ.*

Proposition 2.2.9. *Let $\mathbf{E} \subseteq \mathbf{\Gamma}$ be independent with all its elements having
order at least N. Then $\mathbf{E}$ is weak angular π/N-Kronecker.*

Proof. Let $\langle \mathbf{E} \rangle$ denote the subgroup of $\mathbf{\Gamma}$ generated by $\mathbf{E}$. For each $\gamma \in \mathbf{E}$,
let $\mathbf{\Gamma}_\gamma = \langle \gamma \rangle$ be the cyclic subgroup of $\mathbf{\Gamma}$ generated by γ. The independence
of $\mathbf{E}$ implies that $\langle \mathbf{E} \rangle = \bigoplus_{\gamma \in E} \mathbf{\Gamma}_\gamma$. Its dual group, $\prod_{\gamma \in \mathbf{E}} \widehat{\mathbf{\Gamma}_\gamma}$, is a quotient of
G. By the lemma above, there is no loss of generality in assuming $\mathbf{\Gamma} = \langle \mathbf{E} \rangle$
and that $G = \prod_{\gamma \in \mathbf{E}} \widehat{\mathbf{\Gamma}_\gamma}$.

Let $\varphi : \mathbf{E} \to [0, 2\pi)$. Since each $\gamma \in \mathbf{E}$ has order at least N, $\mathbf{\Gamma}_\gamma$ is either
$\mathbb{Z}$ or $\mathbb{Z}_n$, the cyclic group of order n, for some $n \geq N$. In either case, there is
some $x_\gamma \in \mathbf{\Gamma}_\gamma$ such that

$$d(\arg \gamma(x_\gamma), \varphi(\gamma)) \leq \pi/N,$$

the worst case occurring when $\Gamma_\gamma = \mathbb{Z}_N$. Let $x_\gamma = e$ if $\gamma \notin \mathbf{E}$ and put $x = (x_\gamma) \in G$.

For $\gamma \in \mathbf{E}$, $d(\arg\gamma(x), \varphi(\gamma)) \leq \pi/N$, and thus $\mathbf{E}$ is weak angular π/N-Kronecker. $\qquad\qquad\qquad\qquad\qquad\qquad\qquad\qquad\qquad\qquad\qquad\qquad\qquad \square$

Corollary 2.2.10. *If $\mathbf{E}$ is independent and all elements of $\mathbf{E}$ have order at least three, then $\mathbf{E}$ is weak 1-Kronecker. If all the elements of $\mathbf{E}$ have infinite order, then $\mathbf{E}$ is ε-Kronecker for all $\varepsilon > 0$.*

Remark 2.2.11. If $G = \prod G_\alpha$ is a product of finite cyclic groups, then the set of projections onto the factor groups G_α is an independent set in Γ. No subset of two or more elements in $\mathbb{Z}$ is independent.

Sets with weaker "independence-like" properties can also be ε-Kronecker, as the next example illustrates.

Example 2.2.12. A non-independent weak 1-Kronecker set: Take $G = \mathbb{Z}_2 \oplus \mathbb{Z}_3^{\mathbb{N}}$ and $\mathbf{E} = \{(\chi, \gamma_n)\} \subset \Gamma$ where χ is not the identity of $\mathbb{Z}_2$ and $\{\gamma_n\}_1^\infty$ is an independent subset in the dual of $\mathbb{Z}_3^{\mathbb{N}}$ (such as the set of projections on the factors). Then $\mathbf{E}$ is not independent (for trivial reasons), but $\mathbf{E}$ is weak 1-Kronecker. See Exercise 2.9.3.

2.2.2 ε-Kronecker Sets Are ε-Kronecker(U)

We now give a very useful result.

Theorem 2.2.13. *Let $\varepsilon > 0$ and suppose $\mathbf{E}$ is a weak ε-Kronecker subset of Γ. Then for each open $U \subseteq G$ there exists a finite set $\mathbf{F}$ such that $\mathbf{E} \smallsetminus \mathbf{F}$ is weak ε-Kronecker(U).*

Remark 2.2.14. In the proof of Theorem 2.2.13 we will here, as elsewhere, identify the set of functions $\varphi : \mathbf{E} \to \mathbb{T}$ with the compact space $\mathbb{T}^{\mathbf{E}}$. That product space has the topology of coordinatewise convergence (see p. 209). It is convenient that this topology is "the same" as the weak* topology for (Fourier-Stieltjes transforms of) bounded subsets of measures on G. Each open $U \subseteq \mathbb{T}^{\mathbf{E}}$ contains a subset of the form $\{\psi\} \times \mathbb{T}^{\mathbf{E} \smallsetminus \mathbf{F}}$, where $\mathbf{F} \subseteq \mathbf{E}$ is finite and $\psi \in \mathbb{T}^{\mathbf{F}}$. Similar conclusions hold for the functions mapping $\mathbf{E} \to X$, where X is any of $[0,1]$, $[-1,1]$, $\{-1,1\}$ or the closed unit ball in $\mathbb{C}$.

Proof (of Theorem 2.2.13). By replacing U with a smaller set we may assume U is compact. Since G is compact, there exist $x_1, \ldots, x_N \in G$ such that $G \subseteq \bigcup_{n=1}^{N} x_n U$. For each n let

$$X_n = \{\varphi : \mathbf{E} \to \mathbb{T} : \exists x \in x_n U \text{ such that } |\varphi(\gamma) - \gamma(x)| \leq \varepsilon \; \forall \gamma \in \mathbf{E}\}.$$

We give $\mathbb{T}^{\mathbf{E}}$ the product topology, making it a compact group, and observe that each X_n is closed in $\mathbb{T}^{\mathbf{E}}$. To see this, suppose that $\varphi_\beta \in X_n$ and that

$u_\beta \in x_n U$ satisfy $|\varphi_\beta(\gamma) - \gamma(u_\beta)| \le \varepsilon$ for all $\gamma \in \mathbf{E}$. Suppose also that $\varphi_\beta \to \varphi$ pointwise on $\mathbf{E}$ (this being the topology on $\mathbb{T}^{\mathbf{E}}$). Let $u \in x_n U$ be an accumulation point of the u_β. Passing to a subnet, we may assume $|\varphi(\gamma) - \gamma(u)| = \lim_\beta |\varphi(\gamma) - \gamma(u_\beta)| \le \varepsilon$, so φ is in X_n.

The fact that $\mathbf{E}$ is weak ε-Kronecker ensures that $\mathbb{T}^{\mathbf{E}} = \bigcup_{n=1}^N X_n$, and hence the Baire category theorem for compact Hausdorff spaces implies there exists n such that X_n has interior in $\mathbb{T}^{\mathbf{E}}$.

By Remark 2.2.14, X_n contains a set of the form $\{\psi\} \times \mathbb{T}^{\mathbf{E} \smallsetminus \mathbf{F}}$ for some finite set $\mathbf{F}$ and $\psi \in \mathbb{T}^{\mathbf{F}}$. Therefore, for every $\varphi : (\mathbf{E} \smallsetminus \mathbf{F}) \to \mathbb{T}$, there exists $u \in x_n U$ such that $|\varphi(\gamma) - \gamma(u)| \le \varepsilon$ for all $\gamma \in \mathbf{E} \smallsetminus \mathbf{F}$. Thus, $\mathbf{E} \smallsetminus \mathbf{F}$ is weak ε-Kronecker$(x_n U)$. By Lemma 2.2.4(2), $\mathbf{E} \smallsetminus \mathbf{F}$ is weak ε-Kronecker(U). $\qquad\square$

As remarked earlier, the property of being ε-Kronecker, for $\varepsilon < 2$, is not preserved under translation. However, the following partial translation result is instructive.

Corollary 2.2.15. *Suppose that $\mathbf{E}$ is weak ε-Kronecker and that $\varepsilon' > \varepsilon$. Let $\gamma \in \Gamma$. Then there is a finite set $\mathbf{F} \subseteq \Gamma$ such that $\gamma(\mathbf{E} \smallsetminus \mathbf{F})$ is ε'-Kronecker.*

Proof. Let $U = \{x \in G : |\gamma(x) - 1| < \varepsilon' - \varepsilon\}$. This is an open set so by Theorem 2.2.13 there is a finite set $\mathbf{F}$ such that $\mathbf{E} \smallsetminus \mathbf{F}$ is weak ε-Kronecker(U).

Given $\varphi : \gamma(\mathbf{E} \smallsetminus \mathbf{F}) \to \mathbb{T}$, define $\psi : \mathbf{E} \smallsetminus \mathbf{F} \to \mathbb{T}$ by $\psi(\chi) = \varphi(\gamma\chi)$. Pick $x \in U$ such that $|\psi(\chi) - \chi(x)| \le \varepsilon$ for all $\chi \in \mathbf{E} \smallsetminus \mathbf{F}$. It is easy to see that

$$|\varphi(\gamma\chi) - \gamma\chi(x)| \le |\psi(\chi) - \chi(x)| + |\chi(x) - \gamma\chi(x)| < \varepsilon'. \qquad\square$$

Remark 2.2.16. If γ has finite order, there is even a finite set $\mathbf{F}$ such that $\gamma(\mathbf{E} \smallsetminus \mathbf{F})$ is weak ε-Kronecker. To see this, just take as U the open set $U = \{x : \gamma(x) = 1\}$ (see Exercise C.4.3 (2)) and repeat the argument.

It is easy to see that if $\mathbf{E} \subset \mathbb{N}$ is a Hadamard set and $m \in \mathbb{N}$, then there exists a two-element set $\mathbf{F}$ such that $(\mathbf{E} \smallsetminus \mathbf{F}) \cup \{m\}$ is Hadamard with the same ratio as $\mathbf{E}$. We have the following analogue for ε-Kronecker sets.

Corollary 2.2.17. *Suppose that $\mathbf{E}$ is weak ε-Kronecker. Assume that $\varepsilon' > \varepsilon$ and that γ has infinite order. Then there is a finite set $\mathbf{F}$ such that $(\mathbf{E} \smallsetminus \mathbf{F}) \cup \{\gamma\}$ is ε'-Kronecker.*

Proof. Fix $0 < \tau < \varepsilon' - \varepsilon$. Since γ has infinite order, $\gamma(G)$ is dense in $\mathbb{T}$. We pick a finite subset $X \subseteq G$ such that for every $t \in \mathbb{T}$ there exists some $x \in X$ with $|\gamma(x) - t| < \tau$. For each $x \in X$, choose a neighbourhood U_x of x such that $|\gamma(x) - \gamma(u)| < \varepsilon$ for all $u \in U_x$. By Theorem 2.2.13, for each $x \in X$, there exists a finite subset $\mathbf{F}_x$ such that $\mathbf{E} \smallsetminus \mathbf{F}_x$ is weak ε-Kronecker(U_x). Let $\mathbf{F}$ be the finite set $\mathbf{F} = \bigcup_{x \in X} \mathbf{F}_x$.

Let $\varphi : (\mathbf{E} \smallsetminus \mathbf{F}) \cup \{\gamma\} \to \mathbb{T}$ and pick $x \in X$ such that $|\varphi(\gamma) - \gamma(x)| < \tau$. Select $u \in U_x$ such that $|\varphi(\chi) - \chi(u)| \leq \varepsilon$ for all $\chi \in \mathbf{E} \smallsetminus \mathbf{F}_x$. In particular, this inequality holds for all $\chi \in \mathbf{E} \smallsetminus \mathbf{F}$. Furthermore,

$$|\varphi(\gamma) - \gamma(u)| \leq |\varphi(\gamma) - \gamma(x)| + |\gamma(x) - \gamma(u)| < \tau + \varepsilon < \varepsilon',$$

and so $(\mathbf{E} \smallsetminus \mathbf{F}) \cup \{\gamma\}$ is ε'-Kronecker. $\square$

2.3 The Relationship Between Kronecker Sets and I_0 Sets

2.3.1 ε-Kronecker Sets Are I_0 if $\varepsilon < \sqrt{2}$

The standard iteration argument, Corollary 1.3.4, immediately shows that ε-Kronecker sets with $\varepsilon < 1$ are I_0. In fact, $\varepsilon < \sqrt{2}$ will suffice, as the next result demonstrates. Example 2.3.6 shows that $\sqrt{2}$ is sharp with this property. In contrast, in Example 2.5.6, we construct an I_0 set that is not a finite union of ε-Kronecker sets for any choice of $\varepsilon < \sqrt{2}$. It is not known if such an example exists in $\mathbb{Z}$ [**P 7**].

Recall that Δ is the closed unit ball in $\mathbb{C}$.

Theorem 2.3.1. *Suppose U is a symmetric e-neighbourhood in G. Let $\varepsilon < \sqrt{2}$ and let $\mathbf{E} \subseteq \mathbf{\Gamma}$ be an ε-Kronecker(U) set. Then*

1. *$\mathbf{E}$ is I_0 and the interpolating measure can be chosen to be positive and concentrated on U.*
2. *$\mathbf{E} \cup -\mathbf{E}$ is I_0 and the interpolating measure can be chosen concentrated on U.*
3. *Furthermore, in (1) (resp., (2)) for each $\varepsilon' > 0$ there exists N, depending only on ε and ε', such that for each $\varphi : \mathbf{E} \to \Delta$ (resp., $\varphi : \mathbf{E} \cup \mathbf{E}^{-1} \to \Delta$) there exists $c_n \in [0,1]$ (resp., $c_n \in \Delta$) and $x_n \in U$ such that $|\varphi(\gamma) - \sum_1^N c_n \gamma(x_n)| < \varepsilon'$ for all $\gamma \in \mathbf{E}$ (resp., $\gamma \in \mathbf{E} \cup \mathbf{E}^{-1}$).*

Remark 2.3.2. In the terminology of Chap. 3, (1) implies that an ε-Kronecker set with $\varepsilon < \sqrt{2}$ is an $FZI_0(U)$ set and item (3) implies that it is $FZI_0(U)$ with bounded length depending only on ε.

Proof (of Theorem 2.3.1). (1) Since $\varepsilon < \sqrt{2}$, $\mathbf{E}$ is angular τ-Kronecker for some $\tau < \pi/2$. We approximate the real and imaginary parts of $\varphi : \mathbf{E} \to \Delta$ separately.

To begin, let $\varphi : \mathbf{E} \to \{-1, 1\}$ be given. The angular τ-Kronecker property gives us an $x \in U$ with $d(\arg \varphi(\gamma), \arg \gamma(x)) < \tau$ for all $\gamma \in \mathbf{E}$. Then elementary trigonometry shows $|\varphi(\gamma) - \mathfrak{Re}(\gamma(x))| < 1 - \cos \tau = \varepsilon^2/2 < 1$.

The standard iteration, Corollary 1.3.4, tells us that given any $\varphi : \mathbf{E} \to$ $[-1,1]$ there exists $\mu \in M_d^+(U)$ with $\varphi(\gamma) = \widehat{\mu}(\gamma)$ on $\mathbf{E}$ and norm depending only on τ.

Now consider $\varphi \in \mathrm{Ball}(\ell^\infty(\mathbf{E}))$ and choose $\mu \in M_d^+(U)$ such that $\widehat{\mu} = \mathfrak{Re}\varphi$ on $\mathbf{E}$. To interpolate the imaginary $i\mathfrak{Im}\varphi(\gamma)$ we choose $u_1 \in U$ such that $d(\arg\gamma(u_1), r_\gamma) < \tau$, where $r_\gamma = \pi/2$ if $\mathfrak{Im}\varphi(\gamma) \geq 0$ and $r_\gamma = -\pi/2$ otherwise. Using the first part of the argument, obtain $\nu \in M_d^+(U)$ such that $\widehat{\nu}(\gamma) = -\mathfrak{Re}\gamma(u_1)$ on $\mathbf{E}$. Then $(\nu + \delta_{u_1^{-1}})/2 \in M_d^+(U)$ and for $\gamma \in \mathbf{E}$

$$\left| \frac{\widehat{\nu + \delta_{u_1^{-1}}}(\gamma)}{2} - i\mathfrak{Im}\varphi(\gamma) \right| = \left| \frac{\mathfrak{Im}\gamma(u_1)}{2} - \mathfrak{Im}\varphi(\gamma) \right| \leq 1 - \frac{\cos\tau}{2}$$

since $\mathfrak{Im}\gamma(u_1)/2 \in \pm[(\cos\tau)/2, 1/2]$, depending on the sign of $\mathfrak{Im}\varphi(\gamma)$. The measure $\mu_0 = \mu + (\nu + \delta_{u_1^{-1}})/2 \in M_d^+(U)$ and satisfies $|\varphi(\gamma) - \widehat{\mu_0}(\gamma)|$ $\leq 1 - (\cos\tau)/2$ on $\mathbf{E}$. An application of Proposition 1.3.2 (1) will complete the proof.

(2) Let $\varphi \in \mathrm{Ball}(\ell^\infty(\mathbf{E} \cup \mathbf{E}^{-1}))$. We may assume that φ is real. We use the preceding to find $\mu, \nu \in M_d^+(U)$ such that

$$\widehat{\mu} = \varphi \text{ on } \mathbf{E} \text{ and } \widehat{\nu} = i\varphi \text{ on } \mathbf{E}.$$

Let $\omega_1 = \frac{1}{2}(\mu - i\nu)$. Then $\widehat{\omega_1} = \varphi$ on $\mathbf{E}$, but $\widehat{\omega_1} = 0$ on $\mathbf{E}^{-1}$. Similarly we can find $\omega_2 \in M_d(U)$ such that $\widehat{\omega_2} = \varphi$ on $\mathbf{E}^{-1}$ and $\widehat{\omega_2} = 0$ on $\mathbf{E}$. Then $\omega_1 + \omega_2 \in M_d(U)$ and $\widehat{\omega_1} + \widehat{\omega_2} = \varphi$ on $\mathbf{E} \cup \mathbf{E}^{-1}$.

(3) is Exercise 2.9.7.   □

Remark 2.3.3. The norm of the interpolating measure in both parts of the theorem depends only on ε. This observation, together with Theorem 2.2.13, proves the following corollary:

Corollary 2.3.4. *Suppose $\varepsilon < \sqrt{2}$ and $\mathbf{E}$ is an ε-Kronecker set. There exists $C = C(\varepsilon)$ such that for every non-empty, open $U \subseteq G$ there is a finite set $L = L(U)$ such that every element of $\mathrm{Ball}(\ell^\infty(\mathbf{E} \smallsetminus L))$ can be interpolated by $\mu \in M_d^+(U)$ and $\|\mu\| \leq C(\varepsilon)$.*

2.3.2 An Example of a $\sqrt{2}$-Kronecker Set That Is a Sidon Set but Not I_0

The following lemma shows, in particular, that any two-element set of positive integers is $\sqrt{2}$-Kronecker.

Lemma 2.3.5. *Let $1 \le m < n < \infty$. Let I be a closed interval of length at least $4\pi/m$ and $w \ne z \in \mathbb{T}$. There is a closed subinterval $J \subseteq I$ of length at least $\pi/(8mn)$, such that for all $\theta \in J$ both*

$$|e^{im\theta} - w| < \sqrt{2} \text{ and } |e^{in\theta} - z| < \sqrt{2}.$$

Proof. Since the function $x \mapsto e^{imx}$ is $2\pi/m$-periodic, the length of I ensures that there is some $y \in I$ having $e^{imy} = w$ and $y + \theta \in I$ whenever $\theta \in [-\pi/m, \pi/m]$. Let $\beta = \arg(e^{iny}\bar{z})$.

If $\beta \in [0, \pi/2]$, take $J = [y - \frac{4\pi}{9n}, y - \frac{\pi}{9n}] \subseteq I$. If $x \in J$, then $x = y - \theta$ for some $\theta \in [\frac{\pi}{9n}, \frac{4\pi}{9n}]$, and since $m/n \le 1$ and $n\theta < \pi/2$,

$$|e^{imx} - w| \le |e^{i4\pi/9} - 1| < \sqrt{2}.$$

Also, since $\beta - n\theta \in (-\pi/2, \pi/2)$, it follows that

$$|e^{inx} - z| = |e^{i\beta}e^{-in\theta} - 1| < \sqrt{2}.$$

If $\beta \in [\pi/2, \pi]$, take $J = \{y - \frac{\pi}{2n}(1 + \theta) : \theta \in [\frac{1}{4m}, \frac{1}{2m}]\} \subseteq I$. Thus, if $x = y - \frac{\pi}{2n}(1 + \theta) \in J$,

$$|e^{imx} - w| = |e^{-\frac{im\pi}{2n}(1+\theta)} - 1| < \sqrt{2}.$$

Since $\beta - \frac{\pi}{2}(1+\theta) \in (-\pi/2, \pi/2)$, we have $|e^{inx} - z| = |e^{i\beta}e^{-i\pi(1+\theta)/2} - 1| < \sqrt{2}$. The cases $\beta \in [-\pi, -\pi/2]$ and $[-\pi/2, 0]$ are similar. □

Example 2.3.6. A $\sqrt{2}$-Kronecker set that is Sidon but not I_0: Let $\mathbf{E}$ be the set $\bigcup\limits_{j=1}^{\infty} \{N_j, N_j + j\}$ where $N_1 = 2$ and the N_j are chosen inductively so that

$$\frac{4\pi}{N_{j+1}} < \frac{\pi}{8N_j(N_j + j)}.$$

Let $\varphi : \mathbf{E} \to \mathbb{T}$ be given, say $\varphi(N_j) = w_j$ and $\varphi(N_j + j) = z_j$. According to Lemma 2.3.5, there is a closed interval J_1 of length $\pi/(8N_1(N_1+1)) = \pi/48$ such that for all $x \in J_1$,

$$|e^{iN_1 x} - w_1| < \sqrt{2} \text{ and } |e^{i(N_1+1)x} - z_1| < \sqrt{2}.$$

Because length $J_1 > 4\pi/N_2$, there is a closed subinterval $J_2 \subseteq J_1$ of length at least $\pi/(8N_2(N_2+2))$ and such that for all $x \in J_2$, $|e^{iN_2 x} - w_2| < \sqrt{2}$ and $|e^{i(N_2+2)x} - z_2| < \sqrt{2}$. Proceed inductively to find closed nested intervals J_k such that for all $x \in J_k$,

$$|e^{iN_k x} - w_k| < \sqrt{2} \text{ and } |e^{i(N_k+k)x} - z_k| < \sqrt{2}.$$

Then every $x \in \bigcap_k J_k$ will satisfy $|\varphi(n) - e^{inx}| < \sqrt{2}$ for all $n \in \mathbf{E}$, which shows that $\mathbf{E}$ is at least $\sqrt{2}$-Kronecker.

The two subsets, $\{N_j\}_j$ and $\{N_j + j\}_j$, are both Hadamard and therefore I_0 (Theorem 1.3.9), which implies Sidon. Since a finite union of Sidon sets is Sidon, as we shall see in Corollary 6.3.3, $\mathbf{E}$ is Sidon.

But $\mathbf{E}$ is not I_0 by the same reasoning as given in Example 1.5.2. The disjoint subsets $\{N_j\}$ and $\{N_j + j\}$ do not have disjoint closures in the Bohr compactification of $\mathbb{Z}$, so their union cannot be I_0 (see Corollary 3.4.3). Since it is not I_0, this also shows $\mathbf{E}$ can be no better than $\sqrt{2}$-Kronecker because of Theorem 2.3.1.

2.4 Presence of ε-Kronecker Sets

It is natural to ask which groups Γ contain large ε-Kronecker sets for any (or all) $\varepsilon < \sqrt{2}$. Of course, if Γ contains only elements of order two, then it does not contain an ε-Kronecker set with $\varepsilon \leq \sqrt{2}$ since characters of order two can only take on the values ± 1. Similarly, if Γ contains infinitely many elements of order two, then not all infinite subsets will contain infinite $\sqrt{2}$-Kronecker sets.

In this section it will be shown that provided the subset $\mathbf{E}$ does not have "too many" elements of order 2 (as defined below), then $\mathbf{E}$ will contain a weak 1-Kronecker set of the same cardinality. Algebraic methods will be used to prove this.

Definition 2.4.1. Let Γ_2 be the subgroup generated by the elements of order 2 in Γ and q_2 the quotient mapping $\Gamma \to \Gamma/\Gamma_2$. The set $\mathbf{E} \subseteq \Gamma$ is said to be *2-large* if $|q_2(\mathbf{E})| < |\mathbf{E}|$.

Remark 2.4.2. If $\mathbf{E} \subseteq \Gamma$ is an infinite, independent set that is not 2-large, then the subset of $\mathbf{E}$ consisting of the elements of order 3 or more has the same cardinality as $\mathbf{E}$. It is weak 1-Kronecker by Corollary 2.2.10.

Here is the main existence result.

Theorem 2.4.3. *Let $\mathbf{E} \subseteq \Gamma$ be infinite and not 2-large. Then there exists a weak 1-Kronecker set $\mathbf{F} \subseteq \mathbf{E}$ such that $|\mathbf{F}| = |\mathbf{E}|$.*

To prove Theorem 2.4.3, we will consider several different discrete abelian groups. The proof proper begins on p. 30.

2.4.1 Two Countable Groups: $\mathcal{C}(p^\infty)$ and $\mathbb{Q}$

Let p be a prime and denote by $\mathcal{C}(p^\infty)$ the discrete p-subgroup of $\mathbb{T}$, that is, the group of all $p^n th$-roots of unity. An important classical fact in group theory is that every abelian group is isomorphic to a subgroup of

$$\bigoplus_{\alpha} \mathbb{Q}_{\alpha} \oplus \bigoplus_{\beta} \mathcal{C}(p_{\beta}^{\infty}), \tag{2.4.1}$$

where $\mathbb{Q}_{\alpha}$ are copies of the rationals [165, Theorem 10.30]. We begin with $\mathbb{Q}$ and $\mathcal{C}(p^{\infty})$.

Proposition 2.4.4. *Let $\varepsilon > 0$. Each infinite subset of $\mathcal{C}(p^{\infty})$, or $\mathbb{Q}$, contains an infinite ε-Kronecker set.*

Proof. First, suppose $\mathbf{E}$ is an infinite subset of $\mathcal{C}(p^{\infty})$. Note that each integer, m, defines a character on $\mathcal{C}(p^{\infty})$ by the rule, $e^{2\pi ik/p^n} \to e^{2\pi ikm/p^n}$. Each element of the coset $m + p^n \mathbb{Z}$ acts in the same way on the $p^n th$-roots of unity.

Choose N so large that $|e^{2\pi i/p^N} - 1| < \varepsilon/2$. Take $\gamma_j \in \mathbf{E}$ so that $\gamma_j = e^{2\pi ik_j/p^{n_j}}$ where $n_1 \geq N$, $n_{j+1} - n_j \geq N$ for $j \geq 1$ and $1 \leq k_j < p$. We will prove that the set $\{\gamma_j\}$ is ε-Kronecker.

Let $t_j \in \mathbb{T}$. Since k_1 and p^{n_1} are coprime, the set $\{e^{2\pi ik_1 m/p^{n_1}} : m \in \mathbb{Z}\}$ consists of all the $p^{n_1}th$ roots of unity and consequently there is an integer m_1 such that

$$|e^{2\pi ik_1 m_1/p^{n_1}} - t_1| \leq |e^{2\pi i/p^{n_1}} - 1| < \varepsilon/2.$$

Of course, the same inequality is obtained if m_1 is replaced by any element of the coset $m_1 + p^{n_1}\mathbb{Z}$. By similar arguments, we can choose integer m_2 with

$$|e^{2\pi ik_2(m_1 + m_2 p^{n_1})/p^{n_2}} - t_2| \leq |e^{2\pi ik_2 m_2/p^{n_2 - n_1}} - t_2 e^{-2\pi ik_2 m_1/p^{n_2}}| < \varepsilon/2,$$

as well as

$$|e^{2\pi ik_1(m_1 + m_2 p^{n_1})/p^{n_1}} - t_1| < \varepsilon/2.$$

Again, the same inequalities are obtained if m_2 is replaced by any element of the coset $m_2 + p^{n_2}\mathbb{Z}$. Continuing in this fashion produces a sequence of integers, m_j', with

$$|e^{2\pi ik_\ell m_j'/p^{n_\ell}} - t_\ell| < \varepsilon/2$$

for all $\ell \leq j$. If g is an accumulation point of $\{m_j'\}$ in the compact dual group of $\mathcal{C}(p^{\infty})$, then $|\gamma_j(g) - t_j| \leq \varepsilon/2$, and hence $\{\gamma_j\}$ is ε-Kronecker.

Now consider the case when the discrete dual group is $\mathbb{Q}$. Choose q so large that every Hadamard set in $\mathbb{Q}$ or $\mathbb{R}$, with ratio at least q is $\varepsilon/3$-Kronecker. (See Exercise 2.9.13.)

If $\mathbf{E} \subseteq \mathbb{Q}$ is an unbounded set of real numbers, then we can find a subset $\{\gamma_j\} \subseteq \mathbf{E}$ with $\gamma_{j+1}/\gamma_j \geq q$ for all j, and such a set is ε-Kronecker. Otherwise, $\mathbf{E}$ contains a sequence $\{\gamma_j\}$ with limit $r \in \mathbb{R}$, in the usual topology of $\mathbb{R}$. By passing to a further subsequence, if necessary, we can assume $(\gamma_{j+1} - r)/(\gamma_j - r) \leq 1/q$. Then each finite set $\{\gamma_n - r, \gamma_{n-1} - r, \ldots, \gamma_1 - r\} \subseteq \mathbb{R}$ is Hadamard with ratio at least q, and thus is $\varepsilon/3$-Kronecker. It is easy to see (Exercise 2.9.20) that the Kronecker constant of a set is the supremum of the Kronecker constants of its finite subsets. From that we conclude that $\mathbf{F} = \{\gamma_j - r : j \geq 1\}$ is weak $\varepsilon/3$-Kronecker in the dual, $\widehat{\mathbb{R}_d}$, of $\mathbb{R}$ with the discrete topology. By Corollary 2.2.15, there exists a finite set $\mathbf{F}'$ such that $(\mathbf{F} \setminus \mathbf{F}') + r$ is $\varepsilon/2$-Kronecker. This last set is in $\mathbb{Q}$, so it is $\varepsilon/2$-Kronecker there, as well, by Lemma 2.2.8. □

2.4.2 Proof of Theorem 2.4.3

Let $\mathbf{\Gamma}_0$ denote the torsion subgroup of $\mathbf{\Gamma}$, that is, the largest subgroup of $\mathbf{\Gamma}$ consisting only of elements of finite order, and let $q : \mathbf{\Gamma} \to \mathbf{\Gamma}/\mathbf{\Gamma}_0$ be the natural quotient homomorphism. Then $\mathbf{\Gamma}/\mathbf{\Gamma}_0$ has no non-trivial elements of finite order.

Case I: $\mathfrak{e} := |\mathbf{E}| = |q(\mathbf{E})|$

We may assume that $\mathbf{\Gamma}$ has no elements of finite order since by Lemma 2.2.8, if we find an ε-Kronecker subset of $q(\mathbf{E})$ we may lift it to an ε-Kronecker subset of $\mathbf{E}$. By (2.4.1), we may also assume that $\mathbf{\Gamma}$ is a subgroup of $\bigoplus_{\ell \in B} \mathbb{Q}_\ell$, where the $\mathbb{Q}_\ell$ are copies of the rational numbers and for every ℓ there is an element $\gamma \in \mathbf{E}$ such that the projection, $\Pi_\ell(\gamma)$, onto $\mathbb{Q}_\ell$ is non-trivial. For $\gamma \in \mathbf{E}$, let

$$B(\gamma) = \{\ell \in B : \Pi_\ell(\gamma) \neq 0\}. \tag{2.4.2}$$

Each $B(\gamma)$ is finite and $B = \bigcup_{\gamma \in \mathbf{E}} B(\gamma)$.

Case Ia: Countable E

If some $\Pi_\ell(\mathbf{E})$ is infinite, then Lemma 2.2.8 and Proposition 2.4.4 imply that, for each $\varepsilon > 0$, $\mathbf{E}$ has an ε-Kronecker subset with the same cardinality as $\mathbf{E}$. Otherwise, we may inductively find $\gamma_j \in \mathbf{E}$ such that $B(\gamma_{j+1}) \not\subseteq \bigcup_{k=1}^{j} B(\gamma_k)$. Because the γ_j have infinite order, it is immediate that $\{\gamma_j\}_{j=1}^{\infty}$ is an independent set and hence is ε-Kronecker for all $\varepsilon > 0$ by Corollary 2.2.10.

Case Ib: Uncountable E

Let $\mathcal{S}$ be the set of subsets of $\mathbf{E}$ that are independent and partially order $\mathcal{S}$ by inclusion. Since independence is characterized by properties of finite subsets, it follows that every chain in $\mathcal{S}$ has an upper bound, namely the union of the sets in the chain. Use Zorn's lemma to find a maximal independent subset,

$\mathbf{F}$, of $\mathbf{E}$. Such a set will contain only elements of infinite order and we claim there will be $\mathfrak{c}$ of them.

To count the elements of $\mathbf{F}$, let $\mathbf{H} = \langle \mathbf{F} \rangle$, the group generated by $\mathbf{F}$, so $|\mathbf{H}| \leq \aleph_0 |\mathbf{F}|$. Observe that if $\chi \in \mathbf{E}$ is non-trivial, then by the maximality of $\mathbf{F}$, there will be a positive integer m such that $\chi^m \in \mathbf{H}$. Let $\mathbf{H}_m = \{\chi \in \mathbf{E} : \chi^m \in \mathbf{H}\}$. The map $\chi \to \chi^m$ is one-to-one, and thus $|\mathbf{H}_m| \leq |\mathbf{H}| \leq \aleph_0 |\mathbf{F}|$. Because $\mathbf{E} \subseteq \cup_{m=1}^{\infty} \mathbf{H}_m \cup \{\mathbf{1}\}$, it follows that $|\mathbf{E}| \leq \aleph_0 |\mathbf{F}|$. Thus, $|\mathbf{F}| \geq |\mathbf{E}|$.

Corollary 2.2.10 completes the proof, as in the countable subcase.

Case II: $|q(\mathbf{E})| < \mathfrak{c}$

Without loss of generality, we can assume that $\mathbf{E}$ generates Γ. Let $\mathbf{F}'$ be a maximal independent subset of Γ consisting of elements of infinite order. We claim that $|\mathbf{F}'| < \mathfrak{c}$. Indeed, q maps $\mathbf{F}'$ one-to-one onto an independent set of elements of infinite order since otherwise we would have elements $\gamma_j \in \mathbf{F}'$ and integers $L \geq 1$ and $\ell_j \neq 0$ such that $\prod_1^L \gamma_j^{\ell_j} \in \Gamma_0$, and so for some $m > 0$, we would have $\prod_1^L \gamma_j^{m\ell_j} = 1$, a contradiction. If $|\mathbf{F}'| = \mathfrak{c}$, then $|q(\mathbf{E})| = |q(\Gamma)| \geq |\mathbf{F}'| = \mathfrak{c}$, which we have assumed is not the case.

Then cardinal arithmetic tells us that $|\Gamma/\langle \mathbf{F}' \rangle| = \mathfrak{c}$, and the maximality of $\mathbf{F}'$ implies that $\Gamma/\langle \mathbf{F}' \rangle$ is a torsion group. Using Lemma 2.2.8, we see that it will suffice to find a weak 1-Kronecker subset of $q(\mathbf{E})$, and so we may assume that Γ is a torsion group.

By (2.4.1), we may assume that Γ is a subgroup of $\bigoplus_{\ell \in B} \mathcal{C}(p_\ell^\infty)$, where the index set B has the property that for every $\ell \in B$ there is an element $\gamma \in \mathbf{E}$ such that the projection $\Pi_\ell(\gamma)$ onto $\mathcal{C}(p_\ell^\infty)$ is non-trivial. Furthermore, there must be $\mathfrak{c}$ indices ℓ such that $\Pi_\ell(\mathbf{E})$ contains an element of order at least 3, for otherwise (it is a routine exercise to see that) $\mathbf{E}$ would be 2-large.

We continue to use the notation of (2.4.2). We will use induction if $\mathbf{E}$ is countable and transfinite induction if $\mathbf{E}$ is uncountable. The reader will see that the argument is identical, whether the induction is transfinite or not.

Let $\mathcal{I}$ be a well-ordered index set of cardinality $|B|$, with $1, 2, \ldots$ the first elements of $\mathcal{I}$. Since $\mathbf{E}$ is not 2-large, $\mathbf{E}$ must contain an element of order at least 3. Let $\lambda_1 \in \mathbf{E}$ and $\ell(1) \in B$ be such that the order of $\Pi_{\ell(1)}(\lambda_1)$ is at least 3. That starts the induction.

Suppose $i > 1$ and that we have found $\lambda_{i'} \in \mathbf{E}$ for all $1 \leq i' < i$ such that $B(\lambda_{i'}) \not\subseteq \bigcup_{k<i'} B(\lambda_k)$. If $|\{\lambda_{i'} : 1 \leq i' < i\}| = \mathfrak{c}$, we stop. Otherwise, we note that $A = \bigcup_{i'<i} B(\lambda_{i'})$ also has cardinality less than $\mathfrak{c}$ and that there exist $\lambda(i) \in \mathbf{E}$ and $\ell(i) \in B$ such that $\ell(i) \notin \bigcup_{i'<i} B(\lambda_{\ell(i')})$ and $\Pi_{\ell(i)}(\lambda_i)$ has order at least 3. That completes the inductive step.

Because there are $\mathfrak{c}$ indices ℓ such that $\Pi_\ell(\mathbf{E})$ contains an element of order at least 3, the set $\mathbf{F} = \{\lambda_{\ell(i)} : i \geq 1\}$ must have the same cardinality $\mathfrak{c}$. We now claim that $\mathbf{F}$ is weak 1-Kronecker. Again we use induction, transfinite or not, depending on the cardinality of $\mathbf{E}$. It will be convenient to assume that $G = \prod_{\ell \in B} G_\ell$, where G_ℓ is the dual of $\mathcal{C}(p_\ell^\infty)$, $\ell \in B$. This assumption is justified by Lemma 2.2.8. We shall abuse notation by using $\Pi_\ell(x)$ to denote the ℓ-coordinate of x for $x \in G$.

Let $\varphi : \mathbf{F} \to \mathbb{T}$. Because $\Pi_{\ell(1)}(\lambda_1)$ has order at least 3, it is possible to choose $x_1 \in G_{\ell(1)}$ such that $|\varphi(\lambda_1) - \lambda_1(x_1)| \leq 1$. Suppose now that $i > 1$ and $x_{i'} \in \prod_{k \leq i'} G_{\ell(k)}$ have been chosen for $1 \leq i' < i$ so that $|\varphi(\lambda_k) - \lambda_k(x_{i'})| \leq 1$ and

$$\Pi_{\ell(k)}(x_k) = \Pi_{\ell(k)}(x_{i'}), \tag{2.4.3}$$

whenever $1 \leq k \leq i' < i$.

If i has an immediate predecessor, i', we choose $x \in G_{\ell(i)}$ such that $|\varphi(\lambda_i) - \lambda_i(x_{i'}x)| \leq 1$. Set $x_i = x_{i'}x$. Then (2.4.3) holds for $1 \leq k \leq i' \leq i$.

If i is a limit ordinal, let x_0 be the limit point of the $x_{i'}$ as $i' \to i$. Such a limit point exists because of (2.4.3). This ensures (2.4.3) holds with x_0 in place of $x_{i'}$. Now choose $x \in G_{\ell(i)}$ such that $|\varphi(\lambda_i) - \lambda_i(x_0 x)| \leq 1$ and set $x_i = x_0 x$. It is clear that (2.4.3) now holds with $i' = i$. Finally, let $z = \lim_i x_i$. Then, $|\varphi(\gamma) - \gamma(z)| \leq 1$, so $\mathbf{F}$ is indeed weak 1-Kronecker. $\square$

In Case I, note that more was actually proved.

Corollary 2.4.5. *Suppose Γ_0 is the torsion subgroup of Γ and $q : \Gamma \to \Gamma/\Gamma_0$ is the quotient map. If $\mathbf{E}$ is infinite and $|q(\mathbf{E})| = |\mathbf{E}|$, then for each $\varepsilon > 0$ and neighbourhood $U \subseteq G$ there is an ε-Kronecker(U) set $\mathbf{F} \subseteq \mathbf{E}$, with the same cardinality as $\mathbf{E}$.*

Proof. The proof of the Theorem case I shows the existence of ε-Kronecker subsets, for any specified $\varepsilon > 0$, of the same cardinality as $\mathbf{E}$. To obtain the ε-Kronecker(U) set, just discard a suitable finite subset. $\square$

Corollary 2.4.6. *Suppose the torsion subgroup of Γ is finite. If $\mathbf{E}$ is an infinite set, then, for each $\varepsilon > 0$ and neighbourhood $U \subseteq G$, there is an ε-Kronecker(U) set $\mathbf{F} \subseteq \mathbf{E}$, with the same cardinality as $\mathbf{E}$.*

2.5 Approximating Arbitrary Choices of ±1

It can be seen from the improved standard iteration, Corollary 1.3.3, that each bounded function on $\mathbf{E}$ can be (exactly) interpolated with the Fourier–Stieltjes transform of a discrete measure provided it is possible to approximately interpolate all ±1-valued functions on $\mathbf{E}$. But having this approximation property is clearly not sufficient to ensure that $\mathbf{E}$ is ε-Kronecker for some $\varepsilon < \sqrt{2}$; just consider, for instance, the set of Rademacher functions in $\widehat{\mathbb{D}}$.

One could ask if there are conditions (perhaps on G or Γ) that would ensure that the approximation of arbitrary choices of signs is enough to guarantee the set $\mathbf{E}$ is ε-Kronecker. In this section, we give such criteria and consider related questions. Later, in Sect. 4.4, we consider the case where arbitrary signs can be interpolated exactly but outside of the ε-Kronecker context.

Let G_2 be the annihilator of the subgroup, $\mathbf{\Gamma}^{(2)}$, of all characters whose order is a power of 2. Since G_2 is the dual of the quotient group $\mathbf{\Gamma}/\mathbf{\Gamma}^{(2)}$, which has no elements of order two, Lemma C.1.15 implies every element of G_2 has a square root. This will be significant in what follows.

Theorem 2.5.1. *Let* $\mathbf{E} \subseteq \mathbf{\Gamma}$ *and* $\varepsilon > 0$. *Suppose that for all choices of signs* $\{r_\gamma\}_{\gamma \in \mathbf{E}} \in \mathbb{Z}_2^{\mathbf{E}}$ *there exists an element* $g \in G_2$ *such that* $d(\arg \gamma(g), \arg r_\gamma) < \varepsilon$ *for all* $\gamma \in \mathbf{E}$. *Then* $\mathbf{E}$ *is weak angular* 2ε-*Kronecker.*

Proof. The key step in the proof is to show that for each positive integer k and all choices of angles, $\{s_\gamma\}_{\gamma \in \mathbf{E}}$, which are the arguments of 2^kth roots of unity, there exists some $x_k \in G_2$ such that

$$d(\arg \gamma(x_k), s_\gamma) < (2 - 2^{-k+1})\varepsilon \text{ for all } \gamma \in \mathbf{E}. \tag{2.5.1}$$

This will be proven by an induction argument.

Suppose that (2.5.1) has been established. Fix $\varepsilon' > 2\varepsilon$ and choose k such that $\pi 2^{-k} - \varepsilon 2^{-k+1} + 2\varepsilon < \varepsilon'$. Since the angular distance between two adjacent 2^kth roots of unity is $2\pi/2^k$, for each selection $\{\theta_\gamma\}_{\gamma \in E} \in [0, 2\pi)^{\mathbf{E}}$ we can choose arguments of 2^kth roots of unity, $\{s_\gamma\}_{\gamma \in \mathbf{E}}$, such that $d(\theta_\gamma, s_\gamma) \leq \pi 2^{-k}$ for all γ. With x_k chosen by (2.5.1) and $\gamma \in \mathbf{E}$,

$$d(\arg \gamma(x_k), \theta_\gamma) \leq d(s_\gamma, \theta_\gamma) + d(\arg \gamma(x_k), s_\gamma)$$
$$< \pi 2^{-k} - \varepsilon 2^{-k+1} + 2\varepsilon < \varepsilon'.$$

Consequently, $\mathbf{E}$ is weak angular 2ε-Kronecker.

It only remains to verify (2.5.1). Since ± 1 are the square roots of unity, (2.5.1) holds for $k = 1$ by the hypothesis of the theorem. Proceed by induction and assume the induction assumption is true for k. Let $\{s_\gamma\}_{\gamma \in \mathbf{E}}$ be arguments of the 2^{k+1}th roots of unity and consider $\{2s_\gamma\}$, the arguments of 2^kth roots of unity. By the induction assumption, there is some $x_k \in G_2$ such that for all $\gamma \in \mathbf{E}$, $d(\arg \gamma(x_k), 2s_\gamma) < (2 - 2^{-k+1})\varepsilon$. Since every element of G_2 is a square, there is some $y \in G_2$ such that $y^2 = x_k$. Then $\gamma(y)^2 = \gamma(x_k)$, and hence the argument of $\gamma(y)$ is either equal to $\arg \gamma(x_k)/2$ or $\arg \gamma(x_k)/2 + \pi$. Thus, for all $\gamma \in \mathbf{E}$, either

$$d(\arg \gamma(y), s_\gamma) < (1 - 2^{-k})\varepsilon$$

or

$$d(\arg \gamma(y), s_\gamma + \pi) < (1 - 2^{-k})\varepsilon.$$

In the first case, put $r_\gamma = 1$ and in the second case, put $r_\gamma = -1$. According to the hypothesis of the theorem, there is some $g \in G_2$ such that

$$d(\arg \gamma(g), \arg r_\gamma) < \varepsilon \text{ for all } \gamma \in \mathbf{E}.$$

Let $x_{k+1} = gy \in G_2$. Since $\arg r_\gamma$ is either 0 or π, it follows that either

$$d(\arg \gamma(x_{k+1}), s_\gamma) \leq d(\arg \gamma(y), s_\gamma) + d(\arg \gamma(g), 0)$$
$$< (1 - 2^{-k})\varepsilon + \varepsilon = (2 - 2^{-k})\varepsilon,$$

or

$$d(\arg \gamma(x_{k+1}), s_\gamma) \leq d(\arg \gamma(y), s_\gamma + \pi) + d(\arg \gamma(g), \pi)$$
$$< (1 - 2^{-k})\varepsilon + \varepsilon = (2 - 2^{-k})\varepsilon,$$

depending on whether it is the first or second case, and that completes the induction step. □

Remark 2.5.2. Suppose $G = \mathbb{T} = [-\pi, \pi)$ and assume the choice of $g \in G_2 = \mathbb{T}$ in the hypothesis of the theorem can be chosen from an open interval U that is centred at 0. Then $\mathbf{E}$ is weak angular 2ε-Kronecker$(\overline{2U})$. To see this, argue as in the proof of the theorem but assume in the induction step that the points x_k can be chosen belonging to $(2 - 2^{-k+1})U$. This is true by assumption for $k = 1$. For the induction step, note that we can simply choose $y = x_k/2 \in (1 - 2^{-k})U$, and then $x_{k+1} = y + g \in (1 - 2^{-k})U + U \subseteq (2 - 2^{-k})U \subseteq 2U$.

An easy corollary follows from the theorem.

Corollary 2.5.3. *Assume that Γ has no elements of order two and that for some $\tau < \pi/4$ and all choices of signs $\{r_\gamma\}_{\gamma \in \mathbf{E}}$, there exists $g \in G$ such that $d(\arg \gamma(g), \arg r_\gamma) < \tau$ for all $\gamma \in \mathbf{E}$. Then $\mathbf{E}$ is weak angular ε-Kronecker for some $\varepsilon < \pi/2$.*

Proof. Since Γ has no elements of order two, $G = G_2$. □

In particular, if G is connected, then Γ has no elements of finite order, and hence the corollary applies.

Theorem 2.5.1 can be generalized to the situation where an arbitrary choice of signs is replaced by an arbitrary choice of (fixed) two elements in $\mathbb{T}$ with angular distance π. The proof is asked for in Exercise 2.9.17.

Theorem 2.5.4. *Let $\mathbf{E} \subseteq \Gamma$, $\varepsilon > 0$ and $\theta \in [0, \pi)$. Suppose that for each $\{r_\gamma\}_{\gamma \in \mathbf{E}} \subseteq \{\theta, \theta + \pi\}^{\mathbf{E}}$ there exists an element $g \in G_2$ such that $d(\arg \gamma(g), \arg r_\gamma) < \varepsilon$ for all $\gamma \in \mathbf{E}$. Then $\mathbf{E}$ is weak angular 2ε-Kronecker.*

Theorem 2.5.4 provides a geometric separation condition which ensures a set is ε-Kronecker.

Corollary 2.5.5. *Suppose there are two disjoint intervals, $I, J \subseteq \mathbb{T}$, each with arc length $l < \pi$. Assume that for each $\mathbf{F} \subseteq \mathbf{E}$ there exists an element $g \in G_2$ such that $\gamma(g) \in I$ for all $\gamma \in \mathbf{F}$ and $\gamma(g) \in J$ for all $\gamma \in \mathbf{E} \setminus \mathbf{F}$. Then $\mathbf{E}$ is weak angular $(\pi - m)$-Kronecker, where m is the arc length of the smaller of the two gaps separating I and J.*

Proof. Let θ and $\theta + \pi$ be the two points of distance $\pi/2$ from the centre of the smaller of the two gaps. Since $m \leq \pi$, by symmetry (and without loss of generality), the angular distance from each point in interval I (respectively, interval J) to θ (resp., $\theta + \pi$) is at most $(\pi - m)/2$. The conclusion follows from Theorem 2.5.4. $\qquad\square$

It is not enough for the set $\mathbf{E}$, itself, to contain no elements of order two for the approximation of arbitrary choices of signs (even exactly) to ensure that $\mathbf{E}$ is ε-Kronecker for some $\varepsilon < \sqrt{2}$, as Example 2.5.6 shows. The set there is also I_0, but not a finite union of ε-Kronecker sets for any $\varepsilon < \sqrt{2}$.

Example 2.5.6. An I_0 set that is not a finite union of ε-Kronecker sets: Let $\mathbf{E} = \{(j, \pi_j) : j = 1, 2, \ldots\} \subseteq \mathbb{Z} \times \widehat{\mathbb{D}}$, where $\{\pi_j\}$ is the Rademacher set in $\widehat{\mathbb{D}}$. Notice $\mathbf{E}$ contains no elements of order two, although the subgroup it generates clearly does. Of course, it is possible to interpolate an arbitrary choice of signs, $\{r_j\}$, exactly on $\mathbf{E}$; just take the point $(0, x)$ where $\pi_j(x) = r_j$. By the standard iteration (applied to the real and imaginary parts of candidate φ's), this property is enough to ensure that $\mathbf{E}$ is I_0.

But $\mathbf{E}$ is not a finite union of ε-Kronecker sets for any $\varepsilon < \sqrt{2}$. To prove this, assume that $\mathbf{E}$ were such a union. Then one of the finitely many sets would contain a net $\{(j_\beta, \pi_{j_\beta})\}$ with $j_\beta \to 0$ in the Bohr topology on $\mathbb{Z}$. Furthermore, because the subset $\{(j_\beta, \pi_{j_\beta})\}_\beta$ is (assumed to be) ε-Kronecker, there would be (x, y) such that

$$|(j_\beta, \pi_{j_\beta})(x, y) - i| = |e^{i2\pi j_\beta x} \pi_{j_\beta}(y) - i| < \varepsilon < \sqrt{2} \text{ for all } \beta. \qquad (2.5.2)$$

As $j_\beta \to 0$, given a $\delta > 0$, there is some β such that $|e^{i2\pi j_\beta x} - 1| < \delta$. Since $\pi_{j_\beta}(y)$ is either ± 1, the inequalities (2.5.2) cannot simultaneously hold for small enough δ.

However, $\mathbf{E}$ is $\sqrt{2}$-Kronecker. The argument is similar to the proof that $\mathbb{Z} \smallsetminus \{0\}$ is 2-Kronecker; see Exercise 2.9.2.

2.6 Arithmetic Properties of ε-Kronecker Sets

2.6.1 Are ε-Kronecker Sets Sidon?

All ε-Kronecker sets with $\varepsilon < \sqrt{2}$ are I_0 and hence Sidon. The situation for $\varepsilon \in [\sqrt{2}, 2)$ is less clear. Example 2.3.6 shows that a Sidon set can be $\sqrt{2}$-Kronecker and not I_0. We do not know if *all* $\sqrt{2}$-Kronecker sets are Sidon, much less all ε-Kronecker sets with $\varepsilon \in [\sqrt{2}, 2)$ [**P 4**]. Since the non-Sidon set, $\mathbb{Z} \smallsetminus \{0\}$, is 2-Kronecker (Remarks 2.2.5) the question is settled for $\varepsilon = 2$.

In this section, various arithmetic properties of ε-Kronecker sets, even for $\varepsilon \in [\sqrt{2}, 2)$, are established. In Sect. 6.3.2 it will be seen that Sidon sets also possess these properties, and thus the results that follow can be taken as evidence for an affirmative answer to the question of the section title, for ε in that range.

We now turn to the above-promised arithmetic properties.

Definition 2.6.1. The set $\mathbf{P}$ is called a *parallelepiped of dimension N* if $\mathbf{P} = \prod_{j=1}^{N}\{\chi_j, \gamma_j\}$, where χ_j, $\gamma_j \in \mathbf{\Gamma}$ and $|\mathbf{P}| = 2^N$. The characters χ_j, γ_j, for $j = 1, \ldots, N$ (which need not be distinct), are called the *generators* of $\mathbf{P}$.

An example of a parallelepiped of dimension N is an arithmetic progression in $\mathbb{Z}$ of length 2^N. Indeed, $\{a, a+d, \ldots, a+(2^N-1)d\} = \{a, a+d\} + \sum_{j=1}^{N-1}\{0, 2^j d\}$.

Like a Sidon set (Corollary 6.3.13), an ε-Kronecker set can only contain a small portion of each long arithmetic progression.

Theorem 2.6.2. *Suppose $\mathbf{E}$ is ε-Kronecker for some $\varepsilon < 2$. For every $\tau > 0$ there is a constant $C = C(\tau, \varepsilon)$ such that $|\mathbf{E} \cap \mathbf{P}| \leq C 2^{N\tau}$ for each parallelepiped $\mathbf{P} \subseteq \mathbf{\Gamma}$ of dimension N.*

Proof. We will prove that there exists a constant $N_0 = N_0(\varepsilon, \tau)$ such that if $\mathbf{P}$ is a parallelepiped of dimension N and $\mathbf{E}_1$ is a subset of $\mathbf{E} \cap \mathbf{P}$ of cardinality $2^{N\tau}$, then $N \leq N_0(\varepsilon, \tau)$. That will prove $|\mathbf{E} \cap \mathbf{P}| < 2^{N\tau}$ for all parallelepipeds of dimension N when $N > N_0$ and therefore we will be able to take $C = 2^{N_0}$. Being ε-Kronecker for some $\varepsilon < 2$, $\mathbf{E}$ is angular $(\pi - \delta)$-Kronecker for some $\delta > 0$. Fix an even integer $M > \pi/\delta$ and consider X, the set of all functions mapping $\mathbf{E}_1$ to T_{MN}, the group of MNth roots of unity. There are $(MN)^{|\mathbf{E}_1|}$ such functions. We will call a function in X *multiplicative* if it is the restriction to $\mathbf{E}_1$ of a function defined on the generators of $\mathbf{P}$ and extended by multiplicativity to $\mathbf{P}$. There are at most $(MN)^{2N}$ multiplicative functions in X.

Temporarily fix a multiplicative function Φ. If h is an arbitrary function in X and $|h(\gamma) - \Phi(\gamma)| < 2$ for every $\gamma \in \mathbf{E}_1$, then $h(\gamma) \neq -\Phi(\gamma)$ for all $\gamma \in \mathbf{E}_1$. Since M is even, $-\Phi(\gamma)$ is another MNth root of unity. Hence, there can be at most $(MN - 1)^{|\mathbf{E}_1|}$ functions in X whose distance to the function Φ is strictly less than 2 and a total of at most $(MN-1)^{|\mathbf{E}_1|}(MN)^{2N}$ functions in X with distance less than 2 to some multiplicative function in X.

It can be verified that if N is sufficiently large and $|\mathbf{E}_1| \geq 2^{\tau N}$, then

$$(MN)^{|\mathbf{E}_1|} > (MN - 1)^{|\mathbf{E}_1|}(MN)^{2N}.$$

The strict inequality proves there is some function $h \in X$ whose distance to every multiplicative function in X is equal to 2.

Since $\mathbf{E}$ is angular $(\pi - \delta)$-Kronecker, there is some $x \in G$ such that $d(\arg h(\gamma), \arg \gamma(x)) < \pi - \delta$ for every $\gamma \in \mathbf{E}_1$. Define a multiplicative function Φ on the generators of $\mathbf{P}$ by choosing $\Phi(\gamma)$ to be the MNth root of unity closest to $\gamma(x)$. Since every character in $\mathbf{P}$ is the product of N generators, $d(\arg \Phi(\gamma), \arg \gamma(x)) < \pi/M$ for all $\gamma \in \mathbf{P}$. But then

$$d(\arg h(\gamma), \arg \Phi(\gamma)) \leq d(\arg h(\gamma), \arg \gamma(x)) + d(\arg \Phi(\gamma), \arg \gamma(x))$$
$$< \pi - \delta + \pi/M < \pi.$$

This is a contradiction since the angular distance from h to each multiplicative function in X is π.   □

Corollary 2.6.3. *Suppose* $\mathbf{E} \subseteq \mathbb{Z}$ *is an ε-Kronecker set and* $\mathbf{A}$ *is an arithmetic progression of length N. For each $\tau > 0$ there is a constant C, depending only on ε and τ, such that* $|\mathbf{E} \cap \mathbf{A}| \leq CN^\tau$.

A *square* in Γ is a set of the form $\mathbf{E}_1 \cdot \mathbf{E}_2 \subseteq \Gamma$, where $|\mathbf{E}_1 \cdot \mathbf{E}_2| = |\mathbf{E}_1| |\mathbf{E}_2|$ and $|\mathbf{E}_1| = |\mathbf{E}_2|$. The same argument as above, viewing the characters in $\mathbf{E}_1$ and $\mathbf{E}_2$ as generators of the square, shows that ε-Kronecker sets do not contain arbitrarily large squares. The details are left to Exercise 2.9.10. That is another property possessed by Sidon sets (Proposition 6.3.12).

Proposition 2.6.4. *Suppose* $\mathbf{E}$ *is ε-Kronecker for some $\varepsilon < 2$. There is a constant $N = N(\varepsilon)$ such that $\mathbf{E}$ does not contain a square of cardinality N^2.*

We now turn to the sum of reciprocals of elements of $\mathbf{E}$.

Proposition 2.6.5. *Let* $\{k_j\}_{j=1}^J \subseteq \mathbb{N}$ *be increasing and assume $\gamma \in \Gamma$ has order exceeding k_J. If $\sum_{j=1}^J 1/k_j = s$, then $\alpha(\{\gamma^{k_j}\}_{j=1}^J) \geq \pi(1 - s^{-1})$.*

The following corollary for subsets of $\mathbb{N}$ is immediate from Proposition 2.6.5 and the fact that $\log J \leq \sum_{k=1}^J 1/k$.

Corollary 2.6.6. $\alpha(\{1, \ldots, J\}) \geq \pi(1 - 1/\log J)$.

Proof (of Proposition 2.6.5). Let $\mathbf{E} = \{\gamma^{k_j}\}_{j=1}^J$ and $\delta = \pi/s$. For a positive integer k and $z \in \mathbb{T}$, consider

$$W(k, z) = \bigcup_{\ell=0}^{k-1} \{w \in \mathbb{T} : d(\arg(ze^{2\pi i\ell/k}), \arg w) \leq \delta/k\}.$$

Then $W(k, z)$ consists of k arcs each of (angular) length $2\delta/k$, centred at elements $ze^{2\pi i\ell/k}$, $0 \leq \ell < k$. A sketch of the unit circle may be helpful here.

Choose z_1 so that $W(k_1, z_1) \supseteq \{w : 0 \leq \arg w \leq 2\delta/k_1\}$. Choose $z_2, \ldots, z_J \in \mathbb{T}$ inductively such that

$$W(k_j, z_j) \supseteq \left\{ w : \sum_{\ell=1}^{j-1} 2\delta/k_\ell \leq \arg w \leq \sum_{\ell=1}^{j} 2\delta/k_\ell \right\} \text{ for } 2 \leq j \leq J.$$

Then the hypothesis, $\sum_1^J 1/k_j = s$, ensures $\bigcup_{j=1}^J W(k_j, z_j) = \mathbb{T}$.

Define $\varphi : \mathbf{E} \to \mathbb{T}$ by $\varphi(\gamma^{k_j}) = -z_j^{k_j}$ (the point antipodal to $z_j^{k_j}$) for $1 \leq j \leq J$. We claim $\sup_{\lambda \in \mathbf{E}} d\big(\arg(\varphi(\lambda)), \arg(\lambda(x)) \big) \geq \pi - \delta$ for all $x \in G$. Indeed, for every $x \in G$ there exists $1 \leq j \leq J$ such that $\gamma(x) \in W(k_j, z_j)$. Then $d\big(\arg(z_j), \arg(\gamma(x)) \big) \leq \delta/k_j$, so $d\big(\arg(\gamma^{k_j}(x)), \arg(z_j^{k_j}) \big) \leq \delta$ and

$$d\big(\arg(\varphi(\gamma^{k_j})), \arg(\gamma^{k_j}(x)) \big) \geq \pi - \delta. \qquad \square$$

2.7 Products of ε-Kronecker Sets Are "Small"

In this section we study the small size or "thinness" of products of an ε-Kronecker set, as was done for Hadamard sets in Chap. 1. Not clustering at a continuous character is one form of thinness, addressed in our first section. Another form of thinness is the closure of a sum of copies of $\mathbf{E}$ having zero Haar measure. That is discussed in the second section.

2.7.1 Bohr Cluster Points of Kronecker Sets and Their Products

We begin by proving that an ε-Kronecker set does not cluster in the Bohr topology (see Sect. C.1.3) at a continuous character, a special case of the fact that an I_0 set does not cluster at a continuous character (the Ryll–Nardzewski–Méla–Ramsey Theorem 3.5.1). A more elementary proof can be given for ε-Kronecker sets.

It is unknown if a Sidon set can cluster at a continuous character. That problem will be discussed in more detail in Chap. 8.

Proposition 2.7.1. *An ε-Kronecker set $\mathbf{E}$ does not cluster in the Bohr topology at any $\gamma \in \mathbf{\Gamma}$, if $\varepsilon < 2$.*

Proof. Let $\tau = (2 - \varepsilon)/2 > 0$. Choose an e-neighbourhood $U \subseteq G$ such that $|1 - \gamma(u)| < \tau$ for all $u \in U$. Let $\mathbf{F} \subseteq \mathbf{E}$ be finite such that $\mathbf{E} \smallsetminus \mathbf{F}$ is weak ε-Kronecker(U) and let $u \in U$ be such that $|\chi(u) + 1| < \varepsilon$ for all $\chi \in \mathbf{E} \smallsetminus \mathbf{F}$. Then, $|\gamma(u) - \chi(u)| \geq 2 - \varepsilon - \tau \geq (2 - \varepsilon)/2$ for all $\chi \in \mathbf{E} \smallsetminus \mathbf{F}$, so $\mathbf{E} \smallsetminus \mathbf{F}$ does not cluster at γ in the Bohr topology. $\qquad \square$

The next proposition is an elaboration of the idea in the proof of the Kunen–Rudin Theorem 1.5.1. It applies to Hadamard sets by Theorem 1.5.1, ε-Kronecker sets with $\varepsilon < 2$ by Proposition 2.7.1 and to all I_0 sets by the Ryll–Nardzewski–Méla–Ramsey Theorem 3.5.1.

Proposition 2.7.2. *Suppose $\mathbf{E} \subseteq \mathbf{\Gamma}$ has no Bohr cluster points in $\mathbf{\Gamma}$. The following are equivalent:*

1. *The only element of* $\boldsymbol{\Gamma}$ *which is a Bohr cluster point of* $\mathbf{E} \cdot \mathbf{E}^{-1}$ *is* $\mathbf{1}$.
2. $\gamma \overline{\mathbf{E}} \cap \rho \overline{\mathbf{E}} \subseteq \boldsymbol{\Gamma}$ *for every* $\gamma \neq \rho \in \boldsymbol{\Gamma}$.
3. $\gamma \overline{\mathbf{E}} \cap \rho \overline{\mathbf{E}}$ *is a finite subset of* $\boldsymbol{\Gamma}$ *for every* $\gamma \neq \rho \in \boldsymbol{\Gamma}$.
4. $\big(\gamma (\overline{\mathbf{E}} \smallsetminus \mathbf{E}) \big) \cap (\overline{\mathbf{E}} \smallsetminus \mathbf{E}) = \emptyset$ *for every* $\gamma \neq \mathbf{1} \in \boldsymbol{\Gamma}$.

Similarly, the following are equivalent:

5. *No element of* $\boldsymbol{\Gamma}$ *is a cluster point of* $\mathbf{E} \cdot \mathbf{E}$.
6. $\gamma \overline{\mathbf{E}} \cap \overline{\mathbf{E}}^{-1} \subseteq \boldsymbol{\Gamma}$ *for every* $\gamma \in \boldsymbol{\Gamma}$.
7. $\big(\gamma (\overline{\mathbf{E}} \smallsetminus \mathbf{E}) \big) \cap (\overline{\mathbf{E}} \smallsetminus \mathbf{E})^{-1} = \emptyset$ *for every* $\gamma \in \boldsymbol{\Gamma}$.

Remark 2.7.3. The ρ in (2)–(3) is superfluous, but it will be convenient to have the particular formulation later.

Proof. (1) $\Rightarrow$ (2) Suppose there is some character $\lambda \in \gamma \overline{\mathbf{E}} \cap \rho \overline{\mathbf{E}}$ with $\lambda \in \overline{\boldsymbol{\Gamma}} \smallsetminus \boldsymbol{\Gamma}$. Then λ is a cluster point of nets $\{\gamma \chi_\beta\}$ and $\{\rho \psi_\beta\}$, with $\{\chi_\beta\}, \{\psi_\beta\} \subseteq \mathbf{E}$. Since multiplication is jointly continuous in $\overline{\boldsymbol{\Gamma}}$, $\gamma \rho^{-1} \in \boldsymbol{\Gamma}$ is a cluster point of the net $\{\chi_\beta^{-1} \psi_\beta\}$ and thus of $\mathbf{E} \cdot \mathbf{E}^{-1}$. By (1), $\gamma = \rho$.

(2) $\Rightarrow$ (3) If $\gamma \overline{\mathbf{E}} \cap \rho \overline{\mathbf{E}}$ were an infinite set, it would have a Bohr cluster point. That cluster point could not be a continuous character because of the first assumption of the lemma. Therefore, $\gamma \overline{\mathbf{E}} \cap \rho \overline{\mathbf{E}}$ would not be contained in $\boldsymbol{\Gamma}$, contradicting (2).

(3) $\Rightarrow$ (4) Suppose $\chi \in \gamma (\overline{\mathbf{E}} \smallsetminus \mathbf{E}) \cap (\overline{\mathbf{E}} \smallsetminus \mathbf{E})$. Since χ is a cluster point of $\mathbf{E}$, χ is not in $\boldsymbol{\Gamma}$. That ensures $\gamma \overline{\mathbf{E}} \cap \overline{\mathbf{E}}$ is not a subset of $\boldsymbol{\Gamma}$, and hence (3) implies $\gamma = \mathbf{1}$.

(4) $\Rightarrow$ (1) Suppose $\mathbf{E} \cdot \mathbf{E}^{-1}$ has cluster point $\gamma \in \boldsymbol{\Gamma} \smallsetminus \{\mathbf{1}\}$. Then there are nets $\{\lambda_\beta\}, \{\rho_\beta\} \subseteq \mathbf{E}$ such that $\lambda_\beta \rho_\beta^{-1} \to \gamma$. Without loss of generality we may assume that $\{\rho_\beta\}$ converges to a character $\zeta \in \overline{\boldsymbol{\Gamma}} \smallsetminus \boldsymbol{\Gamma}$. But then $\lambda_\beta \to \gamma \zeta$. Thus, $\zeta \in \gamma^{-1}(\overline{\mathbf{E}} \smallsetminus \mathbf{E}) \cap (\overline{\mathbf{E}} \smallsetminus \mathbf{E})$. Since $\gamma \neq \mathbf{1}$ this contradicts (4).

The proof of the equivalences of (5)–(7) is similar (Exercise 2.9.19).   □

The next result is an ε-Kronecker set form of the Kunen–Rudin Theorem 1.5.1.

Proposition 2.7.4. *Let* $\varepsilon < \sqrt{2}$ *and* $\mathbf{E}$ *be an* ε-Kronecker set. Then

1. $\mathbf{1}$ *is the only continuous character at which* $\mathbf{E} \cdot \mathbf{E}^{-1}$ *clusters.*
2. $\mathbf{E} \cdot \mathbf{E}$ *does not cluster at a continuous character.*

Proof. Let $C > 0$ be given by Corollary 2.3.4.

(1) Let $\gamma \in \boldsymbol{\Gamma}$, $\gamma \neq \mathbf{1}$. Then Proposition 2.7.2(1)–(4) imply it will suffice to show that $\big(\gamma (\overline{\mathbf{E}} \smallsetminus \mathbf{E}) \big) \cap (\overline{\mathbf{E}} \smallsetminus \mathbf{E}) = \emptyset$.

Let $x \in G$ be such that $|1 - \gamma(x)| \geq 1$ and let $U = \{u \in G : |\gamma(u) - \gamma(x)| < 1/(4C)\}$. Now let $L = L(U)$ (whose existence is guaranteed by Corollary 2.3.4) be a finite set such that for every function $\varphi : \mathbf{E} \to \Delta$ there exists $\mu \in M_d^+(U)$ with $\varphi = \hat{\mu}$ on $\mathbf{E} \smallsetminus L$ and $\|\mu\| \leq C$.

Set $\varphi = 1$ on $\mathbf{E}$ and obtain $\mu \in M_d^+(U)$ with $\widehat{\mu} = \varphi$ on $\mathbf{E} \smallsetminus \mathrm{L}$. Since $\widehat{\mu}(\rho) = \varphi(\rho) = 1$ for every $\rho \in \mathbf{E} \smallsetminus \mathrm{L}$ we have

$$|\gamma(x) - \widehat{\mu}(\gamma\rho)| = \left| \int_U \rho(u)(\gamma(x) - \gamma(u))\mathrm{d}\mu \right| \leq \|\mu\|/(4C) = 1/4.$$

This shows $|1 - \widehat{\mu}(\gamma\rho)| \geq 3/4$ and therefore, $\widehat{\mu}(\gamma(\mathbf{E} \smallsetminus \mathrm{L}))$ and $\widehat{\mu}(\mathbf{E} \smallsetminus \mathrm{L})$ have disjoint closures. Consequently, $(\gamma(\overline{\mathbf{E}} \smallsetminus \mathbf{E})) \cap (\overline{\mathbf{E}} \smallsetminus \mathbf{E}) = \emptyset$.

(2) In view of Proposition 2.7.2(5)–(7), it will suffice to show that $(\gamma(\overline{\mathbf{E}} \smallsetminus \mathbf{E})) \cap (\overline{\mathbf{E}} \smallsetminus \mathbf{E})^{-1} = \emptyset$ for all $\gamma \in \Gamma$. Let $U = \{x \in G : |1 - \gamma(x)| < 1/(4C)\}$ and let $\mathrm{L} = \mathrm{L}(U)$. Choose $\varphi = i$ on $\mathbf{E}$ and $\mu \in M_d^+(U)$ with $\widehat{\mu} = \varphi$ on $\mathbf{E} \smallsetminus \mathrm{L}$. Then, $|i - \widehat{\mu}(\gamma\rho)| \leq C/(4C)$ for all $\rho \in \mathbf{E} \smallsetminus \mathrm{L}$. Since $\mu \in M_d^+(U)$, we have $\widehat{\mu} = -i$ on $(\mathbf{E} \smallsetminus \mathrm{L})^{-1}$ and so $\gamma(\overline{\mathbf{E}} \smallsetminus \mathbf{E}) \cap (\overline{\mathbf{E}} \smallsetminus \mathbf{E})^{-1} = \emptyset$.   □

Remarks 2.7.5. (i) With smaller ε, one can have more terms in the products in Proposition 2.7.4. See the section Remarks and Credits for more information.

(ii) It is essential to have $\varepsilon < \sqrt{2}$ in the theorem. Consider the set $\mathbf{E}$ of Example 2.3.6. Then $\mathbf{E} - \mathbf{E} = \mathbb{Z}$, so its Bohr cluster points include all of $\mathbb{Z}$. Replacing the $N_j + j$ terms with $-N_j + j$, one obtains (Exercise 2.9.6) a $\sqrt{2}$-Kronecker set with $\mathbf{E} + \mathbf{E}$ clustering at every point of $\mathbb{Z}$ since it contains the semigroup $\mathbb{N}$ and a compact semigroup is a group (Exercise C.4.16).

In an ordered group, such as $\mathbb{Z}$, we can measure the "step lengths" of an increasing sequence $\{n_j\}$ as the difference $n_{j+1} - n_j$. In the general case, we have the following definition.

Definition 2.7.6. The *step length of* $\mathbf{E} \subseteq \Gamma$ *tends to infinity* if for every finite set $\mathbf{F'} \subseteq \Gamma$, there exists a finite set $\mathbf{F} \subseteq \mathbf{E}$ such that

$$\gamma\chi^{-1} \notin \mathbf{F'} \text{ if } \gamma, \chi \in \mathbf{E} \smallsetminus \mathbf{F}, \gamma \neq \chi.$$

Hadamard sets obviously have step length tending to infinity. On the other hand, any set which is a union of an infinite set and a translate of that set does not. It will be shown later that every Sidon set is a finite union of sets whose step length tends to infinity (Corollary 6.4.7).

With the previous results we can show that ε-Kronecker sets have step length tending to infinity. In fact, a more general result is true.

Proposition 2.7.7. *Suppose* $\mathbf{E} \subseteq \Gamma$ *has no Bohr cluster points in* Γ *and that* $\gamma\overline{\mathbf{E}} \cap \overline{\mathbf{E}} \subseteq \Gamma$ *for all* $\gamma \neq 1$. *Then* $\mathbf{E}$ *has step length tending to infinity.*

Proof. If $\mathbf{E}$ does not have step length tending to infinity, then by definition there is a finite set $\mathbf{F'}$ such that for all finite $\mathbf{F}$, $(\mathbf{E} \smallsetminus \mathbf{F}) \cdot (\mathbf{E} \smallsetminus \mathbf{F})^{-1} \cap \mathbf{F'} \neq \emptyset$. It follows that there exists $\chi \in \mathbf{F'}$ and distinct $\gamma_n, \rho_n \in \mathbf{E}$ such that $\gamma_n \rho_n^{-1} = \chi$

for $n \geq 1$. In particular, $\rho_n \in (\chi^{-1}\mathbf{E}) \cap \mathbf{E}$ for $n \geq 1$. Of course, every cluster point of $\{\rho_n\}$ belongs to $(\chi^{-1}\overline{\mathbf{E}}) \cap \overline{\mathbf{E}}$ and cannot be a continuous character. But that contradicts the second hypothesis of the proposition.   □

A stronger version of the following corollary is in Exercise 2.9.12.

Corollary 2.7.8. *Suppose that $\varepsilon < \sqrt{2}$ and that $\mathbf{E} \subseteq \Gamma$ is ε-Kronecker. Then the step length of $\mathbf{E}$ tends to infinity.*

Proof. Immediate from the preceding results of this section.   □

Corollary 2.7.9. *Let $\varepsilon < \sqrt{2}$ and $\mathbf{E}$ be ε-Kronecker. Then for every finite set $\mathbf{F} \subseteq \Gamma$, $\mathbf{E} \cdot \mathbf{F}$ is I_0.*

Proof. Let $\mathbf{E}, \mathbf{F}$ be as above. Proposition 2.7.2 and Proposition 2.7.4 imply that for each $\gamma \neq \rho \in \Gamma$, $\gamma \overline{\mathbf{E}} \cap \rho \overline{\mathbf{E}}$ is finite. Hence, there is a finite set $\mathbf{H}_{\gamma,\rho}$ such that $\gamma \overline{(\mathbf{E} \smallsetminus \mathbf{H}_{\gamma,\rho})} \cap \rho \overline{(\mathbf{E} \smallsetminus \mathbf{H}_{\gamma,\rho})} = \emptyset$. Let $\mathbf{H} = \bigcup_{\gamma \neq \rho \in \mathbf{F}} \mathbf{H}_{\gamma,\rho}$. Then $\mathbf{H}$ is finite. Since the sets $\gamma \overline{(\mathbf{E} \smallsetminus \mathbf{H})}$ and $\rho \overline{(\mathbf{E} \smallsetminus \mathbf{H})}$ are disjoint when $\gamma \neq \rho$, the local units theorem, Theorem C.1.6, applied to those subsets of $\overline{\Gamma}$ implies that there exists, for each $\gamma \in \mathbf{F}$, a discrete measure $\nu_\gamma \in M_d(G)$ such that

$$\widehat{\nu_\gamma}(\lambda) = \begin{cases} 1 & \text{if } \lambda \in \gamma \overline{(\mathbf{E} \smallsetminus \mathbf{H})} \text{ and} \\ 0 & \text{if } \lambda \in \rho \overline{(\mathbf{E} \smallsetminus \mathbf{H})} \text{ for some } \rho \in \mathbf{F}, \rho \neq \gamma. \end{cases}$$

Now let $\varphi \in \ell^\infty(\mathbf{F} \cdot (\mathbf{E} \smallsetminus \mathbf{H}))$. Since the class of I_0 sets is clearly closed under translation, $\gamma(\mathbf{E} \smallsetminus \mathbf{H})$ is I_0 for each $\gamma \in \mathbf{F}$. Hence, there exists $\mu_\gamma \in M_d(G)$ with $\widehat{\mu_\gamma} = \varphi$ on $\gamma(\mathbf{E} \smallsetminus \mathbf{H})$. Then $\sum_{\gamma \in \mathbf{F}} \nu_\gamma * \mu_\gamma$ has Fourier–Stieltjes transform equal to φ on $\mathbf{F} \cdot (\mathbf{E} \smallsetminus \mathbf{H})$. Since none of the translates of $\mathbf{E} \smallsetminus \mathbf{H}$ cluster in Γ, $\mathbf{F} \cdot (\mathbf{E} \smallsetminus \mathbf{H})$ does not cluster at an element of $\mathbf{F} \cdot \mathbf{H}$. Therefore, we again use the local identity theorem, this time to add the points in $\mathbf{F} \cdot \mathbf{H}$.   □

Remark 2.7.10. Corollary 2.7.9 is an instance of Proposition 5.2.7.

2.7.2 U_0 Sets and the Closure of Products

For a compact set X of a locally compact abelian group, $M_0(X)$ denotes the set of regular, bounded, Borel measures μ supported on X, such that the Fourier–Stieltjes transform of μ vanishes at infinity on the dual group. See Lemma C.1.9 for facts about $M_0(X)$.

Definition 2.7.11. A set X is called a *set of uniqueness in the weak sense* (U_0 *set*) if $M_0(X) = \{0\}$.

The Riemann–Lebesgue lemma shows a U_0 set has zero $\overline{\Gamma}$-Haar measure.

An interpretation of Proposition 1.4.3 is that there exists an angular π/M-Kronecker set $\mathbf{E} \subseteq \mathbb{Z}$ such that $(\mathbf{E} \cup \mathbf{E}^{-1})^{M+2}$ is dense in $\overline{\mathbb{Z}}$. If the Kronecker constant is halved, we obtain a strong converse, namely, the product is U_0.

Theorem 2.7.12. *Suppose that M is a positive integer and $\mathbf{E} \subseteq \Gamma$ is an angular τ-Kronecker set with $\tau < \pi/2M$. Then the closure in the Bohr topology of $(\mathbf{E} \cup \mathbf{E}^{-1})^M$ is a U_0 set.*

Corollary 2.7.13. *Suppose $\mathbf{E}$ is Hadamard with ratio $q > 5$. Then $\overline{\mathbf{E} + \mathbf{E}}$ and $\overline{\mathbf{E} - \mathbf{E}}$ are U_0 sets in $\overline{\mathbb{Z}}$.*

Proof (of Theorem 2.7.12). We proceed by induction on M. To begin, let $\mu \in M_0(\overline{\mathbf{E}} \cup \overline{\mathbf{E}}^{-1})$. Since $\mu|_{\overline{\mathbf{E}}}$ and $\mu|_{\overline{\mathbf{E}}^{-1}}$ are absolutely continuous with respect to μ, Lemma C.1.9 implies it will be enough to show both $\overline{\mathbf{E}}$ and $\overline{\mathbf{E}}^{-1}$ are U_0 sets. Since there are no non-zero point mass measures in M_0, Lemma C.1.9 also implies that $\mu(\mathbf{F}) = 0$ for all finite sets $\mathbf{F}$. Finally, that lemma ensures that there is no loss of generality in assuming $\mu \in M_0(\overline{\mathbf{E}})$ is a probability measure.

For each positive integer k choose a finite set, $\mathbf{F}_k \subseteq \mathbf{E}$, of cardinality k. For each $\varphi \in \{0, \pi\}^{\mathbf{F}_k}$ use the angular τ-Kronecker property of $\mathbf{E}$ to choose $x \in G$ such that $|\arg \gamma(x)| \leq \tau < \pi/2$ for all $\gamma \in \mathbf{E} \smallsetminus \mathbf{F}_k$, or, equivalently,

$$\mathfrak{Re}\gamma(x) \geq \cos\tau > 0 \text{ for all } \gamma \in \mathbf{E} \smallsetminus \mathbf{F}_k \text{ and}$$

$$|\arg \gamma(x) - \varphi(\gamma)| \leq \tau \text{ for all } \gamma \in \mathbf{F}_k.$$

Since $\overline{\mathbf{E}} = (\overline{\mathbf{E} \smallsetminus \mathbf{F}_k}) \bigcup \mathbf{F}_k$ and μ is a probability measure supported on $\overline{\mathbf{E} \smallsetminus \mathbf{F}_k}$, this produces 2^k distinct $x \in G$ such that

$$\mathfrak{Re}\widehat{\mu}(x) = \mathfrak{Re}\left(\int_{\overline{\mathbf{E}}} \gamma(x)\mathrm{d}\mu(\gamma) \right) = \int_{\overline{\mathbf{E} \smallsetminus \mathbf{F}_k}} \mathfrak{Re}\gamma(x)\mathrm{d}\mu(\gamma) \geq \cos\tau > 0.$$

Because k is arbitrary, that proves $\widehat{\mu}$ does not vanish at infinity on G_d, a contradiction.

Now proceed inductively and assume the result holds for some $M \geq 1$. Let $\mathbf{E}$ be an angular τ-Kronecker set for $\tau < \pi/2(M+1)$. Since $(\overline{\mathbf{E}} \bigcup \overline{\mathbf{E}}^{-1})^{M+1}$ is a finite union of sets of the form

$$\overbrace{\overline{\mathbf{E}}^{\pm 1} \cdots \overline{\mathbf{E}}^{\pm 1}}^{M+1},$$

another application of Lemma C.1.9 shows it is enough to prove each of these sets is U_0. For notational convenience, we will assume all signs are $+1$; it will be clear that the choice of signs is essentially irrelevant to the proof.

So let μ be a probability measure in $M_0(\overline{\mathbf{E}}^{M+1})$. Notice that for every finite set $\mathbf{F}$,

$$\overline{\mathbf{E}}^{M+1} = \overline{(\mathbf{E} \smallsetminus \mathbf{F})}^{M+1} \cup \left(\bigcup_{j=1}^{M} \overline{\mathbf{E}}^j \cdot \mathbf{F}^{M-j+1} \right) \cup \mathbf{F}^{M+1}.$$

By the inductive assumption, $\nu(\overline{\mathbf{E}}^j) = 0$ for all $0 \leq j \leq M$ whenever $\nu \in M_0$. Since the translate of a measure in M_0 is again in M_0, it follows that $\mu(\rho\overline{\mathbf{E}}^j) = 0$ for all $0 \leq j \leq M$ and characters ρ. Thus, μ is supported on $\overline{(\mathbf{E} \smallsetminus \mathbf{F})}^{M+1}$.

Now we argue as in the case $M = 1$. For each positive integer k choose a finite set, $\mathbf{F}_k \subseteq \mathbf{E}$, of cardinality k. For each $\varphi \in \{0, \pi\}^{\mathbf{F}_k}$ use the angular τ-Kronecker property of $\mathbf{E}$ to choose $x \in G$ such that

$$|\arg \gamma(x)| \leq \tau < \pi/2(M+1) \text{ for all } \gamma \in \mathbf{E} \smallsetminus \mathbf{F}_k \text{ and}$$
$$|\arg \gamma(x) - \varphi(\gamma)| \leq \tau \text{ for all } \gamma \in \mathbf{F}_k.$$

This produces 2^k distinct $x \in G$ such that

$$|\arg \rho(x)| \leq \tau(M+1) < \pi/2 \text{ for all } \rho \in \left(\overline{\mathbf{E} \smallsetminus \mathbf{F}_k}\right)^{M+1}.$$

Thus

$$\mathfrak{Re}\widehat{\mu}(x) = \mathfrak{Re}\left(\int_{\overline{\mathbf{E}}^{M+1}} \gamma(x) \mathrm{d}\mu(\gamma)\right)$$
$$= \int_{(\overline{\mathbf{E} \smallsetminus \mathbf{F}_k})^{M+1}} \mathfrak{Re}\gamma(x) \mathrm{d}\mu(\gamma) \geq \cos \tau > 0.$$

We derive the same contradiction as before.   □

2.8 Remarks and Credits

Definition and Properties. Unless otherwise indicated, the results in this chapter are from [51–53, 55, 59, 61].

ε-Kronecker sets seem to have first appeared in Kahane's exposition [102, p. 226] of Varopoulos's tensor algebra work, though they are not named. The term "ε-Kronecker set" appears first in Varopoulos [187], and such sets were studied by Givens and Kunen [46], who used the term "ε-free set". Kronecker sets (and, more generally, independent sets) have generated an extensive literature; cf. [56].

ε-Kronecker sets can be defined for non-discrete Γ. Most of the results of this chapter (with obvious modifications) hold when Γ is metrizable (not merely discrete), but difficulties arise when Γ is not metrizable. See Sect. A.1 for details.

Kronecker's classical approximation theorem states that if $h_1, \ldots, h_n$ are rationally independent real numbers and $\theta_1, \ldots \theta_n$ are arbitrary real numbers, then given any $\varepsilon > 0$ there is some real number t such that $|h_j t - \theta_j| < \varepsilon$ mod 1 for all $j = 1, \ldots, n$. Thus, any finite set of rationally independent real numbers is Kronecker. For further discussion, see [87, pp. 435–436].

Example 2.2.7 is a descendent of one in [89, 178, 179], which can also be found in [167, 5.7.6]. Other examples may be found in [59, 64, 66].

A Hadamard set with ratio q is angular τ-Kronecker with $\tau \leq \pi/(q-1)$. This is not much use for small q and we do not know if every Hadamard set is ε-Kronecker for some $\varepsilon < 2$ **[P 3]**, though our examples show that this information, by itself, would be of limited use.

ε-Kronecker and I_0 Sets. That 1-Kronecker sets are I_0 was certainly known to Kahane when he wrote [102], if not to Varopoulos. That ε-Kronecker sets are I_0 for $\varepsilon < \sqrt{2}$ is in [52].

Lemma 2.3.5 is an example of a more general phenomena. Finite sets of positive integers are always ε-Kronecker for some $\varepsilon < 2$. In [76] an extensive investigation is made of the Kronecker constants of finite subsets of $\mathbb{Z}$ and an algorithm is given for calculating these constants. For a two integer set, $\{m, n\}$, the angular Kronecker constant is $\pi \gcd(m, n)/(|m| + |n|)$. For sets of three or more elements the answers are surprisingly complicated. For instance, asymptotically, the angular Kronecker constant of $\{m, m, n+m\}$ is $\pi/3$, but the exact value depends on the congruence mod 3 of $m + 2n$. It is also shown in [76] that the angular Kronecker constant of a finite $\mathbf{E} \subseteq \mathbb{Z}$ is always a rational multiple of π. The exact Kronecker constant of most sets is unknown, in particular, that of $\{1, \ldots, N\}$ **[P 5]**.

Existence of ε-Kronecker Sets. Theorem 2.4.3 is a simplified version of [59, Theorem 2.3]. Theorem 2.4.3 also improves upon [53, Theorem 4.4] (when $\boldsymbol{\Gamma}$ is not 2-large). The 2-large case is addressed in Theorem 4.5.2.

Motivated in part by [184], Galindo and Hernández, in [41] and [42], used topological methods to prove the existence of large ε-Kronecker sets in very abstract settings. A discrete abelian group satisfying their hypothesis is isomorphic to a subgroup of a direct sum of copies of $\mathbb{Q}$ together with a finite group. Thus, the existence of large ε-Kronecker sets for all $\varepsilon > 0$ in their setting follows from Corollary 2.4.5. [59] uses their methods to prove Theorem 2.4.3. The proof is shorter, but perhaps less illuminating, than the one here.

For another approach to the topological method, see Givens and Kunen [46]. Yet another existence theorem for ε-Kronecker sets is [60, Theorem 3.1].

Approximating ±1. The results of Sect. 2.5 are adapted from [55]. They will be used in Sect. 9.3 in establishing a characterization of Sidon sets as proportional ε-Kronecker.

Arithmetic Properties. Historically, arithmetic properties were established first for Sidon sets (see Chap. 6 for more detailed references) and those results motivated the study of arithmetic properties of ε-Kronecker sets.

Products of ε-Kronecker Sets. See Sect. 1.5 and [116, Theorem 2.3] for Hadamard set versions of the results of Sect. 2.7.

A stronger version of Proposition 2.7.4 shows that if $M \geq 1$, $\varepsilon < 2\sin(\pi/4M)$ and $\mathbf{E} \subseteq \boldsymbol{\Gamma}$ is ε-Kronecker, then $\mathbf{E}^{2M}$ has no cluster points in $\boldsymbol{\Gamma}$ and the cluster points of $(\mathbf{E} \cdot \mathbf{E}^{-1})^M$ in $\boldsymbol{\Gamma}$ are exactly the elements of $(\mathbf{E} \cdot \mathbf{E}^{-1})^{M-1}$. We refer the reader to [61, Theorem 4.5] for details and related results.

Corollary 2.7.8 appears in [52] with a different proof. The (longer) proof there gives other conclusions when ε is small. For example, if $\mathbf{E} \subseteq \Gamma$ is 1-Kronecker, then $\mathbf{E}$ cannot contain 50 distinct elements $\gamma_1, \ldots, \gamma_{50}$ such that for some choice of signs,

$$\gamma_1^{\pm 1} \gamma_2^{\pm 1} = \cdots = \gamma_{49}^{\pm 1} \gamma_{50}^{\pm 1}. \tag{2.8.1}$$

See Exercise 2.9.12.

There is a vast literature on sets of uniqueness and multiplicity; see [56] and its references. An approach to U_0 sets different from [56] is through descriptive set theory [111].

Exercises. Exercise 2.9.14 (1) is [46, Lemma 3.8].

2.9 Exercises

Exercise 2.9.1. 1. Show that every finite subset $\mathbf{E} \subseteq \mathbb{Z} \setminus \{0\}$ is ε-Kronecker for some $\varepsilon < 2$.
2. Show that $\mathbb{Z} \setminus \{0\}$ is exactly 2-Kronecker.
3. More generally, show that if $\mathbf{E} \subset \Gamma$ is countable, $\mathbf{E}$ has no elements of finite order and $1 \notin \mathbf{E}$, then $\mathbf{E}$ is 2-Kronecker (or better).
4. Compute $\varepsilon(\mathbb{Z}_3 \setminus \{1\})$.

Exercise 2.9.2. 1. Show that if an ε-Kronecker set contains an element γ of order 2, then $\varepsilon > \sqrt{2}$.
2. Show that the set $\mathbf{E} = \{(j, \pi_j) : j = 1, 2, \ldots\} \subseteq \mathbb{Z} \times \widehat{\mathbb{D}}$, where $\{\pi_j\}$ is the Rademacher set in $\widehat{\mathbb{D}}$, is $\sqrt{2}$-Kronecker.

Exercise 2.9.3. 1. Suppose $\mathbf{E} = \{(\chi, \gamma_n\}_{n=1}^{\infty} \subseteq \Gamma$, where $G = \mathbb{Z}_2 \oplus \mathbb{Z}_k^{\mathbb{N}}$, $\{\gamma_n\}_n$ is a set of independent characters and $k \geq 3$ is odd. Show that $\mathbf{E}$ is weak 1-Kronecker.
2. Suppose $\mathbf{E} = \chi_1 \oplus \cdots \oplus \chi_n \subseteq \Gamma = \bigoplus \widehat{\Gamma}_n$. Show that if each χ_n has order at least 3, then $\mathbf{E}$ is weak 1-Kronecker.

Exercise 2.9.4. 1. Prove Lemma 2.2.8.
2. Give an example of an ε-Kronecker set $\mathbf{E}$ and quotient mapping q that is one-to-one on $\mathbf{E}$, but $q(\mathbf{E})$ is not ε-Kronecker.

Exercise 2.9.5. State and prove an analogue of Corollary 2.2.17 assuming, instead, γ has finite order.

Exercise 2.9.6. Adapt the argument of Example 2.3.6 to show that the set $\{N_j, -N_j + j\}_{j=1}^{\infty}$ is $\sqrt{2}$-Kronecker.

Exercise 2.9.7. Prove Theorem 2.3.1 (3).

Exercise 2.9.8. Suppose $m < n$ are positive integers. Show that if I is an interval of length at least $3\pi/m$, then $\{m, n\}$ is ε-Kronecker(I) for some $\varepsilon < \sqrt{2}$.

Exercise 2.9.9. Let $\mathbf{E} \subseteq \mathbb{Z}$ be ε-Kronecker for some $\varepsilon < 2$. Show that $\mathbf{E}$ has upper density zero, meaning

$$\limsup_{N \to \infty} \frac{|\mathbf{E} \cap [-N, N]|}{2N} = 0.$$

Exercise 2.9.10. Suppose M is an even integer and $\mathbf{E}$ is angular $\pi(1 - 1/M)$-Kronecker. Prove that if $N \log M(M-1) \geq 2 \log M$, then $\mathbf{E}$ does not contain a square of cardinality N^2. (This proves a stronger statement than Proposition 2.6.4.)

Exercise 2.9.11. Let d and N be positive integers. A *d-perturbed arithmetic progression of length* N is a set of the form $\{\gamma_0 \gamma^{k_1}, \gamma_0 \gamma^{k_2}, \ldots, \gamma_0 \gamma^{k_N}\}$ where $k_j \in [(j-1)d, jd)$, $1 \leq j \leq N$, $\gamma_0, \gamma \in \mathbf{\Gamma}$ and γ has order greater than k_N. (When $d = 1$ these are arithmetic progressions.) Let $\tau > 0$. Show that there exists $N = N(d, \tau)$ such that if $\mathbf{E} \subseteq \mathbf{\Gamma}$ is a d-perturbed arithmetic progression of length at least N, then $\alpha(\mathbf{E}) \geq \pi - \tau$.

Exercise 2.9.12. Let $\mathbf{E}$ be ε-Kronecker for $\varepsilon < \sqrt{2}$. Show that there exists an integer $N = N(\varepsilon)$, depending only on ε, such that $\mathbf{E}$ cannot contain $2N$ distinct elements, $\{\gamma_j\}$, satisfying $\gamma_1 \gamma_2^{\pm 1} = \cdots = \gamma_{2N-1} \gamma_{2N}^{\pm 1}$, (whatever the choice of signs).

Exercise 2.9.13. Let $\mathbb{R}_d$ be $\mathbb{R}$ with the discrete topology.

1. Show that if $x_j \in (0, \infty)$ and $x_j/x_{j+1} \geq 4$, then $\{x_j\}$ is weak angular $\pi/(q-1)$-Kronecker in $\mathbb{R}_d$.
2. Let $\varepsilon > 0$ and let $\mathbf{E} = \{\gamma_n\} \subseteq \mathbb{R}_d$ be an infinite sequence. Show that $\mathbf{E}$ has an infinite ε-Kronecker subset.
3. Show that every infinite $\mathbf{E} \subset \mathbb{R}_d$ has a subset $\mathbf{E}'$, with $|\mathbf{E}'| = |\mathbf{E}|$ and having the property that for each $\varepsilon > 0$ there is a finite set $\mathbf{F}$ such that $\mathbf{E}' \smallsetminus \mathbf{F}$ is ε-Kronecker.

Exercise 2.9.14. A discrete abelian group $\mathbf{\Gamma}$ is said to have *infinite exponent* if for every N there is some character in $\mathbf{\Gamma}$ with order at least N.

1. Prove that $\mathbf{\Gamma}$ has infinite exponent if and only if $\mathbf{\Gamma}$ contains an infinite ε-Kronecker set for each $\varepsilon > 0$.
2. Suppose $\mathbf{\Gamma}$ has infinite exponent. Does every infinite subset of $\mathbf{\Gamma}$ contain an ε-Kronecker subset of the same cardinality for each $\varepsilon > 0$? If not, give examples.

Exercise 2.9.15. Let $\mathbf{F} \subseteq \mathbf{\Gamma}$ be a maximal independent subset of $\mathbf{\Gamma}$. Show that $\mathbf{\Gamma}/\langle \mathbf{F} \rangle$ is a torsion group.

Exercise 2.9.16. Let $\mathbf{E} \subseteq \bigoplus_{\ell \in B} \mathcal{C}(p_\ell^\infty)$ be such that for every $\ell \in B$ there is an element $\gamma \in \mathbf{E}$ such that the projection, $\Pi_\ell(\gamma)$, onto $\mathcal{C}(p_\ell^\infty)$ is non-trivial. Show that $\mathbf{E}$ is 2-large if and only if

$$|\{\ell : \Pi_\ell(\mathbf{E}) \text{ contains an element of order } \geq 3\}| < |\mathbf{E}|.$$

Exercise 2.9.17. Generalize Theorem 2.5.1 to the situation where an arbitrary choice of signs is replaced by an arbitrary choice of (fixed) two elements in $\mathbb{T}$ with angular distance π.

Exercise 2.9.18.   1. Let $a \in \mathbb{N}$ and $\mathbf{E} = \{a^k : k \in \mathbb{N}\}$. Show that for each n, the closure (in $\overline{\mathbb{Z}}$) of $\mathbf{E} \pm \cdots \pm \mathbf{E}$ (n terms) is U_0.
   2. Suppose $k_j \in \mathbb{N}$ and $k_j \to \infty$. Let $\mathbf{E} = \{k_1, k_1 k_2, k_1 k_2 k_3, \dots\}$. Show that for each n, the closure (in $\overline{\mathbb{Z}}$) of $\mathbf{E} \pm \cdots \pm \mathbf{E}$ (n terms) is U_0.

Exercise 2.9.19. Prove the equivalences of Proposition 2.7.2(5)–(7).

Exercise 2.9.20. Let $\mathbf{E} \subseteq \Gamma$. Show $\varepsilon(\mathbf{E}) = \sup\{\varepsilon(\mathbf{F}) : |\mathbf{F}| < \infty, \mathbf{F} \subseteq \mathbf{E}\}$.

Chapter 3
I_0 Sets and Their Characterizations

I_0 sets characterized analytically, in terms of function algebras, and topologically. I_0 sets do not cluster at continuous characters.

3.1 Introduction

A subset $\mathbf{E}$ of $\boldsymbol{\Gamma}$ is "I_0" if every bounded function on $\mathbf{E}$ is the restriction of the Fourier–Stieltjes transform of a discrete measure. Hadamard and ε-Kronecker sets with $\varepsilon < \sqrt{2}$ are examples of I_0 sets (Theorem 1.3.9 and Theorem 2.3.1), as are independent sets of characters (Exercise 3.7.11). Many more examples of I_0 sets will be given, particularly in the next chapter where it will be shown that every infinite subset of $\boldsymbol{\Gamma}$ contains an infinite I_0 set. Every I_0 set is obviously Sidon. The converse is not true since the class of Sidon sets is closed under finite unions (Corollary 6.3.3), but the class of I_0 sets is not, as already observed in Example 1.5.2.

In this chapter we establish basic properties of I_0 sets, as well as properties of smaller classes where the interpolating measure is required to be real (RI_0 sets), positive (FZI_0 sets) and/or supported on a small set ($I_0(U)$, $RI_0(U)$ or $FZI_0(U)$).

Most of this chapter is devoted to proving alternate characterizations of I_0 or FZI_0. After formally defining the classes of I_0 sets of interest, we establish characterizations in terms of approximation by trigonometric polynomials (Fourier-Stieltjes transforms of finitely supported measures), the number of whose terms depends only on the error and the I_0 set (Theorems 3.2.5 and 3.2.6).

I_0 sets can be characterized in terms of various function algebras on $\mathbf{E}$, as is shown in Sect. 3.3. A distinguishing and important fact about I_0 sets is that there are also topological characterizations of I_0. For example, the Hartman–Ryll-Nardzewski characterization states that $\mathbf{E}$ is I_0 if and only if disjoint subsets of $\mathbf{E}$ have disjoint closures in the Bohr compactification of $\boldsymbol{\Gamma}$

C.C. Graham and K.E. Hare, *Interpolation and Sidon Sets for Compact Groups*, CMS Books in Mathematics, DOI 10.1007/978-1-4614-5392-5_3, © Springer Science+Business Media New York 2013

(see Sect. 3.4). In the final section of this chapter, it is shown that, like the special cases of Hadamard and ε-Kronecker sets, I_0 sets do not cluster at a continuous character.

3.1.1 Historical Overview

Given $\mathbf{E} \subseteq \mathbf{\Gamma}$, we denote by $B(\mathbf{E})$ the quotient space of $B(\mathbf{\Gamma})$ consisting of restrictions to $\mathbf{E}$ of Fourier transforms of measures on G. This is a Banach space with the quotient norm

$$\|\varphi\|_{B(\mathbf{E})} = \inf\{\|\mu\| : \mu \in M(G) \text{ and } \widehat{\mu}|_{\mathbf{E}} = \varphi\}.$$

The space $B_d(\mathbf{E})$ is defined similarly as a quotient of $B_d(\mathbf{\Gamma})$ and the restricted Fourier algebra, $A(\mathbf{E})$, is the analogous quotient of $A(\mathbf{\Gamma})$.

With this terminology, the definitions of I_0 and Sidon may be restated as: $\mathbf{E}$ is I_0 (Sidon) if and only if $\ell^\infty(\mathbf{E}) = B_d(\mathbf{E})$ (respectively, $B(\mathbf{E})$).

Historically, I_0 sets arose from the study of *almost periodic functions*. They are the complex-valued functions on $\mathbf{\Gamma}$ with continuous extensions to the Bohr compactification, $\overline{\mathbf{\Gamma}}$. The space of almost periodic functions is denoted $AP(\mathbf{\Gamma})$ and by $AP(\mathbf{E})$ we mean the set of restrictions to $\mathbf{E}$ of almost periodic functions. An application of Tietze's extension theorem implies that $AP(\mathbf{E})$ is also the space of restrictions to $\mathbf{E}$ of continuous functions on the Bohr closure, $\overline{\mathbf{E}}$, of $\mathbf{E}$.

The algebra $B_d(\mathbf{E})$ is a subspace of $AP(\mathbf{E})$ (Exercise 3.7.5). By the Stone–Weierstrass theorem (the set of extensions to $\overline{\mathbf{\Gamma}}$ of elements of) $B_d(\mathbf{\Gamma})$ is uniformly dense in $C(\overline{\mathbf{\Gamma}})$. Consequently, the almost periodic functions are uniform limits of Fourier transforms of discrete measures or, equivalently, uniform limits of trigonometric polynomials on $\mathbf{\Gamma}$. Moreover, for every set $\mathbf{E}$, $AP(\mathbf{E})$ is the closure of $B_d(\mathbf{E})$ in the ℓ^∞ norm.

These sets satisfy the following inclusions:

$$B_d(\mathbf{E}) \subseteq AP(\mathbf{E}) \subseteq \ell^\infty(\mathbf{E}) \quad \text{and} \quad B_d(\mathbf{E}) \subseteq B(\mathbf{E}) \subseteq \ell^\infty(\mathbf{E}). \tag{3.1.1}$$

In general, the inclusions are proper and equalities of these function spaces (for a given set $\mathbf{E}$) characterize special sets, such as the characterizations of Sidon and I_0, as noted above.

The original definition of an I_0 set, given by Hartman and Ryll-Nardzewski, was a set $\mathbf{E}$ such that "every bounded function on $\mathbf{E}$ can be extended to an almost periodic function". In contemporary terminology, these are the sets $\mathbf{E}$ for which $AP(\mathbf{E}) = \ell^\infty(\mathbf{E})$. Shortly afterwards, it was realized that the extensions could always be found in $B_d(\mathbf{E})$, and hence the set is I_0 as we have defined it. Much later it was observed that the property $AP(\mathbf{E}) = B(\mathbf{E})$ is equivalent to $\mathbf{E}$ being I_0 and that I_0 is also equivalent to $B(\mathbf{E}) = B_d(\mathbf{E})$; see Sect. 3.3.

This leaves one remaining equality from (3.1.1): $AP(\mathbf{E}) = B_d(\mathbf{E})$. This property is known to characterize the class of *Helsonian sets*. It can be shown that all finite unions of I_0 sets are Helsonian and that all Helsonian sets are Sidon. The classes of Helson and Helsonian sets are defined in Remark 3.5.5. They reappear in Sects. 4.2.2, 9.4.1, and 10.3.2.

3.2 Characterizations by Approximate Interpolation

3.2.1 Subclasses of I_0 Sets

We define some subclasses of I_0 sets and then develop characterizations of them.

Definition 3.2.1. Let $\mathbf{E} \subseteq \mathbf{\Gamma}$. A function $\varphi : \mathbf{E} \to \mathbb{C}$ is *Hermitian* if $\varphi(\gamma) = \overline{\varphi(\gamma^{-1})}$ for all $\gamma \in \mathbf{E} \cap \mathbf{E}^{-1}$.

Definition 3.2.2. Let $U \subseteq G$ be Borel and $\mathbf{E} \subseteq \mathbf{\Gamma}$.

1. The set $\mathbf{E}$ is said to be $I_0(U)$ if whenever $\varphi \in \ell^\infty(\mathbf{E})$ there is a discrete measure, μ, concentrated on U, such that $\widehat{\mu}|_{\mathbf{E}} = \varphi$.
2. The set $\mathbf{E}$ is said to be $RI_0(U)$ (resp., $FZI_0(U)$) if whenever $\varphi \in \ell^\infty(\mathbf{E})$ is a Hermitian function there is a discrete real (resp., positive) measure μ, concentrated on U, such that $\widehat{\mu}|_{\mathbf{E}} = \varphi$.

In either case, when $U = G$, we omit the writing of "(G)".

Remark 3.2.3. We only ask to interpolate Hermitian functions for RI_0 or FZI_0 sets since the Fourier transform of a real measure is Hermitian (see Exercise 1.7.5).

Clearly, every $FZI_0(U)$ set is $RI_0(U)$. In Sect. 4.2.1 it will be shown that all $RI_0(U)$ sets are $I_0(U)$ and that the three classes are distinct.

With this terminology, Theorem 1.3.9(1) states that Hadamard sets are $FZI_0(U)$ for all open $U \subseteq \mathbb{T}$, and Theorem 2.3.1 states that if $\mathbf{E}$ is an ε-Kronecker(U) set for some $\varepsilon < \sqrt{2}$, then $\mathbf{E}$ is $FZI_0(U)$.

We now turn to our characterizations. We begin with a useful observation about translates.

Lemma 3.2.4. *Suppose $U \subseteq G$ and $\mathbf{E} \subseteq \mathbf{\Gamma}$ is $I_0(U)$ (resp., $FZI_0(U)$). Then*

1. *$\gamma \mathbf{E}$ is $I_0(U)$ for every $\gamma \in \mathbf{\Gamma}$.*
2. *$\mathbf{E}$ is $I_0(xU)$ (resp., $FZI_0(xU)$) for every $x \in G$.*

Proof. (1) Let $\varphi \in \mathrm{Ball}(\ell^\infty(\gamma \mathbf{E}))$ and define $\psi \in \mathrm{Ball}(\ell^\infty(\mathbf{E}))$ by $\psi(\chi) = \varphi(\gamma\chi)$. Obtain $\mu = \sum_{j=1}^\infty c_j \delta_{x_j} \in M_d(U)$ with $\widehat{\mu}(\chi) = \psi(\chi)$ for $\chi \in \mathbf{E}$, and consider the measure $\nu = \sum_{j=1}^{\infty} c_j \gamma(x_j) \delta_{x_j} \in M_d(U)$. By taking Fourier transforms, it is easy to see that

$$\widehat{\nu}(\gamma\chi) = \sum_{j=1}^{\infty} c_j \overline{\chi(x_j)} = \widehat{\mu}(\chi) = \psi(\chi) = \varphi(\gamma\chi) \text{ for all } \gamma \in \mathbf{E}.$$

(2) Given $\varphi \in \text{Ball}(\ell^{\infty}(\mathbf{E}))$, obtain $\mu = \sum_{j=1}^{\infty} c_j \delta_{x_j} \in M_d(U)$ such that $\widehat{\mu}(\chi) = \varphi(\chi)\overline{\chi(x)}$ for all $\chi \in \mathbf{E}$. The measure $\nu = \sum_{j=1}^{\infty} c_j \delta_{x_j x} \in M_d(xU)$ has the property that $\widehat{\nu}|_{\mathbf{E}} = \varphi$. The $FZI_0(U)$ case is similar.   □

In particular, to prove $\mathbf{E}$ is $I_0(U)$ for all non-empty, open sets U, it suffices to prove $\mathbf{E}$ is $I_0(U)$ for e-neighbourhoods in G.

Our first characterization was proved by Kalton and will be important in proving other equivalences.

3.2.2 Kalton's Characterization and Immediate Consequences

Theorem 3.2.5 (Kalton). *Let U be a σ-compact subset of G and $\mathbf{E}$ be a subset of Γ. Then the following are equivalent:*

1. *$\mathbf{E}$ is $I_0(U)$.*
2. *There is a constant C such that for all $\varphi \in \text{Ball}(\ell^{\infty}(\mathbf{E}))$, there exists $\mu \in M_d(U)$ with $\|\mu\|_{M(G)} \leq C$ and $\widehat{\mu}(\gamma) = \varphi(\gamma)$ for all $\gamma \in \mathbf{E}$.*
3. *There exist $0 < \varepsilon < 1$ and constant C such that for all $\varphi \in \text{Ball}(\ell^{\infty}(\mathbf{E}))$ there exists $\mu \in M_d(U)$ with $\|\mu\|_{M(G)} \leq C$ and $|\varphi(\gamma) - \widehat{\mu}(\gamma)| \leq \varepsilon$ for all $\gamma \in \mathbf{E}$.*
4. *There exists $0 < \varepsilon < 1$ such that for all $\varphi \in \text{Ball}(\ell^{\infty}(\mathbf{E}))$ there exists $\mu \in M_d(U)$ with $|\varphi(\gamma) - \widehat{\mu}(\gamma)| \leq \varepsilon$ for all $\gamma \in \mathbf{E}$.*
5. *There exist $0 < \varepsilon < 1$ and integer N such that for all $\varphi \in \mathbb{T}^{\mathbf{E}}$ there exists $\mu = \sum_{j=1}^{N} c_j \delta_{x_j} \in M_d(U)$ with $|c_j| \leq 1$ and $|\varphi(\gamma) - \widehat{\mu}(\gamma)| \leq \varepsilon$ for all $\gamma \in \mathbf{E}$.*

In (3) and (4) the statement, "There exist $0 < \varepsilon < 1$", may be replaced by "For every $0 < \varepsilon < 1$", and in (5) "There exists $0 < \varepsilon < 1$ and integer N" may be replaced by "For each $0 < \varepsilon < 1$ there exists an integer N".

For $FZI_0(U)$ sets we have a slight variation. Note that there is no loss of generality in assuming $\mathbf{E}$ is asymmetric because Hermitian functions have unique Hermitian extensions to $\mathbf{E} \cup \mathbf{E}^{-1}$ and real measures have Hermitian Fourier transform.

Theorem 3.2.6. *Let U be a σ-compact subset of G and $\mathbf{E}$ be an asymmetric subset of Γ. Then the following are equivalent:*

1. $\mathbf{E}$ is $FZI_0(U)$.
2. There exists C such that for all Hermitian functions $\varphi \in \mathrm{Ball}(\ell^\infty(\mathbf{E}))$ there exists $\mu \in M_d^+(U)$ with $\|\mu\|_{M(G)} \leq C$ and $\widehat{\mu}(\gamma) = \varphi(\gamma)$ for all $\gamma \in \mathbf{E}$.
3. There exist $0 < \varepsilon < 1$ and constant C such that for all Hermitian $\varphi \in \mathrm{Ball}(\ell^\infty(\mathbf{E}))$ there exists $\mu \in M_d^+(U)$ with $\|\mu\|_{M(G)} \leq C$ and $|\varphi(\gamma) - \widehat{\mu}(\gamma)| \leq \varepsilon$ for all $\gamma \in \mathbf{E}$.
4. There exists $0 < \varepsilon < 1$ such that for all Hermitian $\varphi \in \mathrm{Ball}(\ell^\infty(\mathbf{E}))$ there exists $\mu \in M_d^+(U)$ with $|\varphi(\gamma) - \widehat{\mu}(\gamma)| \leq \varepsilon$ for all $\gamma \in \mathbf{E}$.
5. There exist $0 < \varepsilon < 1$ and integer N such that for all Hermitian $\varphi \in \mathbb{T}^\mathbf{E}$ there exists $\mu = \sum_{j=1}^N c_j \delta_{x_j} \in M_d^+(U)$ with $0 \leq c_j \leq 1$ and $|\varphi(\gamma) - \widehat{\mu}(\gamma)| \leq \varepsilon$ for all $\gamma \in \mathbf{E}$.

In (3) and (4) the statement, "There exist $0 < \varepsilon < 1$", may be replaced by "For every $0 < \varepsilon < 1$", and in (5) "There exists $0 < \varepsilon < 1$ and integer N" may be replaced by "For each $0 < \varepsilon < 1$ there exists an integer N".

A similar set of equivalences can be given for $RI_0(U)$, replacing $M_d^+(U)$ with $M_d^r(U)$.

We begin by introducing terminology and proving preliminary results.

Definition 3.2.7. A discrete measure on G of the form $\sum_{n=1}^N c_n \delta_{x_n}$, with $c_n \in \Delta$, will be said to be of *length* N. The x_n need not be distinct.

For a Borel set $U \subseteq G$, $N \in \mathbb{N}$ and $\varepsilon > 0$, we let $AP(\mathbf{E}, U, N, \varepsilon)$ be the set

$$\{w \in \mathrm{Ball}(\ell^\infty(\mathbf{E})) : \exists \mu \in M_d(U) \text{ of length } N, \ \|\widehat{\mu}|_\mathbf{E} - w\|_\infty \leq \varepsilon\}. \quad (3.2.1)$$

When $\mathbf{E}$ is clear we will omit it. When $U = G$, we omit the U.

By $AP_r(\mathbf{E}, U, N, \varepsilon)$ we mean the subset of $AP(\mathbf{E}, U, N, \varepsilon)$ where the coefficients c_n are restricted to be in $[-1, 1]$. Similarly, $AP_+(\mathbf{E}, U, N, \varepsilon)$ is the subset with the coefficients in $[0, 1]$. When $\mathbf{E}$ is clear we will omit it in this notation as well, and when $U = G$, we omit the U.

We put $\mathbb{H}_\gamma = \{-1, 1\}$ if $\gamma = \gamma^{-1} \in \mathbf{E}$ and $\mathbb{H}_\gamma = \mathbb{T}$ otherwise. Slightly abusing notation, we let

$$\mathbb{H}^\mathbf{E} = \prod_{\gamma \in \mathbf{E}} \mathbb{H}_\gamma.$$

When $\mathbf{E}$ is asymmetric, $\mathbb{H}^\mathbf{E}$ consists of the Hermitian elements of $\mathbb{T}^\mathbf{E}$. We may embed $\mathbb{H}^\mathbf{E}$, $\mathbb{T}^\mathbf{E}$ and $\mathbb{Z}_2^\mathbf{E}$ in the unit ball of $\ell^\infty(\mathbf{E})$ in the natural way.

In terms of this notation, (5) in Theorem 3.2.5 could be restated as (5') *there exist $0 < \varepsilon < 1$ and integer N* (or, equivalently, *for every $0 < \varepsilon < 1$ there exist an integer N*) *such that* $AP(\mathbf{E}, U, N, \varepsilon) \supseteq \mathbb{T}^\mathbf{E}$. Similarly, in Theorem 3.2.6, we may replace (5) by *there exist $0 < \varepsilon < 1$ and integer N* (or, equivalently, *for every $0 < \varepsilon < 1$, there exists an integer N*) with $AP_+(\mathbf{E}, U, N, \varepsilon) \supseteq \mathbb{H}^\mathbf{E}$.

Lemma 3.2.8. *1. For every set $U \subseteq G$ and $\mathbf{E} \subseteq \Gamma$,*

$$AP(U, N, \varepsilon) \cdot AP(U, N, \varepsilon) \subseteq AP(U^2, N^2, 2\varepsilon + \varepsilon^2).$$

2. If U is compact, the sets $AP(U, N, \varepsilon)$ are closed in $\ell^\infty(\mathbf{E})$.

Similar results hold for AP_r and AP_+.

Proof. (1) is an easy exercise.

(2) Suppose the net $\{w^{(\beta)}\} \subseteq AP(U, N, \varepsilon)$ converges to $w \in \ell^\infty(\mathbf{E})$.[1] Since all $\|w^{(\beta)}\|_\infty \leq N$, we have $\|w\|_\infty \leq N$. For each β, let $c_{n,\beta} \in \Delta$ and $x_{n,\beta} \in U$ be such that

$$\left| \sum_{n=1}^{N} c_{n,\beta} \gamma(x_{n,\beta}) - w_\gamma^{(\beta)} \right| \leq \varepsilon \text{ for all } \gamma \in \mathbf{E}.$$

By passing to a subnet, if needed, we may assume that $c_{n,\beta} \to c_n \in \Delta$ and $x_{n,\beta} \to x_n \in U$ for each $n = 1, \ldots, N$. The convergence of the $x_{n,\beta}$ ensures the convergence of each $\gamma(x_{n,\beta})$ to $\gamma(x_n)$ for all $\gamma \in \mathbf{\Gamma}$. Thus, $\sup_{\gamma \in \mathbf{E}} \left| \sum_{n=1}^{N} c_n \gamma(x_n) - w_\gamma \right| \leq \varepsilon$, and hence $w \in AP(U, N, \varepsilon)$. □

Key to the proofs of Theorems 3.2.5 and 3.2.6 is a suitable application of the Baire category theorem.

Proposition 3.2.9. *Suppose $U \subseteq G$ is σ-compact and $\mathbf{E} \subseteq \mathbf{\Gamma}$. If $X = \mathbb{T}$ or $\mathbb{Z}_2$ and $\bigcup_{N=1}^\infty AP(U, N, \varepsilon) \supseteq X^{\mathbf{E}}$, then there exists N such that*

$$AP(U^2, N, 2\varepsilon + \varepsilon) \supseteq X^{\mathbf{E}}.$$

Proof. Let $U = \bigcup_{m=1}^\infty U_m$, where the U_m are compact and $U_{m+1} \supseteq U_m$ for $m \geq 1$. Since $\mathbb{T}^{\mathbf{E}}$ and $\mathbb{Z}_2^{\mathbf{E}}$ are compact groups and each of the sets $X^{\mathbf{E}} \cap AP(U_m, N, \varepsilon)$ is closed, the Baire category theorem implies that for some N_0 and m_0 the set $AP(U_{m_0}, N_0, \varepsilon) \cap X^{\mathbf{E}}$ has non-empty interior, and therefore $AP(U, N_0, \varepsilon) \cap X^{\mathbf{E}}$ will also have non-empty interior. Call this latter intersection Y. A finite number of translates of Y cover the compact group $\mathbb{T}^{\mathbf{E}}$, say $\bigcup_{j=1}^{J} w_j Y = \mathbb{T}^{\mathbf{E}}$.

By assumption each w_j belongs to $AP(U, N_j, \varepsilon) \cap X^{\mathbf{E}}$ for some N_j. Let $\sqrt{N} \geq \max\{N_j : j = 0, \ldots, J\}$. Now apply Lemma 3.2.8 (1). □

Remark 3.2.10. A similar argument shows that if $\bigcup_{N=1}^\infty AP_+(U, N, \varepsilon) \supseteq \mathbb{H}^{\mathbf{E}}$, then for some N, $AP_+(U^2, N, 2\varepsilon + \varepsilon^2) \supseteq \mathbb{H}^{\mathbf{E}}$.

The proofs of Theorems 3.2.5 and 3.2.6 are similar and we will prove them together.

---

[1] We are dropping the "$\mathbf{E}$" from the $AP(\ldots)$ notation, as forewarned.

Proof (of Theorems 3.2.5 and 3.2.6). For both theorems, (1) $\Rightarrow$ (2) follows from the closed graph theorem, and the implications (2) $\Rightarrow$ (3) $\Rightarrow$ (4) are trivial. (5) $\Rightarrow$ (1) follows directly from the basic standard iteration, Proposition 1.3.2.

Theorem 3.2.6, (4) $\Rightarrow$ (5). (Similar, but easier, arguments apply to Theorem 3.2.5.) We first note that an easily formulated and proved variation on the standard iteration argument (using only finitely many iterations) proves the assertion about equivalences of the statements "There exists $0<\varepsilon<1\ldots$" and "For every $0 < \varepsilon < 1\ldots$". Thus, we may assume (4) holds for all $\varepsilon > 0$.

We now prove (5) for every (fixed) $\varepsilon > 0$. Since every non-negative, discrete measure may be approximated in measure norm by a finite, linear and non-negative combination of point mass measures, the hypothesis (4) tells us that $\bigcup_{N=1}^{\infty} AP_+(U, N, \varepsilon/3) \supseteq \mathbb{H}^{\mathbf{E}}$. Therefore, Remark 3.2.10 implies that there exists an integer N such that $AP_+(U^2, N, \varepsilon) \supseteq \mathbb{H}^{\mathbf{E}}$. We quickly deduce a weaker variant of (5), with the approximating measure concentrated on U^2.

To prove the sharper result claimed, we proceed more carefully. The second sentence of the proof of Proposition 3.2.9 (but applied to $AP_+(U, N, \varepsilon/3)$) says that under assumption (4) there is some N such that the set $AP_+(U, N, \varepsilon/3) \cap \mathbb{H}^{\mathbf{E}}$ has non-empty interior in $\mathbb{H}^{\mathbf{E}}$. Hence (see p. 209), there will be a finite set $\mathbf{F} \subseteq \mathbf{E}$ and a $\psi \in \mathbb{H}^{\mathbf{F}}$ such that $\{\psi\} \times \mathbb{H}^{\mathbf{E} \smallsetminus \mathbf{F}} \subseteq AP_+(U, N, \varepsilon/3) \cap \mathbb{H}^{\mathbf{E}}$.

Consider the subset $\mathcal{S}$ of $\ell^{\infty}(\mathbf{E})$ consisting of the Hermitian elements which vanish off $\mathbf{F}$. Because $\mathbf{F}$ is finite, $\mathcal{S}$ is a finite dimensional real subspace. Take a basis of $\mathcal{S}$, say $e_1, \ldots, e_J$, where $e_j \in \mathrm{Ball}(\ell^{\infty}(\mathbf{E}))$. Since all norms are comparable on a finite dimensional space, there is some $c > 0$ such that for all real scalars b_j,

$$\left\| \sum_{j=1}^{J} b_j e_j \right\|_{\ell^{\infty}} \geq c \sum_{j=1}^{J} |b_j|.$$

Each $\pm e_j$ is Hermitian (since they belong to $\mathcal{S}$), so by assumption (4) we may obtain $\mu_j, \nu_j \in M_d^+(U)$ such that for all $\gamma \in \mathbf{E}$,

$$|e_j(\gamma) - \widehat{\mu_j}(\gamma)| < \frac{c\varepsilon}{4N} \text{ and } |e_j(\gamma) + \widehat{\nu_j}(\gamma)| < \frac{c\varepsilon}{4N}. \qquad (3.2.2)$$

By taking suitable, finitely supported, discrete measures, we may assume there is a positive integer M such that the discrete measures μ_j, ν_j have length at most M for all $j = 1, \ldots, J$.

Let $\varphi \in \mathrm{Ball}(\ell^{\infty}(\mathbf{E}))$ be Hermitian. Since φ coincides on $\mathbf{E} \smallsetminus \mathbf{F}$ with an element of $AP_+(U, N, \varepsilon/3) \cap \mathbb{H}^{\mathbf{E}}$, we may find a length N measure $\mu \in M_d^+(U)$ such that

$$|\varphi(\gamma) - \widehat{\mu}(\gamma)| \leq \varepsilon/3 \text{ on } \mathbf{E} \smallsetminus \mathbf{F}. \qquad (3.2.3)$$

Because μ is a positive measure and $\mathbf{E}$ is asymmetric, $(\varphi - \widehat{\mu})|_{\mathbf{F}}$ (extended by 0 on $\mathbf{E} \smallsetminus \mathbf{F}$) belongs to the real vector space $\mathcal{S}$, and therefore

$$\varphi(\gamma) - \widehat{\mu}(\gamma) = \sum_{j=1}^{J} b_j e_j(\gamma)$$

for $\gamma \in \mathbf{F}$ and suitable $b_j \in \mathbb{R}$. Write $b_j = b_j^+ - b_j^-$, where $b_j^\pm \geq 0$. Note that

$$c \sum_{j=1}^{J} |b_j| \leq \|(\varphi - \widehat{\mu})|_{\mathbf{F}}\|_\infty \leq 1 + \|\mu\|_{M(G)} \leq 1 + N. \qquad (3.2.4)$$

Put $\nu = \mu + \sum_{j=1}^{J} b_j^+ \mu_j + \sum_{j=1}^{J} b_j^- \nu_j$. For $\gamma \in \mathbf{F}$, we have

$$|\varphi(\gamma) - \widehat{\nu}(\gamma)| = \left| \varphi - \widehat{\mu} - \sum_{j=1}^{J} b_j^+ \widehat{\mu_j} - \sum_{j=1}^{J} b_j^- \widehat{\nu_j} \right|$$

$$= \left| \sum_{j=1}^{J} b_j e_j - \sum_{j=1}^{J} b_j^+ \widehat{\mu_j} - \sum_{j=1}^{J} b_j^- \widehat{\nu_j} \right|$$

$$= \left| \sum_{j=1}^{J} b_j^+ (e_j - \widehat{\mu_j}) - \sum_{j=1}^{J} b_j^- (e_j + \widehat{\nu_j}) \right|.$$

Combining this observation with (3.2.2) and (3.2.4) gives the estimate

$$|\varphi(\gamma) - \widehat{\nu}(\gamma)| \leq \frac{c\varepsilon}{4N} \sum_{j=1}^{J} |b_j| \leq \frac{(1 + N)\varepsilon}{4N} \leq \frac{\varepsilon}{2} \text{ for every } \gamma \in \mathbf{F}.$$

If $\gamma \in \mathbf{E} \setminus \mathbf{F}$, then $\sum b_j e_j(\gamma) = 0$, so we may write

$$|\varphi(\gamma) - \widehat{\nu}(\gamma)| = \left| \varphi - \widehat{\mu} - \sum_{j=1}^{J} b_j^+ \widehat{\mu_j} - \sum_{j=1}^{J} b_j^- \widehat{\nu_j} \right|$$

$$= \left| \varphi - \widehat{\mu} + \sum_{j=1}^{J} b_j^+ (e_j - \widehat{\mu_j}) - \sum_{j=1}^{J} b_j^- (e_j + \widehat{\nu_j}) \right|.$$

Together with (3.2.2) and (3.2.3), that implies that

$$|\varphi(\gamma) - \widehat{\nu}(\gamma)| < \frac{\varepsilon}{2} + \frac{c\varepsilon}{4N} \sum |b_j| \leq \varepsilon \text{ for every } \gamma \in \mathbf{E} \setminus \mathbf{F}.$$

Furthermore, $\nu \in M_d^+(U)$ has length at most $N + 2JM$. Since N, J, M are all independent of φ, we may take $N + 2JM$ to be the N of (5). $\square$

The least C such that Theorem 3.2.5 (2) holds for all $\varphi \in \mathrm{Ball}(\ell^\infty(\mathbf{E}))$ is known as the $I_0(U)$ *constant of* $\mathbf{E}$. Proposition 1.3.2 shows that if $AP(\mathbf{E}, U, N, \varepsilon) \supseteq \mathbf{T}^{\mathbf{E}}$, then the I_0 constant is at most $N/(1 - \varepsilon)$.

Corollary 3.2.11. *Every finite set is I_0.*

Proof. Suppose $\mathbf{E}$ is a finite set. Given a bounded function, $\varphi : \mathbf{E} \to \mathbb{C}$, let P be the trigonometric polynomial $P(x) = \sum_{\gamma \in \mathbf{E}} \varphi(\gamma)\gamma(x)$ and let ν be the absolutely continuous measure $P dm_G$. Using Riemann sums, approximate ν by a discrete measure μ so that $|\widehat{\nu}(\gamma) - \widehat{\mu}(\gamma)| < \epsilon$ for all $\gamma \in \mathbf{E}$. The details are left to Exercise 3.7.10. It follows from Theorem 3.2.5 (4) that $\mathbf{E}$ is I_0. $\square$

Another proof that finite sets are I_0 will be given later, as outlined in Remark 3.4.2.

Here is a variation on Theorem 3.2.5. A similar variation of Theorem 3.2.6 can be made.

Proposition 3.2.12. *Let $U \subseteq G$ be σ-compact and $\mathbf{E} \subseteq \Gamma$. The following are equivalent:*

1. *The set $\mathbf{E}$ is $I_0(U)$.*
2. *For some $0 < \varepsilon < 1$ and $N \geq 1$, $\mathbb{T}^{\mathbf{E}} \subseteq AP(\mathbf{E}, U, N, \varepsilon)$.*
3. *For some $0 < \varepsilon < 1$ and $N \geq 1$, $\mathrm{Ball}(\ell^{\infty}(\mathbf{E})) \subseteq AP(\mathbf{E}, U, N, \varepsilon)$.*
4. *For some $0 < \varepsilon < 1$, $\mathbb{T}^{\mathbf{E}} \subseteq \bigcup_{N=1}^{\infty} AP(\mathbf{E}, U, N, \varepsilon)$.*
5. *For some $0 < \varepsilon < 1$, $\mathrm{Ball}(\ell^{\infty}(\mathbf{E})) \subseteq \bigcup_{N=1}^{\infty} AP(\mathbf{E}, U, N, \varepsilon)$.*

In each case, the phrase, "For some $0 < \varepsilon < 1$ and $N \geq 1$", may be replaced by "For every $0 < \varepsilon < 1$ there exists a positive integer N".

Proof. Clearly, $(2) \Rightarrow (4)$, and $(3) \Rightarrow (5)$. Item (2) is a restatement of Theorem 3.2.5 (5) and so $(1) \Leftrightarrow (2)$.

Since every element of $\mathrm{Ball}(\ell^{\infty}(\mathbf{E})) = \Delta^{\mathbf{E}}$ is the average of two elements of $\mathbb{T}^{\mathbf{E}}$, (2) implies (3) with double the N of (2), and (4) implies (5). Therefore, we have $(1) \Rightarrow (2) \Rightarrow (3) \Rightarrow (5)$ and $(1) \Rightarrow (2) \Rightarrow (4) \Rightarrow (5)$.

(5) obviously implies that (4) of Theorem 3.2.5 holds, and hence $(5) \Rightarrow (1)$. $\square$

3.2.3 $I_0(U)$ with Bounded Length

Definition 3.2.13. The set $\mathbf{E}$ is $I_0(U, N, \varepsilon)$ if for every $\varphi \in \mathrm{Ball}(\ell^{\infty}(\mathbf{E}))$ there exists a measure $\mu = \sum_{n=1}^{N} c_n \delta_{x_n} \in M_d(U)$ with all $c_n \in \Delta$ and $\|\varphi - \widehat{\mu}|_E\|_{\infty} \leq \varepsilon$. We define $FZI_0(U, N, \varepsilon)$ similarly. If $U = G$ we drop the "U".

Equivalently, $\mathbf{E}$ is $I_0(U, N, \varepsilon)$ if and only if $\mathrm{Ball}(\ell^{\infty}(\mathbf{E})) \subseteq AP(\mathbf{E}, U, N, \varepsilon)$. Kalton's theorem implies that $\mathbf{E}$ is $I_0(U)$ if and only if it is $I_0(U, N, \varepsilon)$ for some N and $\varepsilon < 1$. The standard iteration argument shows that if $\mathbf{E}$ is $I_0(U, N, \varepsilon)$ for some $\varepsilon < 1$, then it is $I_0(U, N', \varepsilon')$ for every $0 < \varepsilon' < 1$, where N' depends only on N, ε and ε'.

Definition 3.2.14. The set $\mathbf{E}$ is $I_0(U)$ *with bounded length* (or $FZI_0(U)$ *with bounded length*) if there is an integer N such that for every non-empty, open set $U \subseteq G$ there is a finite set $\mathbf{F} \subseteq \mathbf{E}$ such that $\mathbf{E} \smallsetminus \mathbf{F}$ is $I_0(U, N, 1/2)$ (resp., $FZI_0(U, N, 1/2)$).

Remarks 3.2.15. (i) The proof that ε-Kronecker(U) sets, with $\varepsilon < \sqrt{2}$ and U a symmetric e-neighbourhood, are $FZI_0(\mathrm{U})$ (Theorem 2.3.1(3)) shows that they are actually $FZI_0(U, N, 1/2)$, where N depends only on ε. Consequently, such sets are $FZI_0(U)$ with bounded length. Similarly, Hadamard sets are $FZI_0(U)$ with bounded length (Theorem 1.3.9(3)).

(ii) It will be shown in Corollary 5.2.6 that if G is connected, then every set in its dual that is $I_0(U)$ with bounded length is $I_0(U)$ for all non-empty, open sets U.

3.2.4 Interpolation of ±1-Valued Functions

To be I_0, it is also sufficient to interpolate all ± 1-valued functions on $\mathbf{E}$.

Proposition 3.2.16. *Let* $U \subseteq G$ *be a compact symmetric e-neighbourhood and suppose there exists* $\varepsilon < 1$ *such that for all* $\varphi \in \mathbb{Z}_2^{\mathbf{E}}$ *there is some* $\mu \in M_d(U)$ *such that* $|\varphi(\gamma) - \widehat{\mu}(\gamma)| \leq \varepsilon$ *for all* $\gamma \in \mathbf{E}$. *Then* $\mathbf{E}$ *is* $I_0(U)$.

Proof. The arguments used in the proof of the improved standard iteration, Corollary 1.3.3, show that under these assumptions, for every $\varphi : \mathbf{E} \to [-1, 1]$, there is a $\mu \in M_d(U)$ with real-valued Fourier transform and such that

$$|\varphi(\gamma) - \widehat{\mu}(\gamma)| \leq (1 + \varepsilon)/2 \text{ for all } \gamma \in \mathbf{E}.$$

Finitely many more applications of the (real) standard iteration show that we can assume $|\varphi(\gamma) - \widehat{\mu}(\gamma)| \leq 1/3$ for all $\gamma \in \mathbf{E}$.

By interpolating the real and imaginary parts of arbitrary $\varphi \in \mathrm{Ball}(\ell^\infty(\mathbf{E}))$ and appealing to Theorem 3.2.5(4), it follows that $\mathbf{E}$ is $I_0(U)$. $\square$

This has an easy corollary.

Corollary 3.2.17. *Suppose* $AP(\mathbf{E}, N, \varepsilon) \supseteq \mathbb{Z}_2^{\mathbf{E}}$ *for some* $\varepsilon < 1$. *Then* $\mathbf{E}$ *is* $I_0(N', 1/2)$ *where* N' *depends only on* N *and* ε.

3.3 Function Algebra Characterizations

It is now easy to see that the original definition of I_0 coincides with the one we have given.

Theorem 3.3.1 (Kahane's AP theorem). *Let $\mathbf{E} \subseteq \Gamma$. The following are equivalent:*

1. $\mathbf{E}$ *is* I_0.
2. $AP(\mathbf{E}) = B(\mathbf{E})$.
3. $AP(\mathbf{E}) = \ell^\infty(\mathbf{E})$.

Proof. $(1) \Rightarrow (2)$ is immediate from the inclusions of (3.1.1).

$(2) \Rightarrow (3)$. Since $\|f\|_\infty \le \|f\|_{B(\mathbf{E})}$, the assumption $AP(\mathbf{E}) = B(\mathbf{E})$ implies that the two Banach spaces have equivalent norms. In particular, there is some constant C such that $C\|f\|_\infty \ge \|f\|_{B(\mathbf{E})}$ for all $f \in AP(\mathbf{E})$.

We claim this statement implies $\mathbf{E}$ is Sidon. To see this, suppose $\varphi \in \ell^\infty(\mathbf{E})$. For every finite subset $\mathbf{F} \subseteq \mathbf{E}$, $\varphi 1_{\mathbf{F}} \in AP(\mathbf{F})$. By assumption, there is a measure $\mu_{\mathbf{F}} \in M(G)$ such that $\widehat{\mu_{\mathbf{F}}} = \varphi 1_{\mathbf{F}}$ on $\mathbf{E}$ and $\|\mu_{\mathbf{F}}\|_{M(G)} \le 2\|\widehat{\mu_{\mathbf{F}}}\|_{B(\mathbf{E})} \le 2C\|\varphi 1_{\mathbf{F}}\|_\infty \le 2C\|\varphi\|_\infty$. Let μ be a weak* cluster point of the set of bounded measures $\mu_{\mathbf{F}}$. Then $\widehat{\mu}|_{\mathbf{E}} = \varphi$, proving $\mathbf{E}$ is Sidon. Hence, $\ell^\infty(\mathbf{E}) = B(\mathbf{E})$. Consequently, $\ell^\infty(\mathbf{E}) = AP(\mathbf{E})$.

$(3) \Rightarrow (1)$ follows directly from Proposition 3.2.12 (5) $\Rightarrow (1)$ because the functions $\sum_{n=1}^N c_n \widehat{\delta_{x_n}}$ are dense in the almost periodic functions. $\qquad \square$

Another characterization was given by Ramsey, Wells and Bourgain.

Theorem 9.4.15 (Ramsey–Wells–Bourgain). *A subset $\mathbf{E}$ of Γ is I_0 if and only if $B_d(\mathbf{E}) = B(\mathbf{E})$.*

This is a deep result whose proof is given in Sect. 9.4. From it one may deduce another variant on the $AP(\mathbf{E}, N, \varepsilon, \mathbb{T}^{\mathbf{E}}) \supseteq \mathrm{Ball}(\ell^\infty(\mathbf{E}))$ criterion for I_0.

Corollary 3.3.2. *Let $\mathbf{E} \subseteq \Gamma$. Then $\mathbf{E}$ is I_0 if and only if for some $0 < \varepsilon < 1$ there is a positive integer N such that for all $f \in B(\mathbf{E})$, with $\|f\|_{B(\mathbf{E})} \le 1$, there exist $c_n \in \Delta$ and $x_n \in G$ with*

$$\left\| \sum_{n=1}^N c_n \widehat{\delta_{x_n}} - f \right\|_{B(\mathbf{E})} \le \varepsilon.$$

Proof. Suppose $\mathbf{E}$ is I_0 with I_0 constant C. Let $\varepsilon > 0$. If $f \in \mathrm{Ball}(\ell^\infty(\mathbf{E}))$, there is some $\mu \in M_d(G)$ such that $\widehat{\mu}|_{\mathbf{E}} = f$ and $\|\mu\| \le C\|f\|_\infty$. Approximating μ by an appropriate finite sum, it follows that for suitable $c_n \in \Delta$ and $x_n \in G$ we have

$$\left\| \sum_{n=1}^N c_n \widehat{\delta_{x_n}}|_{\mathbf{E}} - f \right\|_\infty \le \left\| \sum_{n=1}^N c_n \delta_{x_n} - \mu \right\|_{M(G)} < \varepsilon/C.$$

By considering a discrete measure ν whose transform agrees with $\varphi = \sum_{n=1}^N c_n \widehat{\delta_{x_n}} - f$ on $\mathbf{E}$ and whose measure norm is at most $C\|\varphi\|_\infty$, one can see that the $B_d(\mathbf{E})$-norm of φ is at most ε.

For the other direction, an application of the standard iteration argument shows that given $f \in B(\mathbf{E})$, there exists a discrete measure μ whose transform agrees with f on $\mathbf{E}$. Consequently, $B_d(\mathbf{E}) = B(\mathbf{E})$, and the Ramsey–Wells–Bourgain theorem then implies $\mathbf{E}$ is I_0. □

3.4 Topological Characterizations

The following topological characterization of I_0 is known as the HRN characterization (for Hartman and Ryll-Nardzewski) and is very useful.

Proposition 3.4.1 (HRN characterization). *The following are equivalent:*

1. $\mathbf{E} \subseteq \boldsymbol{\Gamma}$ *is* I_0.
2. *If $\mathbf{E}_0$ and $\mathbf{E}_1$ are disjoint subsets of $\mathbf{E}$, then $\mathbf{E}_1, \mathbf{E}_2$ have disjoint closures in the Bohr compactification of $\boldsymbol{\Gamma}$.*
3. *If $\mathbf{E}_0$ and $\mathbf{E}_1$ are disjoint subsets of $\mathbf{E}$, then there exists a discrete measure μ such that $\widehat{\mu} = 0$ on $\mathbf{E}_0$ and $\widehat{\mu} = 1$ on $\mathbf{E}_1$.*

Proof. (1) $\Rightarrow$ (2). Suppose $\mathbf{E}_0$ and $\mathbf{E}_1$ are disjoint subsets of $\mathbf{E}$. Let $\varphi \in \ell^\infty(\mathbf{E})$ equal 0 on $\mathbf{E}_0$ and 1 on $\mathbf{E} \smallsetminus \mathbf{E}_0$. Because $\mathbf{E}$ is I_0, there is a discrete measure μ such that $\widehat{\mu} = \varphi$ on $\mathbf{E}$. Since $\widehat{\mu}$ is a continuous function on the Bohr group, it must be 0 on $\overline{\mathbf{E}_0}$ and 1 on $\overline{\mathbf{E}_1}$, and hence those closures are disjoint.

(2) $\Rightarrow$ (3). Suppose $\mathbf{E}_0, \mathbf{E}_1$ are disjoint subsets of $\mathbf{E}$. Then their closures are disjoint, so a compactness argument implies there exists an open set $\mathbf{V} \subseteq \overline{\boldsymbol{\Gamma}}$ such that $\overline{\mathbf{E}_0} \cdot \overline{\mathbf{V}} \cdot \overline{\mathbf{V}}^{-1}$ and $\overline{\mathbf{E}_1}$ are disjoint. Choose $g, h \in \ell^2(G_d)$ such that $\widehat{g} = 1_{\overline{\mathbf{V}}}$ and $\widehat{h} = 1_{\overline{\mathbf{E}_0} \cdot \overline{\mathbf{V}}^{-1}}$. Then

$$\mu = \frac{1}{m_{\overline{\boldsymbol{\Gamma}}}(\overline{\mathbf{V}})} g \cdot h \in \ell^1(G_d) = M_d(G).$$

Since $\widehat{\mu} = \frac{1}{m(\overline{\mathbf{V}})}\widehat{g} * \widehat{h}$, one may easily verify that $\widehat{\mu} = 1$ on $\overline{\mathbf{E}_0}$ and 0 off $\overline{\mathbf{E}_0} \cdot \overline{\mathbf{V}} \cdot \overline{\mathbf{V}}^{-1}$. In particular, $\widehat{\mu} = 0$ on $\overline{\mathbf{E}_1}$.

(3) $\Rightarrow$ (1). It suffices to verify that $\bigcup_{N=1}^\infty AP(\mathbf{E}, N, 1/3) \supseteq \mathbb{Z}_2^{\mathbf{E}}$ and then appeal to the standard iteration. The details are left to the reader. □

Remark 3.4.2. Since finite sets are closed in the Bohr compactification, this gives a second proof that finite subsets of $\boldsymbol{\Gamma}$ are I_0. In contrast, even a two-element set need not be $I_0(U)$ for all non-empty, open sets U; take a non-trivial finite group G, the open subset $U = \{e\}$ and $\mathbf{E} = \{\gamma, \mathbf{1}\}$.

Example 1.5.2 shows that the class of I_0 sets is not closed under finite unions. It is easy to deduce from the HRN characterization the following simple criterion for determining whether the union of two I_0 sets is I_0. The proof is left as Exercise 3.7.12(1).

Corollary 3.4.3. *Suppose disjoint sets* **E** *and* **F** *are* I_0. *Then* **E** $\cup$ **F** *is* I_0 *if and only if the Bohr closures of* **E** *and* **F** *are disjoint.*

The HRN criterion may be further refined to show that the size of the separating set can be controlled independently of the partition.

Theorem 3.4.4 (Kahane separation theorem). *Let* **E** $\subset$ Γ. *The following are equivalent:*

1. **E** *is* I_0.
2. *There exists* $\varepsilon > 0$ *such that whenever* $\mathbf{E}_0$ *and* $\mathbf{E}_1$ *are disjoint subsets of* **E**, *then there is a 1-neighbourhood* **V** $\subseteq$ Γ, *with* $\overline{\Gamma}$*-Haar measure at least* ε, *such that* $\mathbf{E}_0 \cdot \mathbf{V}$ *and* $\mathbf{E}_1 \cdot \mathbf{V}$ *are disjoint.*
3. *Whenever* $\mathbf{E}_0$ *and* $\mathbf{E}_1$ *are disjoint subsets of* **E** *there is a Borel subset* **W** $\subseteq$ $\overline{\Gamma}$ *of positive Haar measure and containing the identity, such that* $\mathbf{E}_0 \cdot \mathbf{W}$ *and* $\mathbf{E}_1 \cdot \mathbf{W}$ *are disjoint.*

We begin with a technical lemma.

Lemma 3.4.5. *Let* $\delta > 0$, N *be a positive integer and* $M = \lceil 4\pi/\delta \rceil$. *Let* $x_1, \ldots, x_N \in G$. *There is an open set* **V** $\subseteq$ $\overline{\Gamma}$ *such that* $m_{\overline{\Gamma}}(\mathbf{V}) \geq (1/M)^N$ *and*

$$|\lambda(x_n) - 1| < \delta \text{ for all } n = 1, \ldots, N \text{ and } \lambda \in \mathbf{V}.$$

In the above, $\lceil x \rceil$ is the smallest integer $\geq x$.

Proof (of Lemma 3.4.5). For $1 \leq m \leq M$, let

$$I_m = \Big\{ e^{i\theta} : \frac{(m-1)\pi}{M} < \theta < \frac{(m+1)\pi}{M} \Big\}.$$

These open sets cover $\mathbb{T}$ and hence at least one of the M open preimages,

$$(\widehat{\delta_{x_1}})^{-1}(I_m) = \{\lambda \in \overline{\Gamma} : \overline{\lambda(x_1)} \in I_m\},$$

has $\overline{\Gamma}$-Haar measure at least $1/M$. Choose $\mathbf{V}_1$ to be such a set and suppose $\lambda_1 \in \mathbf{V}_1$. Notice that if $\lambda \in \mathbf{V}_1$, then $|\lambda(x_1) - \lambda_1(x_1)| < 2\pi/M \leq \delta/2$.

Inductively assume there are non-empty, open sets $\mathbf{V}_k \subseteq \mathbf{V}_{k-1}$ for $k = 1, \ldots, K-1$, $k \leq N$, with $m_{\overline{\Gamma}}(\mathbf{V}_k) \geq (1/M)^k$ and having the property that for every (fixed) $\lambda_k \in \mathbf{V}_k$ and arbitrary $\lambda \in \mathbf{V}_k$ we have $|\lambda(x_k) - \lambda_k(x_k)| < \delta$.

The preimages, $(\widehat{\delta_{x_K}})^{-1}(I_m)$ for $m = 1, \ldots, M$, cover $\mathbf{V}_{K-1}$, and hence the intersection of at least one of these sets with $\mathbf{V}_{K-1}$ has measure at least $\frac{1}{M} m_{\overline{\Gamma}}(\mathbf{V}_{K-1}) \geq (1/M)^K$. Select $\mathbf{V}_K$ to be one of those intersections and pick $\lambda_K \in \mathbf{V}_K$.

Take $\mathbf{V} = \lambda_N^{-1}\mathbf{V}_N$. By construction $|\lambda(x_n) - 1| < \delta$ for all $n = 1, \ldots, N$ and $\lambda \in \mathbf{V}$. $\square$

Proof (of Theorem 3.4.4). (1) $\Rightarrow$ (2). Since $\mathbf{E}$ is I_0 there is an integer N such that $AP(\mathbf{E}, N, 1/4) \cap \mathbb{T}^{\mathbf{E}} = \mathbb{T}^{\mathbf{E}}$. We will prove $\varepsilon = (16\pi N)^{-N}$ works in (2).

Let $\mathbf{E}_0$, $\mathbf{E}_1$ be disjoint subsets of $\mathbf{E}$ and choose $\mu = \sum_{n=1}^{N} c_n \delta_{x_n}$, with $c_n \in \Delta$, such that $|\widehat{\mu} - 1| \leq 1/4$ on $\mathbf{E}_0$ and $|\widehat{\mu} + 1| \leq 1/4$ on $\mathbf{E}_1$. Obtain $\mathbf{V}$ from the lemma with $\delta = 1/(4N)$ and these points $x_1, \ldots, x_N$. We will verify that $\mathbf{E}_0 \cdot \mathbf{V} \cap \mathbf{E}_1 \cdot \mathbf{V}$ is empty.

Suppose $\chi = \gamma_0 \beta_0 = \gamma_1 \beta_1 \in \mathbf{E}_0 \cdot \mathbf{V} \cap \mathbf{E}_1 \cdot \mathbf{V}$, with $\gamma_j \in \mathbf{E}_j$ and $\beta_j \in \mathbf{V}$. The choice of $\mathbf{V}$ ensures that $|\beta_j(x_n) - 1| \leq 1/(4N)$ for all n and both j. An easy calculation shows

$$|\widehat{\mu}(\chi) - 1| \leq \left| \sum_{n=1}^{N} c_n \gamma_0(x_n) \left(\beta_0(x_n) - 1\right) \right| + \left| \sum_{n=1}^{N} c_n \gamma_0(x_n) - 1 \right|$$
$$\leq \sum_{n=1}^{N} |c_n| \, |\beta_0(x_n) - 1| + |\widehat{\mu}(\gamma_0) - 1| \leq \frac{1}{2}.$$

Similarly,

$$|\widehat{\mu}(\chi) + 1| \leq \sum_{n=1}^{N} |c_n| \, |\beta_1(x_n) - 1| + |\widehat{\mu}(\gamma_1) + 1| \leq \frac{1}{2},$$

and this is clearly impossible.

(3) $\Rightarrow$ (1). It will be sufficient to prove that disjoint subsets of $\mathbf{E}$ have disjoint closures. So assume $\mathbf{E}_0$, $\mathbf{E}_1 \subseteq \mathbf{E}$ are disjoint. By (3) there is a Borel set $\mathbf{W}$ of positive measure, with $\mathbf{E}_0 \cap (\mathbf{E}_1 \cdot \mathbf{W} \cdot \mathbf{W}^{-1}) = \emptyset$. Since $\mathbf{1}$ is an interior point of $\mathbf{W} \cdot \mathbf{W}^{-1}$, there is a symmetric $\mathbf{1}$-neighbourhood $\mathbf{V}$ such that $\mathbf{V}^2 \subseteq \mathbf{W} \cdot \mathbf{W}^{-1}$. The set $(\mathbf{E}_0 \cdot \mathbf{V}) \cap (\mathbf{E}_1 \cdot \mathbf{V})$ is empty. Because $\overline{\mathbf{E}_j} \subseteq \mathbf{E}_j \cdot \mathbf{V}$, it follows that $\mathbf{E}_0$ and $\mathbf{E}_1$ have disjoint closures.

(2) $\Rightarrow$ (3) is trivial. $\square$

3.5 I_0 Sets Do Not Cluster at a Continuous Character

We conclude this chapter with a proof that general I_0 sets have a property shown already for Hadamard and ε-Kronecker sets: I_0 sets cannot cluster at a continuous character. The proof uses an idea that is important in the proof that any I_0 set is a finite union of $I_0(U)$ sets (Theorem 5.3.1).

Theorem 3.5.1 (Ryll-Nardzewski–Méla–Ramsey). *If $\mathbf{E}$ is an I_0 set, then no continuous character is a cluster point of $\mathbf{E}$.*

Proof. It is enough to show that an I_0 set cannot cluster at $\mathbf{1}$ since if the I_0 set $\mathbf{E}$ clusters at γ, then the I_0 set $\gamma^{-1}\mathbf{E}$ would cluster at $\mathbf{1}$. We argue by contradiction, and suppose that $\mathbf{E}$ did cluster at $\mathbf{1}$.

A Baire category argument (given below) will show that there exist $N \geq 1$, a finite set $\mathbf{F} \subset \mathbf{E}$ and a fixed set $c_1, \ldots, c_N \in \Delta$ such that if $\varphi : \mathbf{E} \to \mathbb{Z}_2$, then there exist $x_1, \ldots, x_N \in G$ with $|\varphi(\gamma) - \sum_1^N c_n \gamma(x_n)| \leq 1/3$ on $\mathbf{E} \smallsetminus \mathbf{F}$.

We claim, assuming the preceding, that $|1 - \sum_1^N c_n| \leq 1/3$. Indeed, let $\varphi \equiv 1$ on $\mathbf{E}$. Choose $x_1, \ldots, x_N$ such that $|\varphi(\gamma) - \sum_1^N c_n \gamma(x_n)| \leq 1/3$ for $\gamma \in \mathbf{E} \smallsetminus \mathbf{F}$. Then $|1 - \sum_1^N c_n| = |1 - \sum_1^N c_n 1(x_n)| \leq 1/3$ since $\mathbf{E}$ clusters at $\mathbf{1}$.

Choosing $\varphi \equiv -1$ on $\mathbf{E}$, we similarly deduce that $|1 + \sum_1^N c_n| \leq 1/3$. But these two inequalities cannot simultaneously hold.

Here is the Baire category argument. For each $M \geq 1$, let D_M be a countable, dense subset of Δ^M. For each $M \geq 1$ and $c = (c_1, \ldots, c_M) \in D_M$ let $\widetilde{AP}(M, c)$ be the set of $\varphi \in \text{Ball}(\ell^\infty(\mathbf{E})) = \Delta^{\mathbf{E}}$ such that there exist $x_1, \ldots, x_M \in G$ with $|\varphi(\gamma) - \sum_1^M c_m \gamma(x_m)| \leq 1/3$ for $\gamma \in \mathbf{E}$. Because G is compact, each $\widetilde{AP}(M, c)$ is closed in $\text{Ball}(\ell^\infty(\mathbf{E}))$. Because $\mathbf{E}$ is I_0, $\bigcup_{M=1}^\infty \bigcup_{c \in D_M} \widetilde{AP}(M, c) \supseteq \text{Ball}(\ell^\infty(\mathbf{E}))$. The Baire category theorem says that some $\widetilde{AP}(M, c)$ has interior. By the definition of the product topology, there is a finite set $\mathbf{F} \subseteq \mathbf{E}$ and open $V \subseteq \Delta^{\mathbf{F}}$ such that $\widetilde{AP}(M, c)$ contains $V \times \Delta^{\mathbf{E} \smallsetminus \mathbf{F}}$. In particular, for every $\varphi : (\mathbf{E} \smallsetminus \mathbf{F}) \to \Delta$, there exist $x_1, \ldots, x_M \in G$ such that $|\varphi(\gamma) - \sum_1^M c_m \gamma(x_m)| \leq 1/3$ for all $\gamma \in \mathbf{E} \smallsetminus \mathbf{F}$. $\square$

Corollary 3.5.2. *If* $\mathbf{E}$ *is an* I_0 *set and* $\mathbf{F}$ *is a finite set, then* $\mathbf{E} \cup \mathbf{F}$ *is* I_0.

Proof. Without loss of generality the sets $\mathbf{E}$ and $\mathbf{F}$ are disjoint. Because $\overline{\mathbf{E}} \smallsetminus \mathbf{E}$ contains no continuous characters and $\mathbf{F}$ is closed in the Bohr topology, Corollary 3.4.3 implies $\mathbf{E} \cup \mathbf{F}$ is I_0. $\square$

In fact, as with ε-Kronecker sets (Theorem 2.7.12), the Bohr closure of an I_0 set does not support a non-zero measure whose transform vanishes at infinity on G_d and thus is a U_0-set (see Definition 2.7.11).

Proposition 3.5.3. *If* $\mathbf{E}$ *is* I_0, *then* $M_0(\overline{\mathbf{E}}) = \{0\}$.

Proof. If $\varphi : \overline{\mathbf{E}} \to \Delta$ is continuous, then, since $\mathbf{E}$ is I_0, there is some $\mu \in M_d(G)$ with $\varphi = \widehat{\mu}|_{\mathbf{E}}$. By continuity, $\varphi = \widehat{\mu}$ on $\overline{\mathbf{E}}$. Therefore, $B_d(\overline{\mathbf{E}}) = A(\overline{\mathbf{E}}) = C(\overline{\mathbf{E}})$. A duality argument implies there is a constant C such that $\|\nu\|_{M(\overline{\mathbf{E}})} \leq C\|\widehat{\mu}\|_\infty$ for all $\nu \in M(\overline{\mathbf{E}})$.

For convenience, let $\Lambda = \widehat{\overline{\Gamma}} = G_d$. Now we proceed by contradiction. Suppose there exists $\mu \neq 0 \in M_0(\overline{\mathbf{E}})$. Since each measure in M_0 is continuous (Lemma C.1.9 (3)), we may find mutually singular probability measures $\nu_1, \nu_2, \cdots \ll \mu$. By Lemma C.1.9 (2), the $\nu_j \in M_0(\overline{\Gamma})$, so the Fourier-Stieltjes transforms of the ν_j's are 1 at $1_\Lambda \in \Lambda$ and tend to 0 away from 1_Λ (recall that Λ has the discrete topology). Inductively choose $\lambda_1, \lambda_2 \in \Lambda$ such that $\|\sum_1^J \widehat{\lambda_j \nu_j}\| \leq 2$ (just translate the humps). Since $\|\sum_1^J \lambda_j \nu_j\| = J$, there can be no finite C with $C\|\nu\|_\infty \geq \|\nu\|$ for all $\nu \in M_0(\overline{\mathbf{E}})$. This contradiction completes the proof. $\square$

Corollary 3.5.4. *If* **E** *is* I_0, *then* $m_{\overline{\Gamma}}(\overline{\mathbf{E}}) = 0$.

Remark 3.5.5. A closed subset, X, of a compact abelian group is called a *Helson* set if $A(X) = C_0(X)$. A subset is *Helsonian* if its closure is a Helson set. The first part of the proof of Proposition 3.5.3 shows that an I_0 set is Helsonian, and the second part of the proof shows that Helson sets are sets of uniqueness in the weak sense.

3.6 Remarks and Credits

Introduction. I_0 sets have been extensively studied since the 1960's and many of the important contributors from that period are listed in what follows. See also Sect. 1.6. The more restricted classes of $I_0(U)$, RI_0 and FZI_0 sets (and their further restrictions, $RI_0(U)$ and $FZI_0(U)$) were introduced in [51, 53]. The name "FZI_0" was suggested by the fact that every Sidon set that does not contain the identity has the Fatou–Zygmund property, meaning the interpolation of bounded Hermitian functions may always be done with a positive measure (see Corollary 6.3.4).

Other I_0 papers from the early years include [79, 81–83, 121, 122, 128, 130, 137, 168, 169, 180, 181, 194].

Characterizations: Approximate Interpolation. Theorem 3.2.5 was proved by Kalton [107] for I_0 sets; our treatment follows closely that of Ramsey [158]. Kalton credits Kahane and Méla for the idea and also gives other characterizations. In place of the Baire category theorem, Kalton's proof, like the one in the appendix of Ramsey's [158], uses the Shields–Kneser–Kemperman theorem, [112, 114, 172], which states, in particular, that if G is a compact, connected group, then $m_G(A \cdot B) \geq m_G(A) + m_G(B)$ for all Borel subsets, A, B, of G, unless $m_G(A) + m_G(B) > 1$, in which case $A \cdot B = G$.

See also [51, 53] for proofs of Theorems 3.2.5 and 3.2.6. Other basic properties of the more restrictive classes of sets are studied in Chap. 4.

Characterizations: Function Algebras and Topological. Theorem 3.3.1(1) $\Leftrightarrow$ (3) is due to Kahane [100]. Theorem 3.3.1(1) $\Leftrightarrow$ (2) and Exercise 3.7.12(2) are from [158], along with a number of other properties, some of which will be discussed in later chapters. The Ramsey–Wells–Bourgain characterization of I_0 sets as those with $B_d(\mathbf{E}) = B(\mathbf{E})$, Theorem 9.4.15, is from Ramsey and Wells [159] and Bourgain [18].

Proposition 3.4.1, the Hartman and Ryll-Nardzewski topological characterization, may be found in [81]. The separation condition of Theorem 3.4.4 is a slight improvement of [100, Proposition 3], which Kahane used in proving Theorem 3.3.1.

Clustering. The fact that I_0 sets cannot cluster at a continuous character, Theorem 3.5.1, was proved by Ryll-Nardzewski [169] for $G = \mathbb{R}$; the general case was proved by Méla [129, pp. 177–8] and (independently) by

Ramsey [156]. Ramsey shows that if $\gamma \in \Gamma$ is a Bohr cluster point of $\mathbf{E} \subseteq \Gamma$, then there exist disjoint sets $\mathbf{E}_1, \mathbf{E}_2 \subseteq \mathbf{E}$ which also cluster at γ, from which the non-clustering of I_0 sets easily follows. The proof here is adapted from Méla.

Extensions to LCA Groups. The definition of I_0 sets may be extended to subsets of locally compact abelian groups, such as $\mathbb{R}$, by requiring (only) the interpolation of continuous, bounded functions on $\mathbf{E}$ by the Fourier–Stieltjes transforms of discrete measures. Many of the papers mentioned above study I_0 sets in this more general setting and many of the results proven in this chapter continue to hold, particularly when Γ is metrizable. This is discussed in more detail in Sect. A.1.

Helson and Helsonian Sets. There is a large literature on Helson sets. See [56]. The term "Helsonian" was introduced by Kahane [100]. That closed Helson sets are sets of uniqueness in the weak sense is due to Helson [84]. The proof here is a variant of [31]. For a contemporary view on Helson's theorem, see [183].

3.7 Exercises

Exercise 3.7.1. Let $\Lambda \subseteq \Gamma$ be a subgroup and $\mathbf{E} \subseteq \Lambda$.

1. Show that $\mathbf{E}$ is I_0 (resp., FZI_0) as a subset of Γ if and only if it is I_0 (resp., FZI_0) as a subset of Λ.
2. Let $U \subseteq G$, $H = \Lambda^\perp$ and U' the image of U in G/H. Show that $\mathbf{E}$ is $I_0(U)$ (resp., $FZI_0(U)$) as a subset of Γ if and only if it is $I_0(U')$ (resp., $FZI_0(U')$) as a subset of Λ.

Exercise 3.7.2. 1. Show that $\mathbf{E}$ is I_0 if and only if there exist N and $\varepsilon < 1$ such that every finite subset $\mathbf{F} \subseteq \mathbf{E}$ is $I_0(G, N, \varepsilon)$.
2. Suppose $\mathbf{E}$ is $I_0(U, N, \varepsilon)$ for some $0 < \varepsilon < 1$ and $\delta > 0$ is given. Determine N' such that $\mathbf{E}$ is $I_0(U, N', \delta)$.

Exercise 3.7.3. Let $U \subseteq G$ be σ-compact. Suppose $AP(\mathbf{E}, U, N, \varepsilon) \supseteq \mathbb{Z}_2^{\mathbf{E}}$ for some $0 < \varepsilon < 1$.

1. Find a bound on the $I_0(U)$ constant of $\mathbf{E}$.
2. Suppose, in addition, that $\mathbf{E}$ is asymmetric and U is a symmetric e-neighbourhood. Show that every real-valued function on $\mathbf{E}$ may be interpolated by a measure $\mu \in M_d^r(U)$ with real-valued Fourier transform.

Exercise 3.7.4. Suppose there exist N and $\varepsilon < 1$ such that for each non-empty, open set U there is some finite set $\mathbf{F} \subset \mathbf{E}$ with $AP(\mathbf{E} \setminus \mathbf{F}, U, N, \varepsilon) \supseteq \mathbb{Z}_2^{\mathbf{E}}$. Show $\mathbf{E}$ is $I_0(U)$ of bounded length.

Exercise 3.7.5. 1. Show that $B_d(\mathbf{E}) \subseteq AP(\mathbf{E})$.
 2. Show that $\mathbf{E}$ is Helsonian if and only if $AP(\mathbf{E}) = B_d(\mathbf{E})$.

Exercise 3.7.6. Show that if $\mathbf{E}$ is I_0 then there exists a constant C such that $\|\widehat{\mu}\|_{B_d(\mathbf{E})} \leq C \|\widehat{\mu}\|_{\ell^\infty(\mathbf{E})}$.

Exercise 3.7.7. Let $U \subseteq G$ be a symmetric neighbourhood of e and suppose that for each $\varphi \in \mathbb{Z}_2^{\mathbf{E}}$ there exists $\mu \in M_d(U)$ such that

$$\sup\{|\varphi(\gamma) - \widehat{\mu}(\gamma)| : \gamma \in \mathbf{E}\} < 1.$$

Show that there is a finite set $\mathbf{F}$ such that $\mathbf{E} \setminus \mathbf{F}$ is $I_0(\overline{U})$.

Exercise 3.7.8. Let $\mathbf{E}$ be an ε-Kronecker set for some $\varepsilon < \sqrt{2}$.

 1. Prove that $\mathbf{E}$ is $FZI_0(N, 1/2)$ with N depending only on ε.
 2. Show that $\mathbf{E}$ is $FZI_0(U)$ with bounded length.

Exercise 3.7.9. Suppose $\mathbf{E}$ is an asymmetric set that contains no elements of order two. Suppose that for each $\varphi_1 : \mathbf{E} \to \mathbb{Z}_2$ and $\varphi_2 : \mathbf{E} \to i\mathbb{Z}_2$ there exist $\mu_1, \mu_2 \in M_d^+(\mathbf{E})$ with $\varphi_j = \widehat{\mu_j}$ on $\mathbf{E}$, $j = 1, 2$. Show that $\mathbf{E}$ is FZI_0.

Exercise 3.7.10. Prove that a finite set is I_0 by using the strategy outlined in Corollary 3.2.11.

Exercise 3.7.11. Show that an independent set is both FZI_0 and $FZI_0(U)$ with bounded length.

Exercise 3.7.12. 1. Prove that the union of two disjoint I_0 sets is I_0 if and only if the Bohr closures of the two sets are disjoint.
 2. Prove that $\mathbf{E}$ is I_0 if and only if whenever $\mathbf{E}_0 \subseteq \mathbf{E}$ there exist a discrete measure μ and closed disjoint subsets $C_0, C_1 \subseteq \mathbb{C}$ such that $\widehat{\mu}(\mathbf{E}_0) \subseteq C_0$ and $\widehat{\mu}(\mathbf{E} \setminus \mathbf{E}_0) \subseteq C_1$

Exercise 3.7.13. Give an example of an ε-Kronecker set $\mathbf{E} \subset \mathbb{Z}$ and a group homomorphism $\Phi : \mathbb{Z} \to \mathbb{Z}$ such that $\mathbf{E} \cup \Phi(\mathbf{E})$ is not I_0.

Exercise 3.7.14. 1. Give the details of the standard iteration argument in Corollary 3.3.2.
 2. Use the weak* density of the unit ball of $A(\mathbf{E})$ in $B(\mathbf{E})$ to prove that Corollary 3.3.2 continues to be true if "$f \in B(\mathbf{E})$" is replaced by "$f \in A(\mathbf{E})$".

Chapter 4
More Restrictive Classes of I_0 Sets

The classes I_0, RI_0 and FZI_0 are distinct. Criteria are given for a set in a larger class to belong to a smaller. Pseudo-Rademacher sets are studied. Every infinite discrete group is shown to contain a large subset that is $FZI_0(U)$ with bounded length. Every infinite subset of Γ is shown to contain a large subset that is $I_0(U)$ with bounded length.

4.1 Introduction

The focus of this chapter will be on more restrictive classes of I_0 sets. In the previous chapter, the $RI_0(U)$ and $FZI_0(U)$ sets were introduced—the sets $\mathbf{E} \subseteq \Gamma$ with the property that every bounded Hermitian function can be interpolated by a real (respectively, positive) discrete measure concentrated on U. Obviously, every $FZI_0(U)$ set is $RI_0(U)$ and it is not difficult to see that every $RI_0(U)$ set is $I_0(U)$. In Sect. 4.2, we will see that these three classes are distinct. Indeed, it will be seen that a set $\mathbf{E}$ is $RI_0(U)$ if and only if $\mathbf{E} \cup \mathbf{E}^{-1}$ is $I_0(U)$. Since $\widehat{\mu}(1) \geq 0$ whenever μ is a positive measure, the singleton $\{1\} \subseteq \Gamma$ is never FZI_0. This will be shown to be the only distinction between RI_0 and FZI_0, however. Topological characterizations of RI_0, similar to the HRN characterization of I_0, will also be given.

An example of an FZI_0 set is an independent set of characters of order 2, such as the set of Rademacher functions in $\widehat{\mathbb{D}}$. More generally, a set $\mathbf{E}$ with the property that every ± 1-valued function defined on $\mathbf{E}$ can be interpolated exactly by some $x \in G$ will be called a *pseudo-Rademacher* set. Pseudo-Rademacher sets are weak $\sqrt{2}$-Kronecker and I_0. In Sect. 4.4, it will be shown that they are $I_0(U)$ with bounded length, but need not be RI_0.

In the final section, the existence of large I_0 sets is investigated. We prove that every infinite subset of Γ contains a subset of the same cardinality that is either weak 1-Kronecker or pseudo-Rademacher and hence a subset that

C.C. Graham and K.E. Hare, *Interpolation and Sidon Sets for Compact Groups*,
CMS Books in Mathematics, DOI 10.1007/978-1-4614-5392-5_4,
© Springer Science+Business Media New York 2013

is $I_0(U)$ of bounded length. Moreover, an infinite Γ always contains a subset that is $FZI_0(U)$ with bounded length and of the same cardinality.

4.2 Distinctiveness of the Classes I_0, RI_0 and FZI_0

4.2.1 Which I_0 Sets Are RI_0?

Proposition 4.2.1. *Let* $\mathbf{E} \subseteq \Gamma$ *and* $U \subseteq G$ *be a symmetric neighbourhood of the identity. Then* $\mathbf{E}$ *is* $RI_0(U)$ *if and only if* $\mathbf{E} \cup \mathbf{E}^{-1}$ *is* $I_0(U)$.

Proof. Assume $\mathbf{E} \cup \mathbf{E}^{-1}$ is $I_0(U)$. Let $\varphi : \mathbf{E} \to \mathbb{C}$ be a bounded Hermitian function and extend φ to $\mathbf{E}^{-1} \setminus \mathbf{E}$ by the rule $\varphi(\gamma^{-1}) = \overline{\varphi(\gamma)}$ for $\gamma \in \mathbf{E}$. Since $\mathbf{E} \cup \mathbf{E}^{-1}$ is $I_0(U)$, we may obtain $\mu \in M_d(U)$ with $\hat{\mu}(\gamma) = \varphi(\gamma)$ for $\gamma \in \mathbf{E} \cup \mathbf{E}^{-1}$. Put $\nu = (\mu + \overline{\mu})/2$. Then $\nu \in M_d^r(U)$ and for $\gamma \in \mathbf{E}$,

$$\varphi(\gamma) = \frac{1}{2}(\hat{\mu}(\gamma) + \overline{\hat{\mu}(\gamma^{-1})}) = \hat{\nu}(\gamma).$$

Therefore, $\mathbf{E}$ is $RI_0(U)$.

Now assume $\mathbf{E}$ is $RI_0(U)$. To show that $\mathbf{E} \cup \mathbf{E}^{-1}$ is $I_0(U)$, we will use the fact that every bounded function $\varphi : \mathbf{E} \cup \mathbf{E}^{-1} \to \mathbb{C}$ is a complex linear combination of bounded Hermitian functions. Indeed, put $\varphi = \psi_1 - i\psi_2$ where $\psi_1(\gamma) = (\varphi(\gamma) + \varphi(\gamma^{-1}))/2$ and $\psi_2(\gamma) = i(\varphi(\gamma) - \varphi(\gamma^{-1}))/2$. Then ψ_1 and ψ_2 are Hermitian functions defined on $\mathbf{E}$, and hence, by assumption, there are measures $\mu_j \in M_d^r(U)$ with $\widehat{\mu_j}(\gamma) = \psi_j(\gamma)$ for $\gamma \in \mathbf{E}$. Let

$$\nu_1 = \frac{1}{2}(\mu_1 + \widetilde{\mu_1}) \text{ and } \nu_2 = \frac{1}{2}(\mu_2 - \widetilde{\mu_2}).$$

Both ν_j belong to $M_d^r(U)$. Since ψ_1 and ψ_2 are purely real- and imaginary-valued, respectively, it is easy to check that $\hat{\nu}_j = \psi_j$ on $\mathbf{E} \cup \mathbf{E}^{-1}$. Therefore, if $\nu = (\nu_1 - i\nu_2)$, then $\nu \in M_d(U)$ and $\varphi = \hat{\nu}$ on $\mathbf{E} \cup \mathbf{E}^{-1}$.   □

Corollary 4.2.2. *1. Finite sets are* RI_0.
2. The union of an RI_0 *set and a finite set is* RI_0.

Proof. These follow from the analogous results for I_0 sets (Corollaries 3.2.11 and 3.5.2) and the previous proposition.   □

Example 4.2.3. An asymmetric subset of $\mathbb{Z}$ that is I_0, but not RI_0: Put $\mathbf{E}_1 = \{8^j + 4j + 1 : j \in \mathbb{N}\}$, $\mathbf{E}_2 = \{8^j + 1 : j \in \mathbb{N}\}$ and $\mathbf{E} = \mathbf{E}_1 \cup -\mathbf{E}_2$. Of course, $\mathbf{E}_1$ and $\mathbf{E}_2$ are both Hadamard sets and are therefore I_0. The set $\mathbf{E} \cup -\mathbf{E}$ is not I_0 since the closures of $\mathbf{E}_1$ and $\mathbf{E}_2$ are not disjoint. Thus, $\mathbf{E}$ is not RI_0.
    For $g = \pi/2$ we have

$$\widehat{\delta_g}(n) = \begin{cases} i = e^{i\pi/2} & \text{for all } n = -8^j - 1 \in -\mathbf{E}_2, \\ -i = e^{-i\pi/2} & \text{for all } n = 8^j + 4j + 1 \in \mathbf{E}_1. \end{cases}$$

Therefore, $\overline{\mathbf{E}}_1 \cap -\overline{\mathbf{E}}_2 = \emptyset$, so $\mathbf{E}_1 \cup -\mathbf{E}_2$ is I_0.

4.2.2 Which RI_0 Sets Are FZI_0?

Interestingly, the only barrier to an RI_0 set being FZI_0 is the inclusion of the identity character. The proof of this is based upon the fact that the closure of an I_0 set in $\overline{\Gamma}$ is a Helson set (see Remark 3.5.5 for the definition) and on the property of Helson sets given by the following theorem. This is known as the Fatou–Zygmund, or FZ, property in the Sidon set context (Definition 6.2.1).

Theorem 4.2.4 (Smith's FZ theorem). *Let $\mathbf{E} \subset \overline{\Gamma}$ be a compact, symmetric Helson set with $1 \notin \mathbf{E}$. Then there exists $C > 0$ such that for all continuous, Hermitian $\varphi : \mathbf{E} \to \Delta$ there exists $\mu \in M^+(G_d)$ such that $\varphi = \widehat{\mu}$ on $\mathbf{E}$ and $\|\mu\| \leq C$.*

We can now characterize which RI_0 sets are FZI_0.

Proposition 4.2.5. *An RI_0 set $\mathbf{E} \subseteq \Gamma$ is an FZI_0 set if and only if $\mathbf{E}$ does not contain the identity element.*

Proof. We have already observed that an FZI_0 set cannot contain the identity, so assume that $\mathbf{E}$ is RI_0 and does not contain the identity. Since $\mathbf{E}$ is I_0, 1 does not belong to $\overline{\mathbf{E}}$, either. As noted above, $\overline{\mathbf{E}}$ is Helson in $\overline{\Gamma}$.

Smith's FZ Theorem 4.2.4 tells us that every Hermitian, continuous $\varphi : \mathbf{E} \to \Delta$ is the restriction of the Fourier-Stieltjes transform of a measure in $M^+(G_d)$. Since $M^+(G_d) = M_d^+(G)$, we see that $\mathbf{E}$ is FZI_0. □

We do not know if there is always a cofinite subset of an $RI_0(U)$ set that is $FZI_0(U)$ **[P 10]**.

4.3 A Topological Characterization of RI_0

It follows directly from the definitions that a set $\mathbf{E}$ is RI_0 if and only if $\mathbf{E} \cup \mathbf{E}^{-1}$ is RI_0, so in studying the property RI_0 there is no loss in working with asymmetric sets.

Proposition 4.3.1. *An asymmetric set $\mathbf{E}$ is RI_0 if and only if both the following conditions are satisfied:*

1. *For every $\mathbf{F} \subseteq \mathbf{E}$ there exists $\sigma \in M_d^r(G)$, with real-valued Fourier transform, such that $\widehat{\sigma}(\mathbf{F})$ and $\widehat{\sigma}(\mathbf{E} \setminus \mathbf{F})$ have disjoint closures in $\mathbb{C}$.*

2. *The closure of $\{\gamma \in \mathbf{E} : \gamma^2 \neq 1\} \subseteq \overline{\Gamma}$ in the Bohr topology does not contain any elements of order two.*

Proof. First, assume $\mathbf{E}$ is RI_0 and let $\mathbf{F} \subseteq \mathbf{E}$. Because $\mathbf{E}$ is asymmetric, the function φ given by $\varphi(\gamma) = 0$ for $\gamma \in \mathbf{F}$ and $\varphi(\gamma) = 1$ for $\gamma \in \mathbf{E} \smallsetminus \mathbf{F}$ is bounded and Hermitian on $\mathbf{E}$. Thus, there is a measure $\mu \in M_d^r(G)$ with $\widehat{\mu} = \varphi$ on $\mathbf{E}$, and, taking $\sigma = \mu + \widetilde{\mu}$, we see that (1) holds.

Let $\mathbf{E}_0 = \{\gamma \in \mathbf{E} : \gamma^2 \neq 1\}$ and suppose there is some $\chi \in \overline{\mathbf{E}_0}$ of order two. Notice that χ also belongs to the closure of $\mathbf{E}_0^{-1}$. The function which is equal to i on $\mathbf{E}_0$ and 0 otherwise on $\mathbf{E}$ is Hermitian and thus may be interpolated by a real, discrete measure μ. By continuity, $\widehat{\mu}(\chi) = i$.

Since μ is a real measure, $\widehat{\mu}(\gamma^{-1}) = \overline{\widehat{\mu}(\gamma)}$ for all $\gamma \in \Gamma$. In particular, $\widehat{\mu}(\gamma) = -i$ for $\gamma \in \mathbf{E}_0^{-1}$. This must also hold on the closure of $\mathbf{E}_0^{-1}$, contradicting the fact that $\widehat{\mu}(\chi) = i$ and $\chi \in \overline{\mathbf{E}_0^{-1}}$. This shows (2) holds.

Conversely, assume properties (1)–(2) hold. To prove $\mathbf{E}$ is RI_0 we will establish that $\mathbf{E} \cup \mathbf{E}^{-1}$ is I_0. By the HRN characterization of I_0, property (1) implies that $\mathbf{E}$ (and similarly $\mathbf{E}^{-1}$) is I_0. Therefore, it will be enough to prove that $\mathbf{E}$ and $\mathbf{E}^{-1} \smallsetminus \mathbf{E}$ have disjoint closures in $\overline{\Gamma}$.

Suppose otherwise. Then there are nets $\{\chi_\alpha\} \subseteq \mathbf{E}$ and $\{\gamma_\beta\} \subseteq \mathbf{E}^{-1} \smallsetminus \mathbf{E}$ that have the same limit ψ. Observe that the characters γ_β are not of order 2 because $\gamma_\beta \in \mathbf{E}^{-1} \smallsetminus \mathbf{E}$. Hence, $\psi \in \overline{\mathbf{E}_0}$ and so property (2) implies ψ is not of order two. That shows that $\{\chi_\alpha\}$ and $\{\gamma_\beta^{-1}\}$ are nets in $\mathbf{E}$ with different limits, ψ, ψ^{-1}, respectively. Since these nets eventually belong to disjoint neighbourhoods of ψ and ψ^{-1}, respectively, there is no loss of generality in assuming the nets are disjoint.

Apply property (1) with $\mathbf{F} = \{\chi_\alpha\}_\alpha$. Since σ is a real measure with real-valued Fourier transform, $\widehat{\sigma}(\psi) = \widehat{\sigma}(\psi^{-1})$. But $\widehat{\sigma}(\psi)$ and $\widehat{\sigma}(\psi^{-1})$ belong to the disjoint sets $\overline{\widehat{\sigma}(\mathbf{F})}$ and $\overline{\widehat{\sigma}(\mathbf{E} \smallsetminus \mathbf{F})}$, respectively, giving a contradiction. □

Corollary 4.3.2. *Suppose Γ has no elements of order two. Then an asymmetric $\mathbf{E} \subset \Gamma$ is RI_0 if and only if for every $\mathbf{F} \subseteq \mathbf{E}$ there exists $\sigma \in M_d^r(G)$, with real-valued Fourier transform, such that $\widehat{\sigma}(\mathbf{F})$ and $\widehat{\sigma}(\mathbf{E} \smallsetminus \mathbf{F})$ have disjoint closures.*

Proof. If Γ has no elements of order two, then the same is true for $\overline{\Gamma}$ (Exercise C.4.11(3)), and hence property (2) is vacuous. □

Remark 4.3.3. Call $\mathbf{E} \subseteq \Gamma$ a *real RI_0 set* (respectively, a *real FZI_0 set*) if for every real-valued, bounded, Hermitian $\varphi : \mathbf{E} \to \mathbb{C}$ there is a real (resp., positive) discrete measure μ with $\widehat{\mu} = \varphi$ on $\mathbf{E}$. Real RI_0 and real FZI_0 sets may be characterized in a similar manner to the main characterization theorems for I_0 sets. See Exercise 4.7.4.

Arguments similar to those used to prove the HRN characterization of I_0 also show that an asymmetric set $\mathbf{E}$ is real RI_0 if and only if for every $\mathbf{F} \subseteq \mathbf{E}$, there exists $\sigma \in M_d^r(G)$, with real-valued Fourier transform, such that $\widehat{\sigma}(\mathbf{F})$ and $\widehat{\sigma}(\mathbf{E} \smallsetminus \mathbf{F})$ have disjoint closures. Thus, Corollary 4.3.2 implies that for groups with no elements of order two, real RI_0 and RI_0 coincide for asymmetric sets.

The set $\mathbf{E}$ of Example 4.2.3 is I_0, but not real RI_0 since it is an asymmetric set in $\mathbb{Z}$ which is not RI_0. The example below shows that real FZI_0 and RI_0 are also distinct classes.

Example 4.3.4. A real FZI_0 set that is not RI_0: Let $\mathbf{E} = \{(j, \pi_j) : j \in \mathbb{N}\} \subseteq \mathbb{Z} \oplus \widehat{\mathbb{D}}$ where the $\{\pi_j\}$ are the Rademacher functions in $\widehat{\mathbb{D}}$. In Example 2.5.6 we showed that this set is not a finite union of ε-Kronecker sets. The independence of the characters $\{\pi_j\}$ ensures that we may interpolate ± 1-valued functions on $\mathbf{E}$ by Fourier transforms of positive discrete measures with real transforms. Of course, this implies that $\mathbf{E}$ is I_0 (see Corollary 3.2.17). That this interpolation property is enough to prove that $\mathbf{E}$ is real FZI_0 is left to Exercise 4.7.4. Choosing a net $(j_\alpha, \pi_{j_\alpha})$ such that $j_\alpha \to 0$ in $\overline{\mathbb{Z}}$, we see that $\mathbf{E}$ and $\mathbf{E}^{-1}$ do not have disjoint closures. Thus, $\mathbf{E}$ is not RI_0.

4.4 Pseudo-Rademacher Sets

By a *Rademacher set* we mean an independent set in Γ consisting only of characters of order two. The prototypical example is the set $\{\pi_n\}_{n=1}^{\infty} \subseteq \widehat{\mathbb{D}}$ of Rademacher functions defined on page xvii. A Rademacher set, $\mathbf{E}$, is FZI_0, since given any choice of $\varphi \in \mathbb{Z}_2^{\mathbf{E}}$, there exists $x \in G$ with $\varphi(\gamma) = \gamma(x)$. We will see later in this section that a Rademacher set is even $FZI_0(U)$ with bounded length. This example motivates the following (weaker) definition.

Definition 4.4.1. The set $\mathbf{E} \subseteq \Gamma$ is called *pseudo-Rademacher(U)* if for every $\varphi : \mathbf{E} \to \mathbb{Z}_2$ there exists $x \in U$ such that $\varphi(\gamma) = \gamma(x)$ for all $\gamma \in \mathbf{E}$.

If a pseudo-Rademacher set contains only elements of order 2, then it is a Rademacher set (Exercise 4.7.6). But there are other pseudo-Rademacher sets: the set of Example 4.3.4, any translate of a Rademacher set by a character independent of that set and any independent set of characters of even orders are all examples of pseudo-Rademacher sets. An infinite pseudo-Rademacher set does not necessarily contain a translate of an infinite Rademacher set.

Pseudo-Rademacher sets are obviously weak $\sqrt{2}$-Kronecker, as well as being I_0. However, they need not be RI_0 (much less FZI_0), as Example 4.3.4 demonstrates, or even contain *any* subsets that are $FZI_0(U)$, as the next example illustrates.

Example 4.4.2. An I_0 set that is pseudo-Rademacher and contains no subset that is $FZI_0(U)$ for some non-empty, open set U: Let $\mathbf{E}_1 = \{\pi_n\}_{n=1}^{\infty} \subset \widehat{\mathbb{D}}$ be the set of Rademacher functions and suppose $\gamma \in \mathbb{Z}_3$ has order three.

Let $\mathbf{\Gamma} = \mathbb{Z}_3 \oplus \widehat{\mathbb{D}}$ and $\mathbf{E} = \{\gamma\} \times \mathbf{E}_1 \subseteq \mathbf{\Gamma}$. No subset of $\mathbf{E}$ is $FZI_0(U)$ for $U = \{x \in G : \gamma(x) = 1\}$ since the Fourier transform of every positive measure concentrated on U takes on only real values on $\mathbf{E}$. The set $\mathbf{E}$ is pseudo-Rademacher since ± 1 can be interpolated exactly and hence I_0. In fact, $\mathbf{E}$ is even FZI_0 (see Exercise 4.7.3).

However, pseudo-Rademacher sets *are* $I_0(U)$ with bounded length. That requires another application of the Baire category theorem, similar to Theorem 2.2.13.

Lemma 4.4.3. *Suppose $\mathbf{E}$ is a pseudo-Rademacher set and $U \subseteq G$ is a neighbourhood of e. Then there is a finite set $\mathbf{F}$ such that $\mathbf{E} \smallsetminus \mathbf{F}$ is pseudo-Rademacher(U).*

Proof. Suppose U is an e-neighbourhood and choose a compact, symmetric e-neighbourhood V such that $V^2 \subseteq U$. By compactness, there are finitely many elements $g_1, \ldots, g_K \in G$ such that $G = \bigcup_{k=1}^{K} g_k V$. Let

$$X_k = \{\varphi \in \mathbb{Z}_2^{\mathbf{E}} : \exists\, x \in g_k V \text{ such that } \varphi(\gamma) = \gamma(x) \,\forall \gamma \in \mathbf{E}\}.$$

The sets X_k are closed in $\mathbb{Z}_2^{\mathbf{E}}$. Since $\mathbf{E}$ is pseudo-Rademacher, their union is all of the compact Hausdorff space $\mathbb{Z}_2^{\mathbf{E}}$.

By the Baire category theorem, one of the sets X_k has non-empty interior. That means there is a finite set $\mathbf{F}$ such that for each $\varphi \in \mathbb{Z}_2^{\mathbf{E}}$ there is an $x \in g_k V$ such that $\varphi(\gamma) = \gamma(x)$ for all $\gamma \in \mathbf{E} \smallsetminus \mathbf{F}$.

Because the trivial character $\mathbf{1}$ is also in $\mathbb{Z}_2^{\mathbf{E}}$, there is some $y \in g_k V$ such that $\gamma(y) = 1$ for all $\gamma \in \mathbf{E} \smallsetminus \mathbf{F}$. But then also $\gamma(y^{-1}) = 1$ for all such γ. Take $z = xy^{-1} \in V^2 \subseteq U$. Then $\gamma(z) = \varphi(\gamma)$ for all $\gamma \in \mathbf{E} \smallsetminus \mathbf{F}$, and that shows $\mathbf{E} \smallsetminus \mathbf{F}$ is pseudo-Rademacher(U). $\quad\square$

Corollary 4.4.4. *Every translate of a pseudo-Rademacher set is $I_0(U)$ with bounded length.*

Proof. Because the property $I_0(U)$ with bounded length is preserved under translation, there is no loss of generality in assuming the translate, $\mathbf{E}$, is actually pseudo-Rademacher.

Use Lemma 4.4.3 to obtain a finite set $\mathbf{F}$ such that $\mathbf{E} \smallsetminus \mathbf{F}$ is pseudo-Rademacher(U). The arguments of Proposition 3.2.16 easily show that $\mathbf{E} \smallsetminus \mathbf{F}$ is $I_0(U, 4, 1/2)$. $\quad\square$

Remark 4.4.5. Similar arguments show that if $\mathbf{E}$ is a Rademacher set, then $\mathbf{E} \smallsetminus \mathbf{F}$ is $FZI_0(U, 1, 1/2)$ since only real-valued functions would need to be interpolated.

Other properties of pseudo-Rademacher sets are derived in the exercises.

4.5 The Existence of Large I_0 and FZI_0 Sets

In Theorem 2.4.3 it was shown that whenever $\mathbf{E} \subseteq \mathbf{\Gamma}$ is not 2-large (see Definition 2.4.1), then $\mathbf{E}$ contains a weak 1-Kronecker subset, $\mathbf{F}$, of the same cardinality as $\mathbf{E}$. As noted in Remark 3.2.15, such a set $\mathbf{F}$ is $FZI_0(U)$ with bounded length. Example 4.4.2 shows this need not be true for a set that is 2-large.

In this section, it will be shown that every 2-large set contains a subset of the same cardinality that is pseudo-Rademacher and hence is $I_0(U)$ with bounded length. It will also be shown that every infinite 2-large group $\mathbf{\Gamma}$ contains a Rademacher set of the same cardinality as $\mathbf{\Gamma}$. Thus, every infinite group contains a subset, of the same cardinality as the group, that is $FZI_0(U)$ with bounded length.

We first prove a lemma.

Lemma 4.5.1. *Every infinite set $\mathbf{E}$ of elements of order 2 contains an independent subset of the same cardinality.*

Proof (of Lemma 4.5.1). Use Zorn's lemma to find a maximal independent set, $\mathbf{A} \subseteq \mathbf{E}$, of elements of order two (see p. 30 for details). We claim $|\mathbf{A}| = |\mathbf{E}|$. If not, let $\mathbf{X}$ be the group generated by $\mathbf{A}$. Since $|\mathbf{A}| < |\mathbf{E}|$, we have $|\mathbf{X}| < |\mathbf{E}|$, and hence there is some $\gamma \in \mathbf{E} \setminus \mathbf{X}$. Because every element of $\mathbf{E}$ has order 2, it follows that $\mathbf{A} \cup \{\gamma\}$ is independent, and that contradicts the maximality of $\mathbf{A}$. $\square$

Theorem 4.5.2. *Let $\mathbf{\Gamma}$ be an infinite discrete group.*

1. *Then $\mathbf{\Gamma}$ contains a subset $\mathbf{F}$ with $|\mathbf{F}| = |\mathbf{\Gamma}|$ that is FZI_0 and $FZI_0(U)$ with bounded length.*
2. *Suppose $\mathbf{E} \subseteq \mathbf{\Gamma}$ is infinite.*

    (i) *If $\mathbf{E}$ is 2-large, then $\mathbf{E}$ contains a subset $\mathbf{F}$ with $|\mathbf{F}| = |\mathbf{E}|$ that is I_0 and $I_0(U)$ with bounded length.*

    (ii) *If $\mathbf{E}$ is not 2-large, then $\mathbf{E}$ contains a subset $\mathbf{F}$ with $|\mathbf{F}| = |\mathbf{E}|$ that is FZI_0 and $FZI_0(U)$ with bounded length.*

In addition, in all cases we may choose the subset $\mathbf{F}$ so that $\mathbf{F} \cdot \mathbf{F}^{-1}$ does not cluster at a non-trivial continuous character.

Corollary 4.5.3 (Hartman–Ryll-Nardzewski existence). *Let $U \subseteq G$ be non-empty and open. Every infinite subset of $\mathbf{\Gamma}$ contains a subset of the same cardinality that is $I_0(U)$.*

Remark 4.5.4. We do not know if there is a set that does not contain an FZI_0 subset of the same cardinality [P 8].

Proof (of Theorem 4.5.2). Part (2ii) holds since Theorem 2.4.3 shows that when **E** is not 2-large one may even choose **F** to be weak 1-Kronecker.

Of course, this fact also establishes (1), except when Γ is 2-large. But then the subgroup of Γ consisting of the characters of order 2 has the same cardinality as Γ. Appealing to Lemma 4.5.1, we may find an independent set of characters of order 2 (hence, a Rademacher set) of the same cardinality as Γ. Such a set is clearly FZI_0 and, as noted in Remark 4.4.5, it is also $FZI_0(U)$ with bounded length.

To prove (2i) suppose **E** is 2-large. As in (2.4.1), we may assume Γ is a subgroup of $\bigoplus_\alpha \Gamma_\alpha$, where each Γ_α is either $\mathbb{C}(p^\infty)$ or $\mathbb{Q}$. Being 2-large, the projection Π of **E** onto the direct sum of the factor groups $\mathbb{C}(2^\infty)$ must contain as many elements of order 2 as there are in **E**.

Appealing to Lemma 4.5.1 again, we obtain an independent subset, **F**, of characters of order two with the same cardinality as **E**. For each $\gamma \in \mathbf{F}$, choose one element $\chi_\gamma \in \mathbf{E}$ such that $\Pi(\chi_\gamma) = \gamma$. It is an easy exercise to check that the resulting subset of **E** is pseudo-Rademacher. By Corollary 4.4.4, it is both I_0 and $I_0(U)$ with bounded constants.

The final comment holds since both weak 1-Kronecker and pseudo-Rademacher sets have the required property; see Proposition 2.7.4 and Exercise 4.7.7. $\square$

The proof just given also establishes the following corollary.

Corollary 4.5.5. *If* **E** *is an infinite subset of* Γ, *then* **E** *contains either a weak 1-Kronecker subset or a pseudo-Rademacher subset of the same cardinality as* **E**. *If* $\mathbf{E} = \Gamma$, *the pseudo-Rademacher set can be chosen to be Rademacher.*

4.6 Remarks and Credits

Distinctiveness of the Classes I_0, RI_0 and FZI_0. Basic properties of the more restricted classes of RI_0 and FZI_0 sets (and their further restrictions, $RI_0(U)$ and $FZI_0(U)$) were studied in [53], including the characterization of which I_0 sets are RI_0, Proposition 4.2.1 and Example 4.2.3.

Theorem 4.2.4 is due to Smith [175].

The proof that all RI_0 sets that do not contain **1** are FZI_0 may be found in [57]. The topological characterization of RI_0, Proposition 4.3.1, may also be found in that paper, as well as basic properties of the classes real RI_0 and real FZI_0 and Example 4.3.4.

Pseudo-Rademacher Sets. Pseudo-Rademacher sets were introduced in [59] and their Kronecker-like properties established there, including the fact that if **E** is pseudo-Rademacher, then $M_0((\mathbf{E} \cup \mathbf{E}^{-1})^k) = \{0\}$ for all positive integers k.

For the classical Rademacher functions and their properties, see [199, I.3 and V.8].

Large I_0 and FZI_0 Sets. The existence of I_0 sets of cardinality $|\Gamma|$ was first proved by Hartman and Ryll-Nardzewski [81, Theorem 5]; another proof was given by Kunen and Rudin [116, Theorem 1.4]. Kalton [75, §4] proved that every infinite subset of Γ contains an infinite I_0 set.

The material on the existence of large I_0 or FZI_0 sets is mainly taken from [59]. More detailed properties of large I_0 or FZI_0 sets are also given there.

4.7 Exercises

Exercise 4.7.1. Exhibit two FZI_0 sets $\mathbf{E}, \mathbf{F} \subset \mathbb{Z}$, whose union is not RI_0, but for which there is a positive, discrete measure σ with the property that $\widehat{\sigma}(\mathbf{E})$ and $\widehat{\sigma}(\mathbf{F})$ have disjoint closures.

Exercise 4.7.2. Show that property (1) of Proposition 4.3.1 need not imply property (2).

Exercise 4.7.3. Let $\mathbf{E} = \{(\gamma, \pi_n)\} \subset \mathbb{Z}_3 \oplus \widehat{\mathbb{D}}$, where $\{\pi_n\}_{n=1}^{\infty}$ is the set of Rademacher functions and γ is an order 3 element of $\mathbb{Z}_3$. (This is the set $\mathbf{E}$ of Example 4.4.2.) Show that $\mathbf{E}$ is FZI_0.

Exercise 4.7.4. 1. Derive the characterizations of real RI_0 sets and real FZI_0 sets mentioned in Remark 4.3.3.
2. Let $\mathbf{E} = \{(j, \pi_j) : j = 1, 2, \dots\} \subseteq \mathbb{Z} \oplus \widehat{\mathbb{D}}$ where $\{\pi_j\} \subset \widehat{\mathbb{D}}$ is the set of Rademacher functions (the set $\mathbf{E}$ of Example 4.3.4). Show that $\mathbf{E}$ is real FZI_0.

Exercise 4.7.5. Show that any translate of a pseudo-Rademacher set is $I_0(U, 4, 1/2)$.

Exercise 4.7.6. Suppose that $\mathbf{E}$ is pseudo-Rademacher and all elements of $\mathbf{E}$ have order 2. Show that $\mathbf{E}$ is a Rademacher set.

Exercise 4.7.7. 1. Prove that pseudo-Rademacher sets do not cluster at a continuous character using the method of Proposition 2.7.1.
2. Show that if $\mathbf{E}$ is a pseudo-Rademacher set, then $\mathbf{E} \cdot \mathbf{E}^{-1}$ does not cluster at a continuous character other than $\mathbf{1}$.
3. Give an example of a pseudo-Rademacher set $\mathbf{E}$ such that $\mathbf{E} \cdot \mathbf{E}$ clusters at a continuous character other than $\mathbf{1}$.

Exercise 4.7.8. 1. Show that if G is infinite, then $|\overline{\Gamma}| = 2^{|G|}$. Hint: Consider a large I_0 set, $\mathbf{E}$, and show that $|\ell^{\infty}(\mathbf{E})| = 2^{|\mathbf{E}|}$.
2. Show that if Γ is infinite, then $|G| = 2^{|\Gamma|}$.

3. A locally compact abelian group H is *self-dual* if $\widehat{H}$ and H are isomorphic as topological groups. Assuming the continuum hypothesis, show that if H is self-dual, then one of the following hold: $|H| < \infty$, $|H| = |\mathbb{R}|$ or $|H| \geq 2^{|\mathbb{R}|}$. Hint: Use the fact that every locally compact abelian group has the form $\mathbb{R}^n \times \Lambda$, where Λ has an open compact subgroup (This is known as the structure theorem; see [167, Theorem 2.4.1] or [87, Theorem 9.8].)

Chapter 5
Unions and Decompositions of $I_0(U)$ Sets

Criteria are given for the union of $I_0(U)$ sets to be $I_0(U)$. Every I_0 set is a finite union of $I_0(U)$ sets with bounded length. Criteria for an I_0 set to be a finite union of RI_0 (or FZI_0) sets are also given.

5.1 Introduction

In the previous chapter we saw that every infinite I_0 set contains a subset of the same cardinality that is $I_0(U)$ with bounded length and even FZI_0 and $FZI_0(U)$ with bounded length if the set is not 2-large (Theorem 4.5.2). In this chapter more will be proven. Every I_0 set will be shown to be a finite union of sets that are $I_0(U)$ with bounded length, and characterizations of I_0 sets that are finite unions of RI_0 or FZI_0 sets will be established. Moreover, every I_0 set in the dual of a connected group will be seen to be $I_0(U)$ for all non-empty, open sets U. The latter fails without the connectedness assumption; just consider a finite group G and the open set $U = \{e\}$. No set of two characters is $I_0(U)$ and yet every subset of characters of this finite Γ is I_0.

Sets that are $I_0(U)$ with bounded length are of interest because of their special properties. For example, like ε-Kronecker sets, their step length tends to infinity, any finite union of translates is I_0, and the product of such a set with its inverse does not cluster at any non-trivial, continuous character.

The proofs of these statements involve a study of the union problem for $I_0(U)$ sets. As remarked above, even the union of two singletons need not be $I_0(U)$. When G is connected there are, however, union results for $I_0(U)$ sets similar to those for I_0 sets. For instance, if two sets with disjoint closures are each $I_0(U)$ for all non-empty, open U, then the same is true for their union. In particular, when G is connected, a set that is $I_0(U)$ with bounded length is $I_0(U)$ for all non-empty, open sets U.

C.C. Graham and K.E. Hare, *Interpolation and Sidon Sets for Compact Groups*,
CMS Books in Mathematics, DOI 10.1007/978-1-4614-5392-5_5,
© Springer Science+Business Media New York 2013

5.2 Unions of $I_0(U)$ Sets

An important consequence of the HRN characterization of I_0 sets is that the union of two disjoint I_0 sets is I_0 if (and only if) the two sets have disjoint closures in $\overline{\Gamma}$ (Corollary 3.4.3). In duals of connected groups, a similar result holds for $I_0(U)$ sets. This will be important in the proof that I_0 sets in duals of connected groups are $I_0(U)$.

We begin with a lemma.

Lemma 5.2.1. *Suppose $U, V, W \subseteq G$ are e-neighbourhoods. Assume $\mathbf{E}$ is $I_0(U)$, $\mathbf{F}$ is $I_0(V)$ and there is a measure $\nu \in M_d(W)$ such that $\widehat{\nu} = 1$ on $\mathbf{E}$ and 0 on $\mathbf{F}$. Then $\mathbf{E} \cup \mathbf{F}$ is $I_0(UW \cup VW)$.*

Proof. The I_0 properties of $\mathbf{E}$ and $\mathbf{F}$ imply that given a $\varphi \in \ell^\infty(\mathbf{E} \cup \mathbf{F})$ there are discrete measures $\mu_1 \in M_d(U)$ and $\mu_2 \in M_d(V)$ such that $\widehat{\mu_1} = \varphi$ on $\mathbf{E}$ and $\widehat{\mu_2} = \varphi$ on $\mathbf{F}$. The discrete measure $\nu * \mu_1 + (\delta_e - \nu) * \mu_2$ is concentrated on $UW \cup VW$ and its transform agrees with φ on $\mathbf{E} \cup \mathbf{F}$. $\square$

Proposition 5.2.2. *Suppose G is a connected group and that $U, V \subseteq G$ are symmetric e-neighbourhoods. Assume that $\mathbf{E}$ is $I_0(U)$, $\mathbf{F}$ is $I_0(V)$ and $\overline{\mathbf{E}} \cap \overline{\mathbf{F}}$ is empty. Then $\mathbf{E} \cup \mathbf{F}$ is $I_0(U^2 V^4 \cup UV^5)$.*

Proof. In view of Lemma 5.2.1, it will suffice to prove that there exists $\nu \in M_d(UV^4)$ with $\widehat{\nu} = 1$ on $\mathbf{E}$ and 0 on $\mathbf{F}$.

Let W be a symmetric e-neighbourhood with $W^2 \subset U \cap V$. Fix a $\gamma \in \overline{\mathbf{E}}$. Because G is connected and $\overline{\mathbf{E}} \cap \overline{\mathbf{F}} = \emptyset$, for every $\lambda \in \overline{\mathbf{F}}$ there is $w \in W$ with $\lambda(w)\gamma^{-1}(w) \neq 1$. Let

$$\nu_\lambda = \frac{1}{|1 - \gamma(w)\lambda^{-1}(w)|^2}(\delta_e - \gamma(w)\delta_w) * (\delta_e - \overline{\gamma(w)}\delta_{w^{-1}}) \in M_d(W^2).$$

Then $\widehat{\nu_\lambda}(\gamma) = 0$, $\widehat{\nu_\lambda}(\lambda) = 1$, $\widehat{\nu_\lambda} \geq 0$ everywhere on $\overline{\Gamma}$ and $\nu_\lambda \in M_d(V)$.

For $\lambda \in \overline{\mathbf{F}}$, set $\Omega_\lambda = \{\rho \in \overline{\Gamma} : \widehat{\nu_\lambda}(\rho) > 1/2\}$. By the compactness of $\overline{\mathbf{F}}$, a finite number of the Ω_λ cover $\overline{\mathbf{F}}$. Hence, there are $\lambda_1, \ldots, \lambda_N$ such that $\tau_1 := \sum_1^N \nu_{\lambda_n} \in M_d(V)$ has $\widehat{\tau_1} \geq 1/2$ on $\overline{\mathbf{F}}$ and 0 at γ. Because $\mathbf{F}$ is $I_0(V)$, there exists $\tau \in M_d(V)$ such that $\widehat{\tau} = 1/\widehat{\tau_1}$ on $\mathbf{F}$. Then $\omega_\gamma := (\delta_e - \tau_1 * \tau) * (\delta_e - \widetilde{\tau_1} * \widetilde{\tau})$ has $\widehat{\omega_\gamma}(\gamma) = 1$, $\widehat{\omega_\gamma} = 0$ on $\mathbf{F}$ and $\widehat{\omega_\gamma} \geq 0$ everywhere. Also, $\omega_\gamma \in M_d(V^4)$.

By the compactness of $\overline{\mathbf{E}}$, there are $\gamma_1, \ldots, \gamma_M$ such that $\tau_1' := \sum_1^M \omega_{\gamma_m}$ has $\widehat{\tau_1'} \geq 1/2$ on $\overline{\mathbf{E}}$ (and 0 on $\mathbf{F}$). Because $\mathbf{E}$ is $I_0(U)$, there exists $\tau' \in M_d(U)$ such that $\widehat{\tau'} = 1/\widehat{\tau_1'}$ on $\mathbf{E}$. Then $\nu = \tau_1' * \tau'$ has $\widehat{\nu} = 1$ on $\mathbf{E}$ and $\widehat{\nu} = 0$ on $\mathbf{F}$. Also, $\nu \in M_d(UV^4)$. $\square$

Corollary 5.2.3. *Suppose G is a connected group.*

1. *Each finite set in Γ is $I_0(U)$ for all non-empty, open sets U.*
2. *If $\mathbf{E}$ is $I_0(U)$ for some non-empty, open set U and $\mathbf{F}$ is a finite set, then $\mathbf{E} \cup \mathbf{F}$ is $I_0(U^2V)$ for each non-empty, open set V.*

Remark 5.2.4. Exercise 5.6.7(2) shows that with the hypotheses of Corollary 5.2.3(2), $\mathbf{E} \cup \mathbf{F}$ is even $I_0(UV)$ for each non-empty, open set V.

Proof (of Corollary 5.2.3). (1) holds because singletons are $I_0(U)$ for all open U.

For (2), note that, since an I_0 set cannot cluster at a continuous character (Ryll–Nardzewski–Méla–Ramsey Theorem 3.5.1), there is no loss of generality in assuming $\overline{\mathbf{E}}$ and $\overline{\mathbf{F}}$ (= $\mathbf{F}$) are disjoint. Now apply Proposition 5.2.2. □

Definition 5.2.5. A set $\mathbf{E}$ is said to be $I_0(U)$ *with bounded constants* if there is a constant K so that for each non-empty, open $U \subseteq G$ there is a finite set $\mathbf{F} \subseteq \mathbf{\Gamma}$ such that for each $\varphi \in \ell^\infty(\mathbf{E} \smallsetminus \mathbf{F})$ there is a measure $\mu \in M_d(U)$ with $\widehat{\mu} = \varphi$ on $\mathbf{E} \smallsetminus \mathbf{F}$ and $\|\mu\| \le K$.

A set that is $I_0(U)$ with bounded length is clearly $I_0(U)$ with bounded constants. It is unknown if these properties coincide [**P 9**].

Corollary 5.2.6. *If G is connected and $\mathbf{E}$ is $I_0(U)$ with bounded constants, then $\mathbf{E}$ is $I_0(U)$ for all U.*

One reason for the interest in sets that are $I_0(U)$ with bounded constants is that the union of a translate of such a set with itself is again I_0, as is true for ε-Kronecker sets with $\varepsilon < \sqrt{2}$ (Corollary 2.7.9).

Proposition 5.2.7. *Suppose $\mathbf{E} \subseteq \mathbf{\Gamma}$ is $I_0(U)$ with bounded constants. Then $\mathbf{E} \cdot \mathbf{F}$ is I_0 for every finite set $\mathbf{F}$.*

Proof. Fix $\gamma \ne \lambda$ in $\mathbf{F}$ and choose $x \in G$ such that $\gamma^{-1}\lambda(x) \ne 1$. Put $z = \gamma^{-1}\lambda(x)$. Suppose U is a neighbourhood of x such that

$$|\gamma^{-1}\lambda(u) - z| < \frac{|1 - z|}{10K} \text{ for all } u \in U,$$

where K is as in Definition 5.2.5. Suppose the finite set $\mathbf{F}_0$ and measure $\nu_0 = \sum a_j \delta_{x_j} \in M_d(U)$ have the property that $\widehat{\nu_0} = 1$ on $\mathbf{E} \smallsetminus \mathbf{F}_0$ and $\|\nu_0\| \le K$. Put $\nu = \sum a_j \gamma(x_j) \delta_{x_j}$. Then $\nu \in M_d(U)$, $\|\nu\| = \|\nu_0\|$ and $\widehat{\nu} = 1$ on $\gamma(\mathbf{E} \smallsetminus \mathbf{F}_0)$. For each $\alpha \in \mathbf{E} \smallsetminus \mathbf{F}_0$,

$$|\widehat{\nu}(\alpha\lambda) - z| = |\widehat{\nu}(\alpha\lambda) - z\widehat{\nu}(\alpha\gamma)| = \left| \int_U (\alpha\lambda - z\alpha\gamma) d\nu \right|$$

$$\le \int_U |\gamma^{-1}\lambda - z| d\nu \le \frac{|1 - z|}{10K} \|\nu\| \le \frac{|1 - z|}{10}.$$

Thus, $|\widehat{\nu} - z| \le |1 - z|/10$ on $\lambda(\mathbf{E} \smallsetminus \mathbf{F}_0)$. That proves $\gamma(\mathbf{E} \smallsetminus \mathbf{F}_0)$ and $\lambda(\mathbf{E} \smallsetminus \mathbf{F}_0)$ have disjoint closures in $\overline{\mathbf{\Gamma}}$, and hence their union is I_0 by Corollary 3.4.3. Since the union of a finite set and an I_0 set is I_0, it follows that $\gamma\mathbf{E} \cup \lambda\mathbf{E}$ is I_0. □

Remarks 5.2.8. (i) A similar result holds for sets that are $FZI_0(U)$ with bounded constants; see Exercise 5.6.3.

(ii) Exercise 5.6.6 asks for an example of a set $\mathbf{E}$ in the dual of $\mathbb{T}^\infty$ that is $I_0(U)$ with bounded constants, such that $\mathbf{E} \cup \mathbf{E}^{-1}$ is not I_0. It is not known if such a set exists in $\mathbb{Z}$ **[P 9]**. It is also not known if $\mathbf{E}$ being $I_0(U)$ with bounded constants with bound close to 1 implies that $\mathbf{E} \cup \mathbf{E}^{-1}$ is I_0.

Another interesting fact about sets that are $I_0(U)$ with bounded constants is that their step length must tend to infinity, again as was previously established for ε-Kronecker sets with $\varepsilon < \sqrt{2}$ (Corollary 2.7.8).

Proposition 5.2.9. *If* $\mathbf{E}$ *is* $I_0(U)$ *with bounded constants, then the step length of* $\mathbf{E}$ *goes to infinity.*

Proof. Suppose $\mathbf{E}$ does not have step length tending to infinity. It is easy to see that then there exist $\gamma \in \Gamma$ and an infinite subset $\mathbf{E}' \subseteq \mathbf{E}$ such that $\gamma \mathbf{E}' \subseteq \mathbf{E}$ and $\mathbf{E}' \cap \gamma \mathbf{E}' = \emptyset$.

Let $\varepsilon > 0$. Let $V = \{x : |1 - \gamma(x)| < \varepsilon\}$. If $V = \{x \in G : \gamma(x) = 1\}$, then no measure concentrated on V can separate the points of $\mathbf{E}'$ from those of $\gamma \mathbf{E}'$, and so $\mathbf{E}' \smallsetminus \mathbf{F}$ cannot be $I_0(V)$, whatever the choice of the finite set $\mathbf{F}$, much less with bounded constants.

Otherwise, choose $\mu \in M_d(V)$ and a finite subset $\mathbf{F} \subseteq \mathbf{E}$ such that $\widehat{\mu} = 1$ on $\mathbf{E}' \smallsetminus \mathbf{F}$ and 0 on $(\gamma \mathbf{E}') \smallsetminus \mathbf{F}$. For all $\lambda \in \mathbf{E}' \smallsetminus (\mathbf{F} \cup \gamma^{-1}\mathbf{F})$,

$$1 = |\widehat{\mu}(\lambda) - \widehat{\mu}(\gamma\lambda)| \leq \|\mu\| \sup\{|1 - \gamma(x)| : x \in V\} \leq \|\mu\|\varepsilon,$$

so $\|\mu\| \geq 1/\varepsilon$. Since ε is arbitrary, $\mathbf{E}$ is not $I_0(U)$ with bounded constants. $\square$

Example 5.2.10. A set that is I_0, but not $I_0(U)$ with bounded constants: Let $\mathbf{E}_1 = \{3^j\}_{j=1}^\infty$ and $\mathbf{E} = \mathbf{E}_1 \cup (\mathbf{E}_1 + 1)$. The set $\mathbf{E}$ is not $I_0(U)$ with bounded constants since its step length does not tend to infinity. However, $\mathbf{E} + \mathbf{F}$ *is* I_0 for all finite sets $\mathbf{F}$ since $\mathbf{E}_1$ is $I_0(U)$ with bounded constants. So there is no converse for Proposition 5.2.7.

Example 5.2.11. A set that is not $I_0(U)$ with bounded constants has step length tending to infinity, and for each non-empty, open U, a cofinite subset is $FZI_0(U)$: The set $\mathbf{E} = \{9^j\} \cup \{9^j + 3j + 1\}$ has the property that given any non-empty, open set, U, there is an integer J such that $\{9^j\}_{j>J} \cup \{9^j + 3j + 1\}_{j>J}$ is $FZI_0(U)$. To see that, assume $[-\pi/9^J, \pi/9^J] \subseteq U$ and let $a = \pi/9^J$. Then $\nu = \delta_0 + \delta_a \in M_d^+(U)$ and $\widehat{\nu} = 0$ on $\{9^j\}_{j>J}$, while $|\widehat{\nu}| \geq |e^{2\pi i/9^J} - 1| > 0$ on $\{9^j + 3j + 1\}_{j>J}$. Exercise 5.6.8 shows $\{9^j\}_{j>J} \cup \{9^j + 3j + 1\}_{j>J}$ is $FZI_0(U^2)$. In particular, $\mathbf{E}$ is I_0. Proposition 5.2.7 implies the set $\mathbf{E}$ is not $I_0(U)$ with bounded constants since $\mathbf{E} \cup (\mathbf{E} + 1)$ is not I_0. The step length of $\mathbf{E}$ tends to infinity, so there is also no converse to Proposition 5.2.9. Later, it will be seen that there is such a converse for Sidon sets (Theorem 7.4.2).

5.3 I_0 Sets Are Finite Unions of Sets That Are $I_0(U)$ with Bounded Length

In this section, we adapt an argument of Méla to prove that every I_0 set in the discrete group Γ is a finite union of sets, each of which is $I_0(U)$ with bounded length. Furthermore, every I_0 set in the dual of a connected group is $I_0(U)$ for all non-empty, open sets U. We also give consequences of this decomposition.

Here is the precise statement.

Theorem 5.3.1 (Méla decomposition theorem). *Let $0 < \tau \le 1/4$, and assume $\mathbf{E}$ is an $I_0(N, \tau/2)$ set. Then there exist $M = M(N, \tau)$ sets, each of which is $I_0(U)$ with bounded length, such that $\mathbf{E}$ is their union.*

The size of M will be clear from the proof.

Remark 5.3.2. In the case when $\mathbf{E}$ is infinite, one of those M subsets will have the same cardinality as $\mathbf{E}$. This provides another proof that every I_0 set contains a subset of the same cardinality that is $I_0(U)$ with bounded length.

Before giving the proof, we will explain how the union results mentioned above follow and give further properties of sets with bounded length.

5.3.1 Consequences of the Decomposition Theorem: Unions and Clustering

Since singletons are $I_0(U)$ for all non-empty, open sets U, it easily follows from the theorem that each I_0 set is a finite union of $I_0(U)$ sets, (though the number of sets may depend on the open set U).

Corollary 5.3.3. *Suppose $\mathbf{E}$ is an I_0 set. Then for each non-empty, open set $U \subseteq G$, there is a finite number of $I_0(U)$ sets $\mathbf{E}_m$ such that $\mathbf{E} = \bigcup_{m=1}^{N} \mathbf{E}_m$.*

More generally, we have the following.

Corollary 5.3.4. *Suppose $\mathbf{E}$ is an I_0 set. There is a constant M with the property that for each non-empty, open set $U \subseteq G$ there is a finite set $\mathbf{F} \subseteq \mathbf{E}$ such that $\mathbf{E} \setminus \mathbf{F}$ is a union of at most M sets that are $I_0(U)$.*

Proof. From Theorem 5.3.1 we know that $\mathbf{E} = \bigcup_{m=1}^{M} \mathbf{E}_m$ where the sets $\mathbf{E}_m$ are $I_0(U)$ with bounded length. In particular, for each m and each open set U, there is a finite set $\mathbf{F}_{m,U}$ such that $\mathbf{E}_m \setminus \mathbf{F}_{m,U}$ is $I_0(U)$. Now take $\mathbf{F} = \bigcup_{m=1}^{M} \mathbf{F}_{m,U}$. $\square$

Combined with Proposition 5.2.9, the theorem gives the following corollary, which will later be shown to be true for all Sidon sets (Corollary 6.4.7).

Corollary 5.3.5. *Suppose* $\mathbf{E}$ *is an* I_0 *set. Then* $\mathbf{E}$ *is a finite union of* I_0 *sets* $\mathbf{E}_m$ *whose step length tends to infinity.*

Coupled with Proposition 5.2.2, one can deduce that when the group G is connected, each I_0 set is actually $I_0(U)$ for all non-empty, open U.

Theorem 5.3.6. *If* G *is a connected group and* $\mathbf{E}$ *is an* I_0 *set, then* $\mathbf{E}$ *is* $I_0(U)$ *for all non-empty, open sets* U.

Proof. Choose the integer M as in Corollary 5.3.4. We may assume $M = 2^K$. Fix a symmetric e-neighbourhood U and choose compact symmetric e-neighbourhoods V, W such that $V^{2 \cdot 6^K + 1} \subseteq U$ and $W^5 \subseteq V$.

According to Corollary 5.3.4, there is a finite set $\mathbf{F}$ with the property that $\mathbf{E} \smallsetminus \mathbf{F} = \bigcup_{m=1}^M \mathbf{E}_m$ where the sets $\mathbf{E}_m$ are each $I_0(V)$ sets. There is no loss of generality in assuming the sets $\mathbf{E}_m$ are disjoint. Being subsets of an I_0 set, they must have disjoint closures in $\overline{\Gamma}$. By repeatedly applying Proposition 5.2.2, we see that $\bigcup_{m=1}^M \mathbf{E}_m$ is $I_0(V^{6^K})$.

Since $\mathbf{F}$ is $I_0(W)$, Proposition 5.2.2 shows $\mathbf{E} = \mathbf{F} \cup \bigcup_{m=1}^M \mathbf{E}_m$ is $I_0(U)$. $\square$

The following corollary is immediate from Theorem 5.3.6 and Proposition 4.2.1.

Corollary 5.3.7. *If* G *is connected, then every* RI_0 *set is* $RI_0(U)$ *for all non-empty, open* U.

While connectedness is essential for Theorem 5.3.6, the following holds when G has only a finite number of connected components.

Corollary 5.3.8. *Suppose the torsion subgroup of* Γ *is finite and that* $\mathbf{E}$ *is* I_0. *Then* $\mathbf{E}$ *is the finite union of sets that are* $I_0(U)$ *for all non-empty, open* U.

Proof. Let Γ_0 be the torsion subgroup of Γ and let G_0 be its annihilator, a connected, open subgroup of G since Γ_0 is finite (Exercise C.4.10). It follows from Corollary 5.3.3 that $\mathbf{E} = \bigcup_{m=1}^N \mathbf{E}_m$ where each $\mathbf{E}_m$ is $I_0(G_0)$.

Temporarily fix m. Since $\mathbf{E}_m$ is an $I_0(G_0)$ set, we may view the elements of $\mathbf{E}_m$, restricted to G_0, as distinct continuous characters on G_0. Viewed in this way, $\mathbf{E}_m$ is an I_0 set in the dual of the connected group G_0. But then Theorem 5.3.6 implies that $\mathbf{E}_m$ is $I_0(V)$ for all open $V \subseteq G_0$. If $U \subseteq G$ is a neighbourhood of e, then $U \cap G_0$ is a neighbourhood of $e \in G_0$, so $\mathbf{E}_m$ is $I_0(U)$ for all non-empty, open U. $\square$

The following is an abstract form of what we already know for Hadamard (Theorem 1.5.1) and ε-Kronecker (Proposition 2.7.4) sets. It is easy to see that Hadamard and ε-Kronecker sets satisfy its hypotheses.

Theorem 5.3.9 (Abstract Kunen–Rudin).

1. *If* $\mathbf{E}$ *is* $I_0(U)$ *with bounded length, then* $\mathbf{E} \cdot \mathbf{E}^{-1}$ *clusters at no continuous character other than* $\mathbf{1}$.

2. *Suppose* **E** *is* $RI_0(U)$ *with bounded length. Then the only continuous character at which* $\mathbf{E} \cdot \mathbf{E}$ *can cluster is* 1.

3. *Suppose* **E** *is* RI_0, *asymmetric and has at most a finite number of elements of order 2. Then* $\mathbf{E} \cdot \mathbf{E}$ *does not cluster at* 1.

Proof. (1). This proof is identical to the proof of Proposition 2.7.4 (1), except that the existence of the finite set $L(U)$ is guaranteed by the hypotheses instead of Corollary 2.3.4.

(2). This follows immediately from the first part since **E** is $RI_0(U)$ with bounded length if and only if $\mathbf{E} \cup \mathbf{E}^{-1}$ is $I_0(U)$ with bounded length.

(3). Since **E** has only a finite number of elements of order 2 and has no cluster points in $\mathbf{\Gamma}$, there is no loss of generality in deleting those points of order 2 from **E**. Because **E** is asymmetric and has no elements of order 2, there exists $\mu \in M_d^r(G)$ such that $\widehat{\mu} = i$ on **E**. But then, $\widehat{\mu} = -i$ on $\mathbf{E}^{-1}$. That proves $\overline{\mathbf{E}} \cap \overline{\mathbf{E}}^{-1} = \emptyset$, and so 1 is not a cluster point of $\mathbf{E} \cdot \mathbf{E}$, by Proposition 2.7.2. □

5.3.2 *Proof of Theorem 5.3.1*

We begin by introducing notation that will be used throughout the proof. The key ideas are the Baire category theorem and the fact that a net in $M(G)$, consisting of discrete measures of length N, clusters in the weak* topology at a length N measure. (Note that it is not true if the length is allowed to grow.)

Given $a = (a_1, \ldots, a_N) \in \Delta^N$ and $g = (g_1, \ldots, g_N) \in G^N$, we denote by $\mu_{a,g}$ the length N, discrete measure on G given by

$$\mu_{a,g} = \sum_{n=1}^{N} a_n \delta_{g_n}.$$

Let $\mathcal{F}$ be the set of all length N, discrete measures μ such that the Fourier transform of μ restricted to **E** has the property that its range is contained in the union of two closed τ-balls in the complex plane, one centred at 1, the other centred at -1. In other words, $\mathcal{F}$ is the set

$$\{\mu \in M_d(G) : \text{length}\, \mu = N \text{ and } \min\left(|\widehat{\mu}(\gamma) - 1|, |\widehat{\mu}(\gamma) + 1|\right) \leq \tau \; \forall \gamma \in \mathbf{E}\}.$$

For each neighbourhood U of e in G, we let

$$\mathcal{F}(a, g, U) = \{\mu_{a,h} \in \mathcal{F} \; : \; g_n h_n^{-1} \in U, \, 1 \leq n \leq N\}.$$

We give $\mathcal{F}$ the weak topology of convergence at each $\gamma \in \mathbf{E}$, so a net $\{\mu_\alpha\}$ converges to μ provided $\widehat{\mu_\alpha}(\gamma) \to \widehat{\mu}(\gamma)$ for each $\gamma \in \mathbf{E}$. Since G is compact and the measures in $\mathcal{F}$ have length N, the set $\mathcal{F}$ is compact.

Corresponding to each $\mu \in \mathcal{F}$ there is a unique function $\varphi \in \mathbb{Z}_2^{\mathbf{E}}$ such that $\sup_{\gamma \in \mathbf{E}} |\varphi(\gamma) - \widehat{\mu}(\gamma)| \leq \tau$. We denote by $T : \mathcal{F} \to \mathbb{Z}_2^{\mathbf{E}}$ the map that sends μ to that function φ. The map T is continuous by the choice of the topologies, and it is onto because $\mathbf{E}$ is $I_0(N, \tau)$.

Finally, we let $\mathcal{E}(a, g, U)$ denote the image under T of $\mathcal{F}(a, g, U)$ and A be a (fixed) countable, dense subset of Δ^N.

In all the following lemmas we assume the hypotheses of Theorem 5.3.1.

Lemma 5.3.10. *Let $g \in G^N$. Then $\mathbb{Z}_2^{\mathbf{E}} = \bigcup_{a \in A} \mathcal{E}(a, g, G)$.*

Proof (of Lemma 5.3.10). Since $\mathbf{E}$ is $I_0(N, \tau/2)$, for each $\varphi \in \mathbb{Z}_2^{\mathbf{E}}$, there is a length N, discrete measure, $\mu = \mu_{b,x} = \sum_{n=1}^{N} b_n \delta_{x_n}$, with $|b_n| \leq 1$ and such that $\sup_{\gamma \in \mathbf{E}} |\varphi(\gamma) - \widehat{\mu}(\gamma)| \leq \tau/2$.

By density we may pick $a = (a_n)_{n=1}^{N} \in A$ such that $\sup_n |a_n - b_n| < \tau/2N$. Then $\mu_{a,g} \in \mathcal{F}(a, g, G)$. Also, $\sup_{\gamma \in \mathbf{E}} |\varphi(\gamma) - \widehat{\mu_{a,g}}(\gamma)| \leq \tau$, and thus $\varphi \in T(\mathcal{F}(a, g, G)) = \mathcal{E}(a, g, G)$. $\qquad\square$

Next, the Baire category theorem implies that one of the sets, $\mathcal{E}(a, g, G)$, has non-zero interior.

Lemma 5.3.11. *There exists a finite set $\mathbf{F} \subseteq \mathbf{E}$, $a \in A$ and $g \in G^N$ such that for each $\varphi \in \mathbb{Z}_2^{\mathbf{E}}$ there is some $\mu \in \mathcal{F}(a, g, G)$ with $|\varphi(\gamma) - \widehat{\mu}(\gamma)| \leq \tau$ for $\gamma \in \mathbf{E} \setminus \mathbf{F}$.*

Proof (of Lemma 5.3.11). The sets $\mathcal{E}(a, g, G)$ are compact being the images of compact sets under a continuous function. Since $\mathbb{Z}_2^{\mathbf{E}}$ is a countable union of such sets, the Baire category theorem implies that at least one of these sets has interior in $\mathbb{Z}_2^{\mathbf{E}}$. Thus, one of the sets must contain all functions φ which have the same (fixed) values on some finite subset $\mathbf{F}$ but are otherwise unrestrained. $\qquad\square$

With more care, we may conclude that the element $g \in G^N$ can be chosen so that for each non-empty, open set U we may obtain good approximations from $\mathcal{F}(a, g, U)$ on a cofinite subset of $\mathbf{E}$.

Lemma 5.3.12. *Let a be as in Lemma 5.3.11. Then there is $g \in G^N$ such that for every e-neighbourhood $U \subseteq G$, there is a finite set $\mathbf{F} \subseteq \mathbf{E}$ such that for each $\varphi \in \mathbb{Z}_2^{\mathbf{E}}$ there is some $\mu \in \mathcal{F}(a, g, U)$ with $|\varphi(\gamma) - \widehat{\mu}(\gamma)| \leq \tau$ for $\gamma \in \mathbf{E} \setminus \mathbf{F}$.*

Proof (of Lemma 5.3.12). Let V be a compact, symmetric neighbourhood of e. Since G is compact, finitely many translates of V cover G, say $G = \bigcup_{j=1}^{J} u_j V$. If $h \in G$, then $h = u_j v$ for some $j = 1, \ldots, J$ and $v \in V$. Consequently, $\mu_{a,h} \in \mathcal{F}(a, u, V)$ where if we put $\mathcal{U} = \{u_1, \ldots, u_J\}$, then $u \in \mathcal{U}^N$. That shows

$$\mathcal{E}(a, g, G) = \bigcup_{u \in \mathcal{U}^N} \mathcal{E}(a, u, V).$$

Since each of the $\mathcal{E}(a, u, V)$ is closed, one of them must have non-empty interior, say $\mathcal{E}(a, g^{(V)}, V)$, with $g^{(V)} \in \mathcal{U}^N \subseteq G^N$.

Thus, for each V, there is a finite set $\mathbf{F} \subseteq \mathbf{E}$ such that for each $\varphi \in \mathbb{Z}_2^{\mathbf{E}}$ there is some $\mu \in \mathcal{F}(a, g^{(V)}, V)$ with $|\varphi(\gamma) - \widehat{\mu}(\gamma)| \leq \tau$ for $\gamma \in \mathbf{E} \smallsetminus \mathbf{F}$.

Let $\mathcal{I}$ be the set of compact, symmetric e-neighbourhoods in G, ordered by reverse inclusion. Then $\mathcal{I}$ is a directed set and $\{g^{(V)} : V \in \mathcal{I}\}$ is a net in the compact set G^N. Let g be a cluster point of that net and let $\mathcal{J}$ be a cofinal subnet of $\mathcal{I}$ such that $\lim_{V \in \mathcal{J}} g^{(V)} = g$.

Finally, suppose U is a compact e-neighbourhood. Take a compact, symmetric e-neighbourhood W such that $W^2 \subseteq U$. Since $\mathcal{J}$ is cofinal, there is a neighbourhood $V \in \mathcal{J}$ with $V \subseteq W$ and $g^{(V)} \in gW$. It follows that

$$\mathcal{F}(a, g^{(V)}, V) \subseteq \mathcal{F}(a, g, U),$$

which gives the desired result. $\qquad\square$

We can now complete the proof of Theorem 5.3.1.

Proof (of Theorem 5.3.1). We partition $\mathbb{T}^N$ into N-dimensional cubes, T_m, with diagonals of length at most τ/N, and let $t^{(m)} \in \mathbb{T}^N$ be the central point of T_m for $1 \leq m \leq M$. For the $g = (g_1, \ldots, g_N) \in G^N$ found in Lemma 5.3.12, let $\mathbf{E}_m = \{\gamma \in \mathbf{E} : (\gamma(g_1), \ldots, \gamma(g_N)) \in T_m\}$.

We claim that each $\mathbf{E}_m$ is $I_0(U)$ with bounded constants. Fix an e-neighbourhood $U \subseteq G$ and let $\mathbf{F}$ be the finite subset also found by Lemma 5.3.12. Suppose $\varphi \in \mathbb{Z}_2^{\mathbf{E}_m}$. By extending φ to be ± 1-valued on $\mathbf{E}$ we may assume $\varphi \in \mathbb{Z}_2^{\mathbf{E}}$. By that lemma there is some $\mu \in \mathcal{F}(a, g, U)$ such that $|\varphi(\gamma) - \widehat{\mu}(\gamma)| \leq \tau$ for all $\gamma \in \mathbf{E} \smallsetminus \mathbf{F}$. Suppose $\mu = \sum_{n=1}^{N} a_n \delta_{x_n}$ where $x_n g_n^{-1} \in U$. Define a second discrete measure ν by

$$\nu = \sum_{n=1}^{N} a_n t_n^{(m)} \delta_{x_n g_n^{-1}} \in M_d(U).$$

For each $\gamma \in \mathbf{E}_m$, $\left|t_n^{(m)} - \gamma(g_n)\right| \leq \tau/N$, and thus $|\widehat{\mu}(\gamma) - \widehat{\nu}(\gamma)| \leq \tau$ on $\mathbf{E}_m \smallsetminus \mathbf{F}$. Hence, $|\varphi(\gamma) - \widehat{\nu}(\gamma)| \leq 2\tau$ for all $\gamma \in \mathbf{E}_m \smallsetminus \mathbf{F}$. Since $2\tau \leq 1/2$, each $\mathbf{E}_m \smallsetminus \mathbf{F}$ is $I_0(U, N, 1/2)$, and thus $\mathbf{E}_m$ is of bounded length, with length at most N. $\qquad\square$

5.4 Union Results for More Restricted Types of I_0 Sets

5.4.1 When an I_0 Set Is a Finite Union of RI_0 Sets

In Example 4.2.3 a construction was given of an I_0 subset of $\mathbb{Z}$, which was not RI_0. The reader may have noticed that the example consisted of a union of two sets that were both ε-Kronecker for some $\varepsilon < 1$, and thus both were even

FZI_0. In this section we will give a topological characterization of the I_0 sets that are finite unions of RI_0 sets. From that characterization it will follow that all I_0 sets in duals of connected groups are finite unions of RI_0 sets. The characterization should be compared with the topological characterization of RI_0 given by Proposition 4.3.1.

Theorem 5.4.1. *An I_0 set $\mathbf{E}$ is a finite union of RI_0 sets if and only if the closure of $\{\gamma \in \mathbf{E} : \gamma^2 \neq 1\}$ in $\overline{\Gamma}$ contains no elements of order 2.*

The following corollary is immediate, with the second sentence coming from the fact that connected groups have torsion-free duals. See Exercise C.4.11 for more information on groups with no elements of order 2.

Corollary 5.4.2. *If Γ has no elements of order two, then each I_0 set in Γ is a finite union of RI_0 sets. In particular, if G is connected, then each I_0 set in Γ is a finite union of RI_0 sets.*

The key idea in the proof of Theorem 5.4.1 is the following simple lemma.

Lemma 5.4.3. *Let $\mathbf{E} \subseteq \Gamma$, $V \subseteq G$ and $0 < \varepsilon < 1$. For $v \in V$, put*

$$\Gamma_v = \{\gamma \in \overline{\Gamma} : \min(|\gamma(v)+1|, |\gamma(v)-1|) > \varepsilon\} \subseteq \overline{\Gamma}$$

and suppose $\bigcup_{v \in V} \Gamma_v \supseteq \overline{\mathbf{E}}$. Then there exist finitely many $v_1, \ldots, v_M \in V$ such that $\bigcup_{m=1}^{M} \Gamma_{v_m} \supseteq \overline{\mathbf{E}}$.

Proof (of Lemma 5.4.3). The sets Γ_v are open in the Bohr topology on $\overline{\Gamma}$ and cover the compact set $\overline{\mathbf{E}}$ by assumption. By compactness, $\overline{\mathbf{E}}$ is covered by finitely many of the sets Γ_{v_m}. $\qquad\square$

Proof (of Theorem 5.4.1). There is no loss of generality in assuming the I_0 set $\mathbf{E}$ is asymmetric, and consequently the necessity follows directly from Proposition 4.3.1.

Since a set of characters of order two is RI_0, we may assume that $\mathbf{E}$ has no elements of order 2. By assumption, the Bohr closure of $\mathbf{E}$ also contains no elements of order two.

Consider $\gamma \in \overline{\mathbf{E}}$. Since the range of a character is a subgroup of $\mathbb{T}$ and γ is not of order two, there must be some $g \in G$ with

$$\min\left(|\gamma(g)+1|, |\gamma(g)-1|\right) \geq \left|1 - e^{i\pi/3}\right| = \varepsilon > 0.$$

By Lemma 5.4.3, there are finitely many $g_1, \ldots, g_M \in G$ with the property that for each $\gamma \in \overline{\mathbf{E}}$ there is some g_m with both $|\gamma(g_m) \pm 1| \geq \varepsilon$.

We first use these g_m to partition $\mathbf{E}$ as $\bigcup_{m=1}^{M} \mathbf{E}_m$, where $\gamma \in \mathbf{E}_m$ if both $|\gamma(g_m) \pm 1| \geq \varepsilon$. Thus, if we put $\mu_m = i(\delta_{g_m} - \delta_{g_m^{-1}})/2$, then $\widehat{\mu_m}$ is real-valued, $\widehat{\mu_m}(\gamma^{-1}) = -\widehat{\mu_m}(\gamma)$ and

$$1 \geq |\widehat{\mu_m}(\gamma)| = |\Im\gamma(g_m)| \geq \sin\frac{\pi}{3} = \frac{\sqrt{3}}{2} \text{ for all } \gamma \in \mathbf{E}_m.$$

Partition each set $\mathbf{E}_m$ as $\mathbf{E}_m^+ \cup \mathbf{E}_m^-$ according to the rule that $\gamma \in \mathbf{E}_m^\pm$ provided $\widehat{\mu_m}(\gamma) \in \pm[\sqrt{3}/2, 1]$.

We claim the $\mathbf{E}_m^\pm$ are RI_0. This will be proven by establishing that $\mathbf{E}_m^\pm \cup (\mathbf{E}_m^\pm)^{-1}$ is an I_0 set and then applying Proposition 4.2.1. Of course, each set $\mathbf{E}_m^\pm$ is I_0, being a subset of an I_0 set.

We introduce further measures,

$$\mu_m^\pm = \frac{1}{2}\left(\pm\frac{2\mu_m}{\sqrt{3}} + \delta_e\right).$$

For $\gamma \in \mathbf{E}_m^+$, we have

$$\widehat{\mu_m^+}(\gamma) = \frac{1}{2}\left(\frac{2\widehat{\mu_m}(\gamma)}{\sqrt{3}} + 1\right) \in [1, 3/2]$$

$$\widehat{\mu_m^+}(\gamma^{-1}) \in [-1/2, 0].$$

The calculations are similar for μ_m^- on $\mathbf{E}_m^-$. That shows that the sets $\mathbf{E}_m^+$ and $(\mathbf{E}_m^+)^{-1}$ (and similarly, $\mathbf{E}_m^-$ and $(\mathbf{E}_m^-)^{-1}$) have disjoint closures in $\overline{\boldsymbol{\Gamma}}$. By the HRN union theorem (specifically, Corollary 3.4.3) for I_0 sets with disjoint closures, each $\mathbf{E}_m^\pm \cup (\mathbf{E}_m^\pm)^{-1}$ is I_0 and therefore each $\mathbf{E}_m^\pm$ is RI_0. $\qquad\square$

Corollary 5.4.4. *If $\boldsymbol{\Gamma}$ contains only finitely many elements of order two, then every I_0 set is a finite union of RI_0 sets.*

Proof. If $\gamma \in \overline{\mathbf{E}} \setminus \mathbf{E}$ is of order two, then, by Corollary C.1.18, γ is continuous. Because an I_0 set does not cluster at elements of $\boldsymbol{\Gamma}$, Theorem 5.4.1 applies. $\square$

Corollary 5.4.5. *If G is connected, every I_0 set is a finite union of sets that are $RI_0(U)$ for all non-empty, open U.*

Proof. Combine Corollary 5.4.4 and Corollary 5.3.7. $\qquad\square$

5.4.2 When $\mathbf{E}$ Is a Finite Union of FZI_0 Sets

Since RI_0 sets that do not contain $\mathbf{1}$ are FZI_0 (Proposition 4.2.5), the following is immediate from Theorem 5.4.1.

Corollary 5.4.6. *If $\mathbf{E} \subseteq \boldsymbol{\Gamma} \setminus \{\mathbf{1}\}$ is I_0, then $\mathbf{E}$ is a finite union of FZI_0 sets if and only if the closure of $\{\gamma \in \mathbf{E} : \gamma^2 \neq \mathbf{1}\}$ contains no elements of order two.*

It is unknown if for every $RI_0(U)$ set $\mathbf{E}$ there is a finite set $\mathbf{F}$ such that $\mathbf{E} \setminus \mathbf{F}$ is $FZI_0(U)$ [**P 10**]. However, there is a finite union result.

Theorem 5.4.7. *Suppose* **E** *is an* $RI_0(U)$ *set. Given an e-neighbourhood* $W \subseteq G$, *there is a finite subset* **F** *such that* $\mathbf{E} \smallsetminus \mathbf{F}$ *is a finite union of* $FZI_0(UW)$ *sets.*

Since a real, discrete measure concentrated on U is the difference of two positive, discrete measures concentrated on U, the main task in proving Theorem 5.4.7 is to show that a cofinite subset of **E** may be decomposed into finitely many sets, $\mathbf{E}_j$, each with the property that the function which is identically -1 on $\mathbf{E}_j$ can be approximated by the Fourier transform of positive W-measures. This can be done via two lemmas.

Lemma 5.4.8. *Let* $V \subseteq G$ *be an e-neighbourhood and let* γ *be a discontinuous character. Given* $0 < \delta < \sqrt{3}$, *there exists* $v \in V$ *such that* $|\gamma(v) - 1| > \delta$.

Proof (of Lemma 5.4.8). Since γ is discontinuous, there is a net $x_\alpha \to e$ and $\tau > 0$ such that

$$|\gamma(x_\alpha) - 1| > \tau.$$

If $\tau > \delta$, just pick $v = x_\alpha \in V$. Otherwise, we argue as follows. Let $z \in \mathbb{T}$ have $|1 - z| = \tau$. Since $\tau \leq \delta$, there exists a minimal $k \geq 2$ such that $|z^k - 1| > \delta$. (Here we use the fact that $\delta < \sqrt{3}$.) Choose an e-neighbourhood W such that $W^k \subset V$. Let $x_\alpha \in W$. Then $|1 - \gamma(x_\alpha)| > \tau$, and one of the powers x_α^j satisfies $|1 - \gamma(x_\alpha^j)| > \delta$, $1 \leq j \leq k$. $\qquad\square$

Lemma 5.4.9. *Let* **E** *be an infinite* I_0 *set and suppose* $V \subseteq G$ *is an e-neighbourhood. Let* $0 < \delta < \sqrt{3}$. *There is a finite set* **F** *and finitely many* $v_1, \dots, v_M$ $\in V$ *such that for all* $\gamma \in \mathbf{E} \smallsetminus \mathbf{F}$ *there is some* v_m *with* $|\gamma(v_m) - 1| > \delta$.

Proof (of Lemma 5.4.9). First we prove that there are only finitely many continuous characters $\gamma \in \Gamma$ satisfying $|\gamma(v) - 1| \leq \delta$ for all $v \in V$. That set of characters will be the finite set **F** which we exclude.

So assume otherwise, say the set $\mathbf{E}' = \{\gamma_\alpha\}$ satisfies $|\gamma_\alpha(v) - 1| \leq \delta$ for all $v \in V$. Since every infinite set contains an infinite I_0 set (Corollary 4.5.3), there is no loss of generality in assuming $\mathbf{E}'$ is an I_0 set. Then any cluster point, χ, of $\mathbf{E}'$ is discontinuous. The definition of convergence in the Bohr topology means $|\chi(v) - 1| \leq \delta$ for all $v \in V$, and that contradicts Lemma 5.4.8. Thus, there are only finitely many continuous characters, say those in the set **F**, which satisfy $|\gamma(v) - 1| \leq \delta$ for all $v \in V$.

Since **E** is I_0, if $\chi \in \overline{\mathbf{E} \smallsetminus \mathbf{F}}$, then either $\chi \in \mathbf{E} \smallsetminus \mathbf{F}$ or χ is discontinuous. In either case, $\sup_{v \in V} |\chi(v) - 1| > \delta$ for all such χ. A compactness argument now completes the proof. $\qquad\square$

Proof (of Theorem 5.4.7). Suppose **E** is $RI_0(U, N, \varepsilon)$ for some $\varepsilon > 0$ and integer N. Pick an integer M such that $(1 - \sin \pi/8)^M < (1 - \varepsilon)/2N$. Let $V \subseteq G$ be a symmetric e-neighbourhood with $V^{2M} \subseteq W$.

Obtain the points $v_1, \dots, v_M \in V$, from Lemma 5.4.9, with $\delta = |1 - e^{\pi i/3}|$. Partition $\mathbf{E} = \bigcup_{m=1}^{M} \mathbf{E}_m$ by the rule that $\gamma \in \mathbf{E}_m$ if $|\gamma(v_m) - 1| > \delta$. Partition the sets $\mathbf{E}_m$ further as $\mathbf{E}_{m,1} \cup \mathbf{E}_{m,2}$ depending on whether $\arg \gamma(v_m)$ belongs to $\pm[\pi/3, \pi/2 + \pi/8]$ or $\pm(\pi/2 + \pi/8, \pi]$.

If $\gamma \in \mathbf{E}_{m,1}$, then $\arg \gamma(v_m^2) \in \pm[2\pi/3, \pi + \pi/4]$, so

$$\mathfrak{Re}\gamma(v_m^2) \in [-1, -\cos \pi/3].$$

If $\gamma \in \mathbf{E}_{m,2}$, then $\mathfrak{Re}\gamma(v_m) \in [-1, -\sin \pi/8]$. In either case, there is a choice of $u_{m,k} \in V^2$, for $k = 1, 2$, so that if $\mu_{m,k} = (\delta_{u_{m,k}} + \delta_{u_{m,k}^{-1}})/2$, then $\left| \widehat{\mu_{m,k}}(\gamma) + 1 \right| \leq 1 - \sin \pi/8$ for $\gamma \in \mathbf{E}_{m,k}$. Now put

$$\sigma_{m,k} = \sum_{\ell=1}^{M} \binom{M}{\ell} \mu_{m,k}^\ell \in M_d^+(V^{2M}) \subseteq M_d^+(W).$$

Then $\left| \widehat{\sigma_{m,k}}(\gamma) + 1 \right| = \left| \widehat{\mu_{m,k}}(\gamma) + 1 \right|^M \leq (1 - \varepsilon)/2N$ for $\gamma \in \mathbf{E}_{m,k}$.

This measure provides us with a suitable approximation of -1 and will be used to show that each set $\mathbf{E}_{m,k}$ is $FZI_0(UW)$. To see this, suppose φ is a bounded Hermitian function defined on $\mathbf{E}_{m,k}$, and let $\omega = \sum_{n=1}^{N} a_n \delta_{g_n} \in M_d^r(U)$, with all $-1 \leq a_n \leq 1$, have the property that $|\varphi(\gamma) - \widehat{\omega}(\gamma)| \leq \varepsilon$ for $\gamma \in \mathbf{E}_{m,k}$. Write $a_n = a_n^+ - a_n^-$ with $a_n^+, a_n^- \geq 0$ and consider

$$\omega' = \sum_{n=1}^{N} a_n^+ \delta_{g_n} + \sigma_{m,k} * \sum_{n=1}^{N} a_n^- \delta_{g_n}.$$

Clearly, $\omega' \in M_d^+(UW)$. It is a routine exercise to show that $|\varphi(\gamma) - \widehat{\omega'}(\gamma)| \leq (1 + \varepsilon)/2$. Thus, $\mathbf{E}_{m,k}$ is $FZI_0(UW)$. $\qquad\square$

Corollary 5.4.10. *If G is connected and $\mathbf{E}$ is I_0, then for each non-empty, open set U there is a finite set $\mathbf{F}$ such that $\mathbf{E} \setminus \mathbf{F}$ is a finite union of $FZI_0(U)$ sets.*

Proof. Immediate from Corollary 5.4.5 and Theorem 5.4.7. $\qquad\square$

5.5 Remarks and Credits

Except for Example 5.2.11, found in [53], and the abstract Kunen–Rudin Theorem 5.3.9, which is new, the remaining results of this chapter are adapted from [62, 63]. Lemma 5.3.12 corrects an error in [62, Lemma 3.4]. Theorem 5.3.1 is a general version of Méla's [130, Theorem 6]. The proof here is essentially his. See Corollary 7.4.3 for the corresponding result for Sidon sets.

5.6 Exercises

Exercise 5.6.1. Show that a set can be $I_0(U)$ for all non-empty, open sets U, but not have step length tending to infinity.

Exercise 5.6.2. Let $G = \mathbb{Z}_3^{\mathbb{N}}$ and let $\mathbf{E} = \{\pi_k : 1 \leq k < \infty\}$, where π_k is the projection onto the kth factor group $\mathbb{Z}_3$. Let

$$U_\ell = \overbrace{\{0\} \times \cdots \times \{0\}}^{\ell} \times \prod_{\ell+1}^{\infty} \mathbb{Z}_3.$$

1. Show that $\mathbf{E}$ is the union of ℓ sets sets each of which is $I_0(U_\ell)$.
2. Show that the number ℓ cannot be reduced.

Exercise 5.6.3. Suppose $\mathbf{E}$ is asymmetric and $FZI_0(U)$ with bounded constants and that $\mathbf{F}$ is an asymmetric finite set. Show that if $\mathbf{F}^{-1} \cap \mathbf{E}$ is empty, then $\mathbf{E} \cdot \mathbf{F}$ is FZI_0.

Exercise 5.6.4. Construct an example of an I_0 set that is not a finite union of $RI_0(U)$ sets for all non-empty, open U. This shows there is no analogue to Corollary 5.3.8 for $RI_0(U)$ sets. Hint: Let $G = \mathbb{Z}_3 \times \mathbb{T}^{\mathbb{N}}$.

Exercise 5.6.5. Give examples of the following:

1. A set $\mathbf{E}$ that is $I_0(U)$ with bounded length, but $\mathbf{E} \cdot \mathbf{E}$ clusters at $\gamma \in \Gamma$, $\gamma \neq \mathbf{1}$.
2. A set $\mathbf{E}$ that is RI_0 and asymmetric, but $\mathbf{E} \cdot \mathbf{E}$ clusters at $\mathbf{1}$.
3. A set $\mathbf{E}$ that is RI_0 and contains no elements of order two, but $\mathbf{E} \cdot \mathbf{E}$ clusters at $\mathbf{1}$.
4. A set $\mathbf{E}$ that is $I_0(U)$ with bounded length, asymmetric, contains no elements of order two, but $\mathbf{E} \cdot \mathbf{E}$ clusters at $\mathbf{1}$.

These examples illustrate the necessity of the hypotheses in Theorem 5.3.9(2), (3).

Exercise 5.6.6. Let $G = \mathbb{T}^\infty$. Construct $\mathbf{E} \subset \Gamma$ that is $I_0(U)$ with bounded constants but such that $\mathbf{E} \cup \mathbf{E}^{-1}$ is not I_0.

Exercise 5.6.7. 1. Suppose $U, V, W \subseteq G$ are e-neighbourhoods. Assume $\mathbf{E}$ is $I_0(U)$, $\mathbf{F}$ is $I_0(V)$ and there is a measure $\nu \in M_d(W)$ such that $\hat{\nu} = 1$ on $\mathbf{E}$ and 0 on $\mathbf{F}$. Prove $\mathbf{E} \cup \mathbf{F}$ is $I_0(UW \cup V)$.
2. Suppose G is a connected group, $\mathbf{E}$ is $I_0(U)$ for some open set U and $\mathbf{F}$ is a finite set. Show that $\mathbf{E} \cup \mathbf{F}$ is $I_0(UV)$ for each non-empty, open set V.

Exercise 5.6.8. Prove the following $FZI_0(U)$ variation of Lemma 5.2.1: Assume that U, V, W are open subsets of G. Suppose $\mathbf{E}$ is $FZI_0(U)$, $\mathbf{F}$ is $FZI_0(V)$ and that there is a measure $\nu \in M_d^+(W)$ such that $\hat{\nu} = 0$ on $\mathbf{F}$ and $|\hat{\nu}| \geq \delta > 0$ on $\mathbf{E}$. Show that $\mathbf{E} \cup \mathbf{F}$ is $FZI_0(UW \cup V)$.

Exercise 5.6.9. Suppose Γ contains only finitely many elements of order two. Show that if $U \subseteq G$ is a non-empty, open set and $\mathbf{E} \subseteq \Gamma$ is an I_0 set, then there is a finite subset $\mathbf{F}$ such that $\mathbf{E} \smallsetminus \mathbf{F}$ is a finite union of $RI_0(U)$ sets.

Chapter 6
Sidon Sets: Introduction and Decomposition Properties

Sidon sets defined. Basic properties and characterizations. Examples. Quasi-independent and Rider sets. $\Lambda(p)$ sets. Decompositions of Sidon sets.

6.1 Introduction

Recall that a set $\mathbf{E} \subseteq \mathbf{\Gamma}$ is called a *Sidon set* if for every $\varphi \in \ell^\infty(\mathbf{E})$, there is a measure $\mu \in M(G)$ such that $\widehat{\mu}(\gamma) = \varphi(\gamma)$ for all $\gamma \in \mathbf{E}$. If the measure can always be chosen to be concentrated on a subset $U \subseteq G$, then $\mathbf{E}$ is called a *Sidon(U) set*. Obviously, $I_0(U)$ sets are Sidon(U); in this chapter the converse will be shown to be false.

Sidon sets have been extensively studied since the 1920s, when Sidon proved that Hadamard sets have the property of Corollary 6.2.5, and thus are what we now call "Sidon". Excellent expositions of what was known up to the late 1970s can be found in [88, Chap. 37] and [123].

A proof that Hadamard sets (with ratio ≥ 3) are Sidon, using the notion of Riesz products, was outlined in Exercise 1.7.9. That Riesz product construction can be generalized to all discrete abelian groups to prove that sets possessing various independence-type properties, such as dissociateness or quasi-independence, are Sidon, providing many new examples of Sidon sets. These examples illustrate a connection between arithmetic properties of a set and its (possible) Sidonicity. That culminates in Chap. 7, where Pisier's characterization of Sidonicity in terms of quasi-independence is established. The Riesz product construction, along with various equivalent characterizations of Sidonicity, can be found in Sect. 6.2 of this chapter.

In Sect. 6.3 we give a brief overview of the other major results of the pre-1980 era: a finite union of Sidon sets is Sidon; each Sidon set in the dual of a connected group is Sidon(U) for all non-empty, open U; each Sidon set not containing the identity has the property that the interpolating measure can be chosen to be positive; and (like Hadamard sets) each Sidon set has the

C.C. Graham and K.E. Hare, *Interpolation and Sidon Sets for Compact Groups*, CMS Books in Mathematics, DOI 10.1007/978-1-4614-5392-5_6, © Springer Science+Business Media New York 2013

property that every L^2 function, with Fourier transform supported on the Sidon set, belongs to L^p for every $p < \infty$. In Chap. 7 it will be shown that this latter property can also be used to characterize Sidonicity. Arithmetic properties of Sidon sets are summarized, as well. Proofs will generally not be given if they can be found in [123], unless they are needed later.

An important open problem in the study of Sidon sets is to determine which interesting classes of sets have the property that every Sidon set is a finite union of sets from the class. In Sect. 6.4 probabilistic methods are used to prove that every Sidon set is a finite union of sets with a weak independence property and hence a finite union of sets with step length tending to infinity. We return to this problem, as well, in Chaps. 7 and 9.

6.2 Characterizations and Examples

6.2.1 Characterizations

If $\mathbf{E}$ is a Sidon set, then an application of the closed graph theorem shows that there is a constant C such that for each $\varphi \in \ell^\infty(\mathbf{E})$, there is some "interpolating" $\mu \in M(G)$ such that $\widehat{\mu}(\gamma) = \varphi(\gamma)$ for all $\gamma \in \mathbf{E}$ and $\|\mu\|_{M(G)} \leq C\|\varphi\|_\infty$. The infimum of such C is called the *Sidon constant*, $S(\mathbf{E})$, of $\mathbf{E}$.

An equivalent formulation, easily seen by a dual space argument, is that $\mathbf{E}$ is Sidon if and only if there is a constant C such that for every $\mathbf{E}$-polynomial f we have $\sum_{\gamma \in \mathbf{E}} |\widehat{f}(\gamma)| \leq C\|f\|_\infty$. That can be restated: for every $\mu \in M_d(\mathbf{E})$ we have $\|\mu\|_{M_d(\mathbf{E})} \leq C\|\widehat{\mu}\|_\infty$. The infimum of those C is equal to $S(\mathbf{E})$.

We now formalize definitions.

Definition 6.2.1. 1. The set $\mathbf{E} \subseteq \Gamma$ is said to have the *Fatou–Zygmund property (FZ)* if there is a constant, $S_0(\mathbf{E})$, such that for each Hermitian $\varphi \in \ell^\infty(\mathbf{E})$, there is some positive measure $\mu \in M(G)$ such that $\widehat{\mu}(\gamma) = \varphi(\gamma)$ for all $\gamma \in \mathbf{E}$ and $\|\mu\|_{M(G)} \leq S_0(\mathbf{E})\|\varphi\|_\infty$.

2. Suppose $U \subseteq G$ is Borel. The set $\mathbf{E} \subseteq \Gamma$ is said to be a *Sidon(U)set* if an interpolating measure μ can always be chosen to belong to $M(U)$. We denote the constant of interpolation in this case by $S(\mathbf{E}, U)$.

Since each bounded function is a linear combination of two bounded Hermitian functions, each set with the Fatou–Zygmund property is a Sidon set with the Sidon constant at most $2S_0$. Of course, a Sidon(G) set is simply a Sidon set, an $I_0(U)$ set is Sidon(U) and a FZI_0 set has the Fatou–Zygmund property. In particular, Hadamard and ε-Kronecker sets with $\varepsilon < \sqrt{2}$ are Sidon. Every infinite set in Γ contains an infinite Sidon set since it contains an infinite I_0 set (Corollary 4.5.3).

There are many characterizations of Sidonicity. Below we state some of the important early ones.

Theorem 6.2.2. *For* $\mathbf{E} \subseteq \Gamma$ *the following are equivalent:*

1. $\mathbf{E}$ *is a Sidon set with Sidon constant* $S(\mathbf{E})$.
2. *For each* $\varphi : \mathbf{E} \to \mathrm{Ball}(\ell^{\infty}(\mathbf{E}))$, *there exists* $\mu \in M(G)$ *such that*

$$\sup\{|\varphi(\gamma) - \widehat{\mu}(\gamma)| : \gamma \in \mathbf{E}\} < 1. \tag{6.2.1}$$

3. *For each* $\varphi : \mathbf{E} \to \{-1, 1\}$, *there exists* $\mu \in M(G)$ *such that (6.2.1) holds.*
4. *There is a constant* $S(\mathbf{E})$ *such that*

$$\sum_{\gamma} |\widehat{f}(\gamma)| \leq S(\mathbf{E}) \|f\|_{\infty} \text{ for all } f \in \mathrm{Trig}_{\mathbf{E}}(G).$$

The proof of this theorem may be found in many books.

The characterization of Sidon(U) sets is more complicated in that we do not know if "$\delta < 1$" is essential in Theorem 6.2.3(2) and (3).

Theorem 6.2.3. *For* $\mathbf{E} \subseteq \Gamma$ *and* U *a symmetric, compact subset of* G *the following are equivalent:*

1. $\mathbf{E}$ *is a Sidon(U) set with Sidon(U) constant* $S(\mathbf{E}, U)$.
2. *There exists* $0 < \delta < 1$ *such that for each* $\varphi : \mathbf{E} \to \Delta$, *there exists* $\mu \in M(U)$ *such that*

$$\sup\{|\varphi(\gamma) - \widehat{\mu}(\gamma)| : \gamma \in \mathbf{E}\} < \delta. \tag{6.2.2}$$

3. *There exists* $0 < \delta < 1$ *such that for each* $\varphi : \mathbf{E} \to \{-1, 1\}$ *there exists* $\mu \in M(U)$ *such that (6.2.2) holds.*
4. *There is a constant* $S(\mathbf{E}, U)$ *such that*

$$\sum_{\gamma} |\widehat{f}(\gamma)| \leq S(\mathbf{E}, U) \|f|_{U}\|_{\infty} \text{ for all } f \in \mathrm{Trig}_{\mathbf{E}}(G).$$

To obtain Theorem 6.2.3 (4) from (3) we will use a Baire category theorem argument, whose technique has already appeared several times; see, for example, Theorem 3.2.5.

Lemma 6.2.4. *Let* $U \subseteq G$ *be symmetric and compact. Suppose there exists* $0 < \delta < 1$ *such that for each* $\varphi : \mathbf{E} \to \{-1, 1\}$, *there exists* $\mu \in M(U)$ *with*

$$\sup\{|\varphi(\gamma) - \widehat{\mu}(\gamma)| : \gamma \in \mathbf{E}\} \leq \delta. \tag{6.2.3}$$

Then for every $\varepsilon > 0$ *there exists a constant* $C = C(\delta, \varepsilon)$ *such that for each* $\psi : \mathbf{E} \to [-1, 1]$, *there is a measure* $\mu \in M(U)$ *with* $\|\mu\| \leq C$ *and*

$$\sup\{|\psi(\gamma) - \widehat{\mu}(\gamma)| : \gamma \in \mathbf{E}\} \leq \varepsilon. \tag{6.2.4}$$

Proof (of Lemma 6.2.4). For each integer $n \geq 1$, let Ω_n be the set of $\varphi \in \mathbb{Z}_2^{\mathbf{E}}$ such that there exists $\mu \in M(G)$ with $\|\mu\| \leq n$ and $\sup_{\gamma \in \mathbf{E}} |\varphi(\gamma) - \widehat{\mu}(\gamma)| \leq \delta$. It is easy to verify that each Ω_n is closed in $\mathbb{Z}_2^{\mathbf{E}}$ (Exercise 6.6.3).

The hypothesis of the lemma tells us that $\mathbb{Z}_2^{\mathbf{E}} = \bigcup_{n=1}^{\infty} \Omega_n$. The Baire category theorem implies that some Ω_n has non-empty interior in $\mathbb{Z}_2^{\mathbf{E}}$ and hence contains a set of the form $\Omega' = \{\psi\} \times \mathbb{Z}_2^{\mathbf{E} \setminus \mathbf{F}}$, where $\mathbf{F}$ is a finite set and $\psi \in \mathbb{Z}_2^{\mathbf{F}}$. In particular, for every $\varphi : \mathbf{E} \to \{-1, 1\}$ there exists $\mu \in M(U)$ with

$$|\varphi - \widehat{\mu}| \leq \delta \text{ on } \mathbf{E} \setminus \mathbf{F} \text{ and } \|\mu\| \leq n. \tag{6.2.5}$$

By replacing μ with $\frac{1}{2}(\mu + \widetilde{\mu})$ (which still belongs to $M(U)$ as U is symmetric), we may assume that $\widehat{\mu}$ is real-valued.

The standard iteration, Proposition 1.3.2, but applied to arbitrary measures rather than discrete measures, tells us that for every $\varphi : \mathbf{E} \setminus \mathbf{F} \to [-1, 1]$ there exists $\mu' \in M(U)$ with $\widehat{\mu'} = \varphi$ on $\mathbf{E} \setminus \mathbf{F}$ and $\|\mu'\| \leq C_1 = n/(1-\delta)$. By interpolating real and imaginary parts, we see that every $\varphi : \mathbf{E} \setminus \mathbf{F} \to \Delta$ may be interpolated exactly by $\mu'' \in M(U)$ with $\|\mu''\| \leq 2C_1$.

If $U = G$, we add an appropriate trigonometric polynomial of at most $|\mathbf{F}|$ terms and coefficients bounded by $2C_1$ in absolute value, to elements of Ω', to see that φ can be interpolated exactly by a measure in $M(G)$ with norm at most $C = 2C_1 + (2C_1 + 1)|\mathbf{F}|$. The lemma thus holds, even with $\varepsilon = 0$.

When $U \neq G$ we must be more careful. Let $\mu \in M(U)$ have norm at most $2C_1$ and have $\widehat{\mu} = \varphi$ on $\mathbf{E} \setminus \mathbf{F}$. Enumerate the elements of $\mathbf{F}$ as $\lambda_1, \ldots, \lambda_J$. The hypotheses of the lemma and the standard iteration tell us that we may find $\nu_j \in M(U)$ such that

$$|1_{\{\lambda_j\}} - \widehat{\nu_j}| < \varepsilon/(8C_1 J) \text{ for } 1 \leq j \leq J \text{ on } \mathbf{E}.$$

(We have no control of the norms of the ν_j.) We now add a linear combination of the ν_j to μ to obtain μ' of norm at most $2C_1 + (1 + 2C_1) \sum_1^J \|\nu_j\|$, having $|\varphi - \widehat{\mu'}| < \varepsilon$ on $\mathbf{E}$. □

Proof (of Theorem 6.2.3). (1) $\Rightarrow$ (2) and (2) $\Rightarrow$ (3) are obvious.

(3) $\Rightarrow$ (4). Let δ be as in (3) and let C be given by Lemma 6.2.4. For $f \in \text{Trig}_{\mathbf{E}}(G)$ we put $f_1 = (f + \widetilde{f})/2$ and $if_2 = (f - \widetilde{f})/2$. Then $f = f_1 + if_2$ and $\widehat{f_1}, \widehat{f_2}$ are real-valued $\mathbf{E}$-polynomials, so there is no loss of generality in assuming $\widehat{f}$ is real.

Put $\varphi(\gamma) = \text{sgn} \overline{\widehat{f}(\gamma)}$ for $\gamma \in \mathbf{E}$. By assumption (3) and Lemma 6.2.4 there is a measure $\mu \in M(U)$ with $\|\mu\| \leq C$ and $|\varphi(\gamma) - \widehat{\mu}(\gamma)| \leq \delta < 1$ for all $\gamma \in \mathbf{E}$. Parseval's formula and the fact that μ is concentrated on U imply

$$\sum_{\gamma \in \mathbf{E}} |\widehat{f}(\gamma)| = \sum_{\gamma \in \mathbf{E}} \overline{\widehat{f}(\gamma)} \varphi(\gamma) = \sum_{\gamma \in \mathbf{E}} \overline{\widehat{f}(\gamma)} \left(\varphi(\gamma) - \widehat{\mu}(\gamma)\right) + \sum_{\gamma \in \mathbf{E}} \overline{\widehat{f}(\gamma)} \widehat{\mu}(\gamma)$$

$$= \sum_{\gamma \in \mathbf{E}} \overline{\widehat{f}(\gamma)} \left(\varphi(\gamma) - \widehat{\mu}(\gamma)\right) + \int_U \overline{f} d\mu.$$

Thus, $\sum_{\gamma\in\mathbf{E}}|\widehat{f}(\gamma)| \leq \sum_{\gamma\in\mathbf{E}}|\widehat{f}(\gamma)|\delta + \|f|_U\|_\infty\|\mu\|_{M(U)}$. Subtracting, we see that $\sum_{\gamma\in\mathbf{E}}|\widehat{f}(\gamma)| \leq \|f|_U\|_\infty C/(1-\delta)$.

(4) $\Rightarrow$ (1). Fix $\varphi \in \ell^\infty(\mathbf{E})$. Notice that assumption (4) ensures that if two $\mathbf{E}$-polynomials agree on U, then they have the same Fourier transform and hence coincide. Thus, we may define a linear functional

$$T : \mathrm{Trig}_\mathbf{E}(G)|_U \to \mathbb{C}$$

by $T(f|_U) = \sum_{\gamma\in\mathbf{E}}\widehat{f}(\gamma)\varphi(\gamma)$. Assumption (4) also ensures that

$$\|T\| \leq S(\mathbf{E},U)\|\varphi\|_\infty.$$

The operator T extends to a bounded linear functional on all of $C(U)$ by the Hahn–Banach theorem, and hence, by the Riesz representation theorem, there is a measure $\mu \in M(U)$ with $\|\mu\|_{M(G)} = \|T\|$ and $T(f) = \int_U \overline{f}d\mu$ for all $f \in C(U)$. In particular, if f is the character $\gamma \in \mathbf{E}$, then $T(f|_U) = \varphi(\gamma) = \int_U \overline{\gamma}d\mu = \widehat{\mu}(\gamma)$ since μ is supported on U. $\qquad\square$

Corollary 6.2.5. $\mathbf{E} \subseteq \Gamma$ *is Sidon if and only if whenever* $f \in L_\mathbf{E}^\infty(G)$, *then* $\widehat{f} \in \ell^1(\mathbf{E})$.

Proof. Suppose $\mathbf{E}$ is Sidon and $f \in L_\mathbf{E}^\infty(G)$. Take a bounded approximate identity of trigonometric polynomials, $\{K_\rho\}$. By (4) of the theorem, there is a constant $S(\mathbf{E})$ such that

$$\sum_{\gamma\in\mathbf{E}}|\widehat{f}(\gamma)\widehat{K_\rho}(\gamma)| \leq S(\mathbf{E})\|f * K_\rho\|_\infty \leq CS\|f\|_\infty,$$

where $\|K_\rho\|_1 \leq C$ for all ρ. Since $\widehat{K_\rho}(\gamma) \to 1$ this proves necessity. Sufficiency follows by an application of the closed graph theorem to deduce Theorem 6.2.3(4). $\qquad\square$

6.2.2 Weak Independence Properties and Riesz Products

Generalizations of independent sets are very important in the study of Sidon sets. Important classes include the dissociate sets, quasi-independent sets and k-independent sets.

Definition 6.2.6. 1. The set $\mathbf{E} \subseteq \Gamma \smallsetminus \{1\}$ is called *dissociate* if whenever $\gamma_1,\ldots,\gamma_n$ are distinct elements in $\mathbf{E}$, $m_j \in \{0,\pm1,\pm2\}$ and $\prod_{j=1}^n \gamma_j^{m_j} = 1$, then $\gamma_j^{m_j} = 1$ for all j.
2. The set $\mathbf{E} \subseteq \Gamma \smallsetminus \{1\}$ is called *quasi-independent* if the same statement holds for $m_j \in \{0,\pm1\}$.

3. Let k be a positive integer. The set $\mathbf{E} \subseteq \mathbf{\Gamma} \smallsetminus \{1\}$ is called *k-independent* if whenever $\gamma_1, \ldots, \gamma_n$ are distinct elements in $\mathbf{E}$, $m_j \in \{0, \pm1\}$, $\sum |m_j| \leq k$ and $\prod_{j=1}^{n} \gamma_j^{m_j} = 1$, then $\gamma_j^{m_j} = 1$ for all j.

Independent sets are dissociate, dissociate sets are quasi-independent and a set is independent if and only if it is k-independent for all positive integers k. All three classes of sets are asymmetric. A Hadamard set with ratio at least 3 is dissociate, and every Hadamard set is a finite union of dissociate sets; see Exercise 1.7.4. However, there exist quasi-independent sets that are not finite unions of Hadamard sets (Exercise 6.6.4).

A very important observation is that the Riesz product construction of Theorem 1.2.4 extends to dissociate sets. Indeed, suppose $\mathbf{E}$ is a dissociate subset of $\mathbf{\Gamma}$ and $\varphi : \mathbf{E} \to \mathbb{C}$ is bounded and Hermitian. Then φ is real-valued on the order two elements and has a unique extension to a bounded Hermitian function on $\mathbf{E} \cup \mathbf{E}^{-1}$, still called φ, by setting $\varphi(\gamma^{-1}) = \overline{\varphi(\gamma)}$.

In addition, suppose $\|\varphi\|_\infty \leq 1/2$.

For each $\gamma \in \mathbf{E}$, define the trigonometric polynomial $p_\gamma = p_{\gamma,\varphi}$ by

$$p_\gamma = p_{\gamma,\varphi} = \begin{cases} 1 + 2\mathfrak{Re}\big(\varphi(\gamma)\gamma\big) & \text{if } \gamma^2 \neq 1, \\ 1 + \varphi(\gamma)\gamma & \text{if } \gamma^2 = 1. \end{cases} \tag{6.2.6}$$

For a finite subset $\mathbf{F} \subseteq \mathbf{E}$, let

$$P_\mathbf{F} = P_{\mathbf{F},\varphi} = \prod_{\gamma \in \mathbf{F}} p_\gamma. \tag{6.2.7}$$

Since each $p_\gamma \geq 0$, the functions $P_\mathbf{F}$ are non-negative, trigonometric polynomials. The dissociate property of $\mathbf{E}$ makes it easy to see that

$$\|P_\mathbf{F}\|_1 = \int P_\mathbf{F} = \widehat{P_\mathbf{F}}(1) = 1 \text{ and that } \widehat{P_\mathbf{F}}(\gamma) = \varphi(\gamma) \text{ if } \gamma \in \mathbf{F} \cup \mathbf{F}^{-1}.$$

Furthermore,

$$\widehat{P_\mathbf{F}}(\lambda) = \begin{cases} \prod_{\gamma \in \mathbf{F}'} \varphi(\gamma) & \text{if } \lambda = \prod_{\gamma \in \mathbf{F}'} \gamma \text{ for asymmetric } \mathbf{F}' \subseteq \mathbf{F} \cup \mathbf{F}^{-1}, \\ 0 & \text{otherwise.} \end{cases}$$
$$\tag{6.2.8}$$

The finite subsets of $\mathbf{E}$ form a directed set under inclusion, and hence we can view $\{P_\mathbf{F} : \mathbf{F} \subseteq \mathbf{E} \text{ is finite}\}$ as a net in the closed unit ball of $M(G)$. Since that ball is weak* compact, the net has a cluster point in the weak* topology. Consideration of Fourier–Stieltjes transforms shows that this cluster point is unique. That unique measure, μ, is known as the *Riesz product* associated with $\mathbf{E}$ and φ. The measure μ is positive and $\|\mu\|_{M(G)} = 1 = \widehat{\mu}(1)$. Off of $\mathbf{E} \cup \mathbf{E}^{-1} \cup \{1\}$, the Fourier transform of μ is at most $1/4$ in modulus and on $\mathbf{E}$ it agrees with φ. These comments prove the following corollary.

Corollary 6.2.7. *Each dissociate set is a Fatou–Zygmund set with constant at most 2 and a Sidon set with constant at most 4.*

This corollary may be used to show every infinite subset of $\mathbf{\Gamma}$ contains a Fatou–Zygmund subset of the same cardinality (Exercise 6.6.13).

Remark 6.2.8. If none of the elements of $\mathbf{E}$ has order two, then we can apply the above reasoning directly to each $\varphi : \mathbf{E} \to \mathbb{C}$ that has supremum at most $1/2$. In this case, the Sidon constant is at most 2. If all the elements of $\mathbf{E}$ have order two, we can apply the same reasoning with each Hermitian $\varphi : \mathbf{E} \to \mathbb{C}$ of norm at most one. Thus, in the order two case, the Fatou–Zygmund constant is 1 and the Sidon constant is at most 2.

A similar Riesz product argument shows that a quasi-independent set is Sidon. For that it is convenient to introduce the following notations.

Definition 6.2.9. For a subset $\mathbf{F} \subseteq \mathbf{\Gamma}$ and positive integer k, let

$$Q_k(\mathbf{F}) = \left\{ \prod_{\gamma \in \mathbf{F}} \gamma^{\varepsilon_\gamma} : \gamma \in \mathbf{F}, \varepsilon_\gamma \in \{0, -1, 1\} \text{ and } \sum_\gamma |\varepsilon_\gamma| = k \right\} \text{ and}$$

$$W_k(\mathbf{F}) = \left\{ \prod_{\gamma \in \mathbf{F}} \gamma^{\varepsilon_\gamma} : \gamma \in \mathbf{F}, \varepsilon_\gamma \in \mathbb{Z} \text{ and } \sum_\gamma |\varepsilon_\gamma| \leq k \right\}.$$

The elements of $Q_k(\mathbf{F})$ will be called *quasi-words from* $\mathbf{F}$ *of presentation length* k and the elements of $W_k(\mathbf{F})$ will be called *words from* $\mathbf{F}$ *of length at most* k. Of course, $Q_k(\mathbf{F}) \subseteq W_k(\mathbf{F})$.

We shall only use $Q_k(\mathbf{F})$ here, but it is convenient to introduce both notations now since $W_k(\mathbf{F})$ will reappear implicitly in (6.3.2).

Definition 6.2.10. Given a non-negative integer k and $\chi \in \mathbf{\Gamma}$, let $R_k(\mathbf{E}, \chi)$ denote the number of asymmetric subsets $\mathbf{S} \subseteq \mathbf{E} \cup \mathbf{E}^{-1}$ of cardinality k satisfying $\prod_{\gamma \in \mathbf{S}} \gamma = \chi$. In particular, $R_k(\mathbf{E}, \gamma)$ is the number of ways to represent γ using elements of $Q_k(\mathbf{E})$.

An asymmetric set $\mathbf{E} \subseteq \mathbf{\Gamma}$ is said to be a *Rider set* if there is a constant $C > 0$ such that $R_k(\mathbf{E}, 1) \leq C^k$ for all k. The smallest such C is called the *Rider constant* of $\mathbf{E}$.

Since a quasi-independent set has $R_k(\mathbf{E}, 1) = 1$ for all k, every quasi-independent set is a Rider set.[1] Thus, the Sidonicity of quasi-independent sets follows from the next theorem. Exercise 6.6.5 asks for a direct proof that quasi-independence implies Sidon.

---

[1] The only possible $\mathbf{S}$ is the empty set, though of course it is debatable whether $\emptyset$ is to be included. If not, all $R_k(\mathbf{E}, 1) = 0$.

Theorem 6.2.11 (Rider). *Let* $\mathbf{E}$ *be a Rider set. Then* $\mathbf{E}$ *is a Sidon set with Sidon constant bounded by a function of the Rider constant.*

Proof. Choose $C \geq 1$ such that $R_k(\mathbf{E}, 1) \leq C^k$ for all k and let $\varphi(\chi) = 1/(2C)$ for all $\chi \in \mathbf{E} \cup \mathbf{E}^{-1}$. Given $\mathbf{F}$, a finite subset of $\mathbf{E}$, we define the Riesz product (trigonometric polynomial) $P_{\mathbf{F}}$ exactly as in (6.2.6) (and following), noting that we do not have the same information about the values of $\widehat{P_{\mathbf{F}}}$ when $\mathbf{F}$ is not dissociate. Then that Riesz product has

$$\widehat{P_{\mathbf{F}}}(1) = 1 + \sum_{k=1}^{\infty}(2C)^{-k}R_k(\mathbf{F}, 1) \leq 1 + \sum_{k=1}^{\infty}2^{-k} = 2. \qquad (6.2.9)$$

Therefore, $\|P_{\mathbf{F}}\|_1 \leq 2$ and, for $\chi \in \mathbf{F} \cup \mathbf{F}^{-1}$,

$$\sum_{k=1}^{\infty}(2C)^{-k}R_k(\mathbf{F}, \chi) = \widehat{P_{\mathbf{F}}}(\chi) \leq \|P_{\mathbf{F}}\|_1 \leq 2. \qquad (6.2.10)$$

Let $\psi : \mathbf{E} \to \{-1, 1\}$ and extend it to be Hermitian on $\mathbf{E} \cup \mathbf{E}^{-1}$. Let $0 < a < 1/(4C)$ and set $\varphi(\gamma) = a\psi(\gamma)/(2C)$ for $\gamma \in \mathbf{E}$. Also, let $\mu_{\mathbf{F}} = \mu_{\mathbf{F},a}$ be the resulting Riesz product (again noting that we will not have the same information about $\widehat{\mu_{\mathbf{F}}}$ as in the dissociate case). Then

$$\mu_{\mathbf{F}} = 1 + \sum_{k=1}^{\infty} \sum_{\substack{|\mathbf{S}|=k \\ \mathbf{S} \subset \mathbf{F} \cup \mathbf{F}^{-1} \\ \mathbf{S} \text{ asymmetric}}} \prod_{\rho \in \mathbf{S}} \frac{a\psi(\rho)}{2C}\,\rho. \qquad (6.2.11)$$

As in (6.2.9), $\|\mu_{\mathbf{F}}\|_{M(G)} \leq 2$. Put $\frac{2C}{a}\widehat{\mu_{\mathbf{F}}}(\chi) = \psi(\chi) + \beta_{\mathbf{F}}(\chi)$. Since for all $\gamma \in \mathbf{F} \cup \mathbf{F}^{-1}$, $\sum_{k=1}^{\infty}(2C)^{-k}R_k(\mathbf{F}, \chi) \leq 2$, we have

$$|\beta_{\mathbf{F}}(\chi)| \leq \frac{2C}{a} \sum_{k \geq 2} a^k(2C)^{-k}R_k(\mathbf{F}, \chi) \leq 4Ca < 1. \qquad (6.2.12)$$

Thus, $\left|\psi(\chi) - \frac{2C}{a}\widehat{\mu_{\mathbf{F}}}(\chi)\right| \leq 4Ca$ for all $\chi \in \mathbf{E}$. The same estimate on the Fourier transform for $\chi \in \mathbf{E}$ holds for any weak* cluster point of the net of measures $\{\mu_{\mathbf{F}}\}$, indexed by the finite subsets of $\mathbf{E}$. Applying Theorem 6.2.2 (3) proves that $\mathbf{E}$ is Sidon with Sidon constant bounded by a function of the Rider constant. $\qquad \square$

Corollary 6.2.12. *Every quasi-independent set is Sidon with Sidon constant bounded by a constant independent of the set.*

Remarks 6.2.13. (i) If $\mathbf{E}$ is quasi-independent, then $\widehat{P_{\mathbf{F}}}(1) = 1$, and so $\|\mu_{\mathbf{F}}\|_{M(G)} = 1$. If, in addition, the elements of $\mathbf{E}$ all have order two, then

E is actually an independent set. In particular, it is a dissociate set, and hence the Sidon constant is at most 2.

(ii) Exercise 6.6.5 gives an improvement in the bound for the Sidon constant of Rider sets and shows that the Sidon constant of a quasi-independent set is at most $6\sqrt{6}$. In the quasi-independent case, as in Remark 6.2.8, if none of the elements of **E** have order two, the argument of Exercise 6.6.5 may be applied directly, giving a Sidon constant of $3\sqrt{6}$.

We do not know if every dissociate set (much less every quasi-independent set) is I_0 [**P 6**]. However, our next result shows that each Rider set is a finite union of quasi-independent sets.

Proposition 6.2.14. *Every Rider set* **E** *is a finite union of quasi-independent sets.*

Proof. We may assume **E** does not contain 1 and the Rider constant is an integer $C \geq 3$. Randomly partition **E** into C^2 sets, where any element of **E** belongs to any one of the sets with equal probability and the choices are independent. Specifically, let $\xi_\gamma(\omega)$, $\gamma \in \mathbf{E}$, be independent random variables, each taking on the values $1, \ldots, C^2$ with equal probability. Let

$$\mathbf{E}_\ell(\omega) = \{\gamma : \xi_\gamma(\omega) = \ell\}, 1 \leq \ell \leq C^2.$$

We claim that with positive probability there is a random partition such that each of the C^2 subsets, $\mathbf{E}_\ell(\omega)$, is quasi-independent. To see this, note that a subset $\mathbf{E}_\ell(\omega)$ is not quasi-independent if and only there exist an integer k, characters $\gamma_1, \ldots, \gamma_k \in \mathbf{E}_\ell(\omega)$ and $\varepsilon_1, \ldots, \varepsilon_k \in \{-1, 1\}$ such that $1 = \prod_{j=1}^k \gamma_j^{\varepsilon_j}$, that is, **1** is a quasi-word of length k using letters in $\mathbf{E}_\ell(\omega)$. Because $\xi_\gamma(\omega) = \ell$ with probability $1/C^2$ and the random variables $\xi_{\gamma_1}, \ldots, \xi_{\gamma_k}$ are independent, the probability that a given collection of k letters from **E** all belong to $\mathbf{E}_\ell(\omega)$ is C^{-2k}. Since there are at most C^k ways of writing **1** as a quasi-word of length k with letters from **E** and C^2 subsets $\mathbf{E}_\ell(\omega)$, the probability that **1** is a quasi-word of length k with letters from one of the subsets $\mathbf{E}_\ell(\omega)$ is at most $C^2 \times C^{-2k} \times C^k = C^{2-k}$.

Since **1** is not a quasi-word from **E** of length one or two, it follows that the probability that one of the subsets is not quasi-independent is at most

$$\sum_{k \geq 3} C^{2-k} = 1/(C-1) < 1.$$

Hence, with positive probability, there is a choice ω such that all the sets $\mathbf{E}_\ell(\omega)$ are quasi-independent. This proves **E** is a union of at most C^2 quasi-independent sets. $\square$

6.3 Important Properties

6.3.1 The Union Theorem, the Fatou–Zygmund Property and Sidon(U)

An important distinction between Sidon and I_0 sets is that the class of Sidon sets is closed under finite unions. In particular, each finite union of quasi-independent sets is Sidon; the converse is, in general, an open problem [P 1], though when k is a product of distinct primes, then every Sidon set in the dual of $\mathbb{Z}_k^{\mathbb{N}}$ is the finite union of quasi-independent sets.

The union theorem was proven first by Drury, who used Riesz product measures in a very clever manner. A major idea in Drury's proof is the following result from which one can also deduce that Sidon sets without **1** have the Fatou–Zygmund property.

Theorem 6.3.1 (Drury's union theorem). *Suppose* $\mathbf{E} \subseteq \mathbf{\Gamma} \smallsetminus \{\mathbf{1}\}$ *is a symmetric Sidon set with Sidon constant* S. *Given* $0 < \varepsilon < 1$ *and Hermitian function* $\varphi \in Ball(\ell^\infty(\mathbf{E}))$, *there is a positive measure* $\mu \in M(G)$ *with* $\widehat{\mu}|_{\mathbf{E}} = \varphi$, $|\widehat{\mu}(\gamma)| \leq \varepsilon$ *for all* $\gamma \notin \mathbf{E} \cup \{\mathbf{1}\}$ *and* $\|\mu\|_{M(G)} \leq 32S^4/\varepsilon$.

Corollary 6.3.2. *Let* $\mathbf{E}$ *be a Sidon set with Sidon constant* S. *Given* $0 < \varepsilon < 1$ *and* $\varphi \in Ball(\ell^\infty(\mathbf{E}))$, *there is some* $\mu \in M(G)$ *with* $\widehat{\mu}|_{\mathbf{E}} = \varphi$, $|\widehat{\mu}(\gamma)| \leq \varepsilon$ *for all* $\gamma \notin \mathbf{E}$ *and* $\|\mu\|_{M(G)} \leq 512S^4/\varepsilon$.

The important union theorem for Sidon sets is an easy consequence of this corollary. Later in the book, we will give an alternate proof of the union theorem using the Pisier characterization of Sidonicity (see Remark 7.2.3).

Corollary 6.3.3 (Union theorem). *The union of two Sidon sets is Sidon.*

This shows that there are Sidon sets that are not I_0, since the non-I_0 set of Example 1.5.2 is the union of two I_0 sets (hence Sidon).

Corollary 6.3.4. *Each Sidon set that does not contain the identity has the Fatou–Zygmund property.*

Proof. If $\mathbf{E}$ is Sidon, so is $\mathbf{E}^{-1}$ and hence also $\mathbf{E} \cup \mathbf{E}^{-1}$. Since $\mathbf{E} \cup \mathbf{E}^{-1}$ is symmetric we can appeal to Theorem 6.3.1.   □

Because every measure has a unique decomposition as a sum of discrete and continuous measures, it is natural to study not only the class of sets where the interpolation can be done with discrete measures (the I_0 sets), but, also, the class where the interpolation can be done with continuous measures. Interestingly, the latter class coincides with the Sidon sets. This can be deduced from Drury's ideas and the fact that if $\mathbf{E}$ is a Sidon subset of an infinite group $\mathbf{\Gamma}$, then for each measure ν with discrete part ν_d,

$$\|\widehat{\nu_d}\|_\infty \leq \sup\{|\widehat{\nu}(\gamma)| : \gamma \in \mathbf{\Gamma} \smallsetminus \mathbf{E}\}.$$

Theorem 6.3.5. *Suppose* **E** *is a Sidon set. Given* $\varphi \in \ell^\infty(\mathbf{E})$ *there is a continuous measure* μ *such that* $\widehat{\mu}|_\mathbf{E} = \varphi$.

Drury's ideas were also used in Déchamps-Gondim's characterization of the Sidon sets that are Sidon(U) for all non-empty, open U.

Theorem 6.3.6 (Déchamps-Gondim). *A Sidon set* $\mathbf{E} \subseteq \mathbf{\Gamma}$ *is Sidon(U) for all non-empty, open sets* $U \subseteq G$ *if and only if for each finite subgroup* $\mathbf{X} \subseteq \mathbf{\Gamma}$, **E** *is the union of a finite set and a set which has the property that it intersects each coset of* **X** *in at most one point.*

Corollary 6.3.7. *If* G *is connected, then every Sidon set in the dual of* G *is Sidon(U) for all open* $U \subseteq G$.

Proof. The dual of a connected group has no non-trivial finite subgroups. □

6.3.2 $\Lambda(p)$ Sets and Arithmetic Properties

One of the classical results about Hadamard sequences, $\{n_k\}$, is that if $f = \sum a_k e^{in_k x}$ is integrable, then $f \in L^p$ for all $p < \infty$ (Theorem 1.2.7). This fact holds for a larger class of sets, known as $\Lambda(p)$ sets, which not only includes Sidon sets but also sets such as $\{3^j + 3^k : j, k \in \mathbb{N}\}$.

Definition 6.3.8. Suppose $1 < p < \infty$. A set $\mathbf{E} \subseteq \mathbf{\Gamma}$ is called a $\Lambda(p)$ *set* if whenever $f \in L^1_\mathbf{E}(G)$, then $f \in L^p(G)$.

An application of the closed graph theorem shows that **E** is a $\Lambda(p)$ set if and only if there is a constant B_p such that $\|f\|_p \le B_p \|f\|_1$ for all $f \in \mathrm{Trig}_\mathbf{E}(G)$. If $p > 2$, this is also equivalent to the existence of a constant C_p such that $\|f\|_p \le C_p \|f\|_2$ for all $f \in \mathrm{Trig}_\mathbf{E}(G)$. The least such constant C_p is called the $\Lambda(p)$ *constant of* **E**.

Obviously, if **E** is $\Lambda(p)$, then **E** is $\Lambda(q)$ for each $q < p$. It is known that if $1 < p < 2$ and **E** is a $\Lambda(p)$ set, then **E** is also a $\Lambda(q)$ set for some $q > p$, while if $p > 2$ there are sets that are $\Lambda(p)$ but not $\Lambda(q)$ for any $q > p$.

The union of two $\Lambda(p)$ sets for $p > 2$ is clearly $\Lambda(p)$. The union problem is open for $\Lambda(p)$ sets with $p \le 2$.

In this terminology, Theorem 1.2.7 states that Hadamard sets are $\Lambda(p)$ for all $p < \infty$. More generally, Sidon sets are $\Lambda(p)$.

Theorem 6.3.9 (Zygmund–Rudin). *Suppose* **E** *is a Sidon set with Sidon constant* $S(\mathbf{E})$. *Then* **E** *is a* $\Lambda(p)$ *set for all* $p < \infty$ *with* $\Lambda(p)$ *constant at most* $2S(\mathbf{E})\sqrt{p}$ *for all* $p > 2$.

The Pisier characterization of Sidon sets, Theorem 7.2.1, establishes (among other things) that the property of being $\Lambda(p)$ for all $p > 2$, with $\Lambda(p)$ constant $O(\sqrt{p})$, characterizes Sidon sets. That Sidon sets have this property is classic and is proved in standard references on Sidon sets, such as [123]. A proof is included here because it forms part of the proof of Theorem 7.2.1.

Proof (of Theorem 6.3.9). We will first show that if $\mathbf{F}$ is the Rademacher set in the dual of $\mathbb{D}' = \mathbb{Z}_2^{\mathbf{E}}$ and $h \in \mathrm{Trig}_{\mathbf{F}}(\mathbb{D}')$, then for all $k \in \mathbb{N}$,

$$\|h\|_{2k} \le 2\sqrt{k}\|h\|_2. \qquad (6.3.1)$$

That is, of course, a special case of the theorem.

To begin, let $h = \sum_1^J a_j \pi_j \in \mathrm{Trig}_{\mathbf{F}}(\mathbb{D}')$ have real coefficients, a_j. Then

$$\int_{\mathbb{D}'} |h(y)|^{2k}\, dy = \int \left(\sum_j a_j \pi_j \right)^{2k} dy$$

$$= \sum_{\substack{m_j \in \mathbb{N} \\ \sum_j m_j = 2k}} \frac{(2k)!}{m_1! \cdots m_J!} \int_{\mathbb{D}'} \prod_{j=1}^J (a_j^{m_j} \pi_j^{m_j})\, dy.$$

Since $\mathbf{F}$ is independent and all π_j are of order two, each integral on the right side of the above is zero unless all m_j in that term are even. When all m_j are even, the integral of that term is $\prod a_j^{m_j}$. Hence,

$$\|h\|_{2k}^{2k} = \int_{\mathbb{D}'} |h(y)|^{2k}\, dy = \sum_{\substack{m_j \in \mathbb{N} \\ \sum m_j = k}} \frac{(2k)!}{(2m_1)! \cdots (2m_J)!} \prod_{j=1}^J a_j^{2m_j}.$$

On the other hand,

$$\|h\|_2^{2k} = \left(\sum_j a_j^2 \right)^k = \sum_{\substack{m_j \in \mathbb{N} \\ \sum m_j = k}} \frac{k!}{m_1! \cdots m_J!} \prod a_j^{2m_j}.$$

But if $\sum m_j = k$, then

$$\frac{(2k)!}{(2m_1)! \cdots (2m_J)!} \frac{m_1! \cdots m_J!}{k!} = \frac{(k+1) \cdots (2k)}{\prod_{j=1}^J ((m_j+1) \cdots (2m_j))}$$

$$\le \frac{(k+1) \cdots (2k)}{2^k} \le k^k.$$

Thus, $\|h\|_{2k} \le \sqrt{k}\|h\|_2$.

For complex coefficients, an easy calculation using the triangle inequality shows that $\sqrt{2k}$ will do.

Now assume G is an arbitrary compact abelian group and $\mathbf{E}$ is a Sidon subset of the dual group $\mathbf{\Gamma}$.

Let $f = \sum_{\gamma \in \mathbf{E}} a_\gamma \gamma \in \mathrm{Trig}_{\mathbf{E}}(G)$. Let $\mathbf{F} = \{\pi_\gamma : \widehat{f}(\gamma) \neq 0\}$ be a set of Rademacher functions on $\mathbb{D}'$, so (6.3.1) applies to all $\mathbf{F}$-polynomials (i.e., elements of $\mathrm{Trig}_{\mathbf{F}}(\mathbb{D}')$). Also put $g_y(x) = g(x,y) = \sum_{\gamma \in \mathbf{E}} a_\gamma \pi_\gamma(y) \gamma(x)$ for $x \in G$ and $y \in \mathbb{D}'$. For each $y \in \mathbb{D}'$, consider the bounded function $\gamma \mapsto \pi_\gamma(y)$ for $\gamma \in \mathbf{E}$. Since $\mathbf{E}$ is Sidon, there exists a measure $\mu_y \in M(G)$ such that $\widehat{\mu_y}(\gamma) = \pi_\gamma(y)$ for all $\gamma \in \mathbf{E}$ and $\|\mu_y\| \leq S(\mathbf{E})$. Because $\pi_\gamma(y) = \pm 1$, we have $f(x) = \mu_y * g_y(x)$. Therefore,

$$\|f\|_p \leq \|g_y\|_p \|\mu_y\|_{M(G)} \leq S(\mathbf{E})\|g_y\|_p.$$

Also, $g_y = \mu_y * f$, so $\|g_y\|_p \leq S(\mathbf{E})\|f\|_p$.

Then, for each positive integer k and $y \in \mathbb{D}'$,

$$\int_G |f(x)|^{2k}\, \mathrm{d}x \leq S(\mathbf{E})^{2k} \int_G |g(x,y)|^{2k}\, \mathrm{d}x.$$

Now define the $\mathbb{D}'$-polynomial $h_x(y) = g(x,y)$ for each $x \in G$. By Plancherel's theorem, $\|h_x\|_2 = \|f\|_2$. Furthermore, Fubini's theorem and (6.3.1) applied to each function h_x and even integer $2k$ give

$$\int_{\mathbb{D}'} \int_G |f(x)|^{2k}\, \mathrm{d}x\, \mathrm{d}y \leq S(\mathbf{E})^{2k} \int_G \int_{\mathbb{D}'} |h_x(y)|^{2k}\, \mathrm{d}y\, \mathrm{d}x$$

$$\leq S(\mathbf{E})^{2k} \int_G (\sqrt{2k}\|h_x\|_2)^{2k}\, \mathrm{d}x$$

$$= S(\mathbf{E})^{2k}(\sqrt{2k}\|f\|_2)^{2k}.$$

Therefore, $\|f\|_{2k} \leq S(\mathbf{E})\sqrt{2k}\|f\|_2$. Since $\|f\|_p$ is increasing in p, $\|f\|_p \leq \|f\|_{2k} \leq S(\mathbf{E})\sqrt{2k}\|f\|_2$ whenever $2k \geq p$. Taking $2k$ to be the smallest even integer dominating p gives $\|f\|_p \leq S(\mathbf{E})2\sqrt{p}\|f\|_2$. □

Another class of $\Lambda(p)$ sets are the k-fold products of Rider sets.

Proposition 6.3.10. *Suppose $\mathbf{E}$ is a Rider set. For all $2 < p < \infty$, $Q_k(\mathbf{E})$ is a $\Lambda(p)$ set with $\Lambda(p)$ constant at most $A_k p^{k/2}$, where A_k depends only on k and the Rider constant of $\mathbf{E}$.*

For still other examples of $\Lambda(p)$ sets we introduce the function which counts the number of words in $W_k(E)$ whose product is γ,

$$r_k(\mathbf{E}, \gamma) = \left| \left\{ (\chi_1, \ldots, \chi_k) \in \mathbf{E}^k : \prod_{j=1}^{k} \chi_j = \gamma \right\} \right|. \qquad (6.3.2)$$

Proposition 6.3.11. *Suppose there is a constant C such that for some integer $k \geq 2$, $\sup_\gamma r_k(\mathbf{E}, \gamma) \leq C$. Then $\mathbf{E}$ is a $\Lambda(2k)$ set with $\Lambda(2k)$ constant at most $C^{1/k}$.*

Like Hadamard and ε-Kronecker sets, Sidon and $\Lambda(p)$ sets are severely restricted in terms of the arithmetic structures they can contain. For instance, Sidon sets cannot contain a product of two infinite sets. $\Lambda(p)$ sets cannot contain arithmetic progressions of length k, (meaning a subset of k distinct elements of the form $\{\gamma, \gamma\rho, \ldots, \gamma\rho^k\}$) for arbitrarily large k or even parallelepipeds of large dimension (see Definition 2.6.1).

Proposition 6.3.12.   *1. If $\mathbf{E}$ is a Sidon set, then*

$$\sup\{\min(|\mathbf{A}|, |\mathbf{B}|) : \mathbf{A} \cdot \mathbf{B} \subseteq \mathbf{E}\} < \infty.$$

2. If $\mathbf{E}$ is a $\Lambda(p)$ set for some $p > 2$, with $\Lambda(p)$ constant C_p, and $\mathbf{F}$ is an arithmetic progression with N terms, then $|\mathbf{E} \cap \mathbf{F}| \le 3C_p^2 N^{2/p}$.
3. If $\mathbf{E}$ is a $\Lambda(p)$ set for some $p > 1$, then there are constants C and $\tau > 0$ such that if $\mathbf{P}$ is a parallelepiped of dimension N, then $|\mathbf{E} \cap \mathbf{P}| \le C2^{N\tau}$.

Corollary 6.3.13. *If $\mathbf{E}$ is a Sidon set and $\mathbf{F}$ is an arithmetic progression of length N, then $|\mathbf{E} \cap \mathbf{F}| \le 24\, S(\mathbf{E})^2 \log N$.*

Proof. Apply Proposition 6.3.12 with $p = 2 \log N$. $\square$

Up to the constant this is sharp, as the example $\mathbf{E} = \{3^k\}$ demonstrates.

Example 6.3.14. A $\Lambda(p)$ set that is not Sidon: It follows from these facts that if $k \ge 2$, then $\{3^{j_1} + 3^{j_2} + \cdots + 3^{j_k} : j_1 < j_2 < \ldots < j_k\}$ is a $\Lambda(p)$ set for all $p < \infty$ but not a Sidon set (Exercise 6.6.6).

As a partial converse to Proposition 6.3.11, it is known that when $\mathbf{E} \subseteq \mathbb{N}$ is $\mathbf{\Lambda}(2k)$, then on average the numbers $r_k(\mathbf{E}, n)$ are bounded.

Proposition 6.3.15. *If $\mathbf{E} \subseteq \mathbb{N}$ is $\mathbf{\Lambda}(2k)$, then there is a constant C such that $\sum_{n=1}^{N} r_k^2(\mathbf{E}, n) \le CN$ for all N.*

If $\mathbf{F}$ is a finite set, we can view the sets $Q_k(\mathbf{F})$ as generalized arithmetic progressions. $\Lambda(p)$ sets can also contain only small portions of these sets.

Proposition 6.3.16.   *1. Suppose $\mathbf{E}$ is a $\Lambda(p)$ set for some $p > 2$, with $\Lambda(p)$ constant C_p. If $\mathbf{F}$ is a finite set, then*

$$|Q_k(\mathbf{F}) \cap \mathbf{E}| \le 2e^2 C_p^2 k^{2|\mathbf{F}|/p}.$$

2. If $\mathbf{E}$ is a Sidon set and $\mathbf{F}$ is a finite set, then

$$|Q_k(\mathbf{F}) \cap \mathbf{E}| \le C|\mathbf{F}| \log k,$$

where C depends only on the Sidon constant of $\mathbf{E}$.

This bound will be important in the next section in proving that each Sidon set is a finite union of k-independent sets.

6.4 Decompositions of Sidon Sets

A fundamental and difficult problem in the study of Sidon sets is to determine which more restricted classes of sets have the property that every Sidon set is a finite union of sets from the class. Hadamard sets do not have this property, for instance: Sidon sets, and even ε-Kronecker sets, need not be finite unions of Hadamard sets since not even all quasi-independent sets are finite unions of Hadamard sets. See Exercise 6.6.4(2) for the assertion about quasi-independent sets and Example 2.2.7 for the assertion about ε-Kronecker sets. For many other interesting classes of sets this problem is open. In particular, it is unknown **[P 1]** if every Sidon set in $\Gamma \smallsetminus \{1\}$ is:

- A finite union of quasi-independent sets
- A finite union of ε-Kronecker sets for some $\varepsilon < \sqrt{2}$
- A finite union of I_0 sets.

These problems are all open for $\Gamma = \mathbb{Z}$ and for most other groups, though there are answers in two special cases (see the section Remarks and credits). However, every Sidon set not containing **1** *is* a finite union of k-independent sets. This striking fact was proven by Bourgain; his proof is given in the first subsection. From this characterization one can also deduce that every Sidon set is a finite union of sets whose step length tends to infinity and consequently a finite union of sets that are Sidon(U) for all non-empty, open sets U. That is proved in the second subsection.

6.4.1 Sidon Sets Are Finite Unions of k-Independent Sets

The proof that Sidon sets in $\Gamma \smallsetminus \{1\}$ are finite unions of k-independent sets is based on an investigation of the sets, $Q_k(\mathbf{F})$ (see Definition 6.2.9), of quasi-words of presentation length k. This union property actually holds, more generally, for sets $\mathbf{E}$ with the property that if $\mathbf{F} \subseteq \Gamma$ is finite, then $|\mathbf{E} \cap Q_k(\mathbf{F})| \leq C_k |\mathbf{F}|$, where C_k depends only on k and the set $\mathbf{E}$.

Theorem 6.4.1 (Bourgain's k-independent set theorem). *Suppose that for some integer k there is a constant C_k such that every finite subset $\mathbf{F} \subseteq \mathbf{E}$ satisfies the condition*

$$|\mathbf{E} \cap Q_k(\mathbf{F})| \leq C_k |\mathbf{F}|.$$

Then $\mathbf{E}$ is a union of N sets, each of which is k-independent, where N depends only on k and C_k.

Appealing to Proposition 6.3.16 gives the decomposition result for Sidon sets.

Corollary 6.4.2. *For each k, a Sidon set $\mathbf{E} \subseteq \Gamma \smallsetminus \{1\}$ is a finite union of k-independent sets.*

The proof of the theorem relies upon combinatorial arguments. For the duration of the proof, given $\mathbf{A} \subseteq \boldsymbol{\Gamma}$, let

$$\mathcal{P}_k(\mathbf{A}) = \{\mathbf{A}' \subseteq \mathbf{A} : |\mathbf{A}'| = k\}.$$

Lemma 6.4.3. *Suppose $\mathcal{I}$ and $\mathbf{A} \subset \boldsymbol{\Gamma}$ are finite sets and $k, N \in \mathbb{N}$. Let $\delta = N^{-1/2k}$ and assume $\delta|\mathbf{A}|$ is much greater than k. For each $i \in \mathcal{I}$, let B_i be a subset of $\mathcal{P}_k(\mathbf{A})$. Suppose that for each $i \in \mathcal{I}$, $|B_i| = N$ and that, if $\boldsymbol{\pi}, \boldsymbol{\pi}' \in B_i$, then either $\boldsymbol{\pi} = \boldsymbol{\pi}'$ or $\boldsymbol{\pi} \cap \boldsymbol{\pi}'$ is empty. Then there are subsets $\mathcal{I}' \subseteq \mathcal{I}$ and $\mathbf{A}' \subseteq \mathbf{A}$ such that:*

1. $|\mathcal{I}'| \geq |\mathcal{I}|/2$.
2. $\delta|\mathbf{A}|/2 \leq |\mathbf{A}'| \leq \delta|\mathbf{A}|$.
3. *For each $i \in \mathcal{I}'$, there is at least one element $\boldsymbol{\pi} \in B_i$ with $\boldsymbol{\pi} \subseteq \mathbf{A}'$.*

Proof. Let $n = \lfloor \delta|\mathbf{A}| \rfloor$. Then every $\mathbf{A}' \in \mathcal{P}_n(\mathbf{A})$ satisfies (2).

For each $i \in \mathcal{I}$, define a function F_i on $\mathcal{P}_n(\mathbf{A})$ by setting

$$F_i(\mathbf{Y}) = \sum_{\boldsymbol{\pi} \in B_i} \prod_{\rho \in \boldsymbol{\pi}} 1_{\mathbf{Y}}(\rho) \text{ for } \mathbf{Y} \in \mathcal{P}_n(\mathbf{A}).$$

Of course, $\prod_{\rho \in \boldsymbol{\pi}} 1_{\mathbf{Y}}(\rho) = 0$ or 1 and equals 1 if and only if $\boldsymbol{\pi} \subseteq \mathbf{Y}$. Thus, $F_i(\mathbf{Y}) \geq 1$ (equivalently, $F_i(\mathbf{Y}) > 0$) if and only if there is some $\boldsymbol{\pi} \in B_i$ with $\boldsymbol{\pi} \subseteq \mathbf{Y}$.

The number of sets $\mathbf{Y} \in \mathcal{P}_n(\mathbf{A})$ that contain a given subset π of cardinality k is $\binom{|\mathbf{A}|-k}{n-k}$, and since the cardinality of $\mathcal{P}_n(\mathbf{A})$ is $\binom{|\mathbf{A}|}{n}$, the average (or expected) value of F_i, denoted $\mathbb{E}(F_i)$, is given by

$$\mathbb{E}(F_i) = \sum_{\boldsymbol{\pi} \in B_i} \frac{\binom{|\mathbf{A}|-k}{n-k}}{\binom{|\mathbf{A}|}{n}} \stackrel{|\mathbf{A}|}{\sim} N\delta^k.$$

Here $f \stackrel{|\mathbf{A}|}{\sim} g$ means $f/g \to 1$ as $|\mathbf{A}| \to \infty$.

If $\boldsymbol{\pi}, \boldsymbol{\pi}' \in B_i$, then $\prod_{\rho \in \boldsymbol{\pi}} 1_{\mathbf{Y}}(\rho) \prod_{\rho' \in \boldsymbol{\pi}'} 1_{\mathbf{Y}}(\rho') \neq 0$ if and only if $\mathbf{Y}$ contains $\boldsymbol{\pi} \cup \boldsymbol{\pi}'$. As these sets are disjoint if $\boldsymbol{\pi} \neq \boldsymbol{\pi}'$, the number of sets $\mathbf{Y} \in \mathcal{P}_n(\mathbf{A})$ with this property is $\binom{|\mathbf{A}|-2k}{n-2k}$ or approximately $\delta^{2k}|\mathcal{P}_n(\mathbf{A})|$. Thus, the average value of F_i^2 is

$$\mathbb{E}(F_i^2) = \mathbb{E}\left(\sum_{\boldsymbol{\pi} \in B_i}\left(\prod_{\rho \in \boldsymbol{\pi}} 1_{\mathbf{Y}}(\rho)\right)^2\right) + \mathbb{E}\left(\sum_{\boldsymbol{\pi} \neq \boldsymbol{\pi}' \in B_i} \prod_{\rho \in \boldsymbol{\pi}} 1_{\mathbf{Y}}(\rho) \prod_{\rho' \in \boldsymbol{\pi}'} 1_{\mathbf{Y}}(\rho')\right)$$

$$\stackrel{|\mathbf{A}|}{\sim} \sum_{\boldsymbol{\pi} \in B_i} \delta^k + \sum_{\boldsymbol{\pi} \neq \boldsymbol{\pi}' \in B_i} \delta^{2k} = N\delta^k + N(N-1)\delta^{2k}.$$

Given the choice of $\delta = N^{-1/2k}$ we deduce that

$$\mathbb{E}(F_i) \stackrel{|\mathbf{A}|}{\sim} \sqrt{N} \text{ and } \mathbb{E}(F_i^2) \stackrel{|\mathbf{A}|}{\sim} N(1 + \varepsilon_N),$$

where $\varepsilon_N \to 0$ as $N \to \infty$.

We claim the probability that $F_i(\mathbf{Y}) \geq 1$ (equivalently, $F_i(\mathbf{Y}) > 0$) is at least $1/2$ for all $i \in \mathcal{I}$. To see this, note that Hölder's inequality implies

$$N \overset{|\mathbf{A}|}{\sim} \left(\mathbb{E}(F_i)\right)^2 \leq \mathbb{P}\{F_i \geq 1\} \mathbb{E}(F_i^2) \overset{|\mathbf{A}|}{\sim} \mathbb{P}\{F_i \geq 1\} N(1 + \varepsilon_N).$$

Thus, $\mathbb{P}\{F_i \geq 1\} \geq (1 + \varepsilon_N)^{-1} \geq 1/2$ for large enough N.

For $\mathbf{Y} \in \mathcal{P}_n(\mathbf{A})$, let $q(\mathbf{Y})$ be the number of indices $i \in \mathcal{I}$ such that $F_i(\mathbf{Y}) \geq 1$. Since the expected value of q equals

$$\mathbb{E}(q) = \mathbb{E}\left(\sum_{i \in \mathcal{I}} 1_{F_i^{-1}[1,\infty)}\right) = \sum_{i \in \mathcal{I}} \mathbb{P}\{F_i(\mathbf{Y}) \geq 1\} \geq |\mathcal{I}|/2,$$

there must be some $\mathbf{Y} \in \mathcal{P}_n(\mathbf{A})$ with $F_i(\mathbf{Y}) \geq 1$ for at least half the indices in $\mathcal{I}$. Let $\mathbf{A}'$ be such a choice of $\mathbf{Y}$ and let $\mathcal{I}' = \{i : F_i(\mathbf{Y}) \geq 1\}$. Then properties (1) and (3) hold by definition.   □

Lemma 6.4.4. *Assume* $\mathbf{E} \subseteq \Gamma$ *has the property that* $|\mathbf{E} \cap Q_k(\mathbf{F})| \leq C_k|\mathbf{F}|$ *for all finite subsets* $\mathbf{F} \subseteq \mathbf{E}$ *and some (fixed)* k. *Let* $\mathbf{A} \subseteq \mathbf{E}$ *be finite,* N *be a positive integer and*

$$\mathcal{I} = \{\gamma \in \mathbf{E} : \gamma \in Q_k(\boldsymbol{\pi}) \text{ for at least } N \text{ disjoint sets } \boldsymbol{\pi} \subseteq \mathbf{A}\}.$$

Then $|\mathcal{I}| \leq 2C_k N^{-1/2k}|\mathbf{A}|$.

Proof. For each $\gamma \in \mathcal{I}$, put $\mathbf{B}_\gamma = \{\boldsymbol{\pi} \in \mathcal{P}_k(\mathbf{A}) : \gamma \in Q_k(\boldsymbol{\pi})\}$. The definition of $\mathcal{I}$ implies that by reducing $\mathbf{B}_\gamma$, if necessary, we can suppose the sets $\mathbf{B}_\gamma$ all have N elements and if $\boldsymbol{\pi}, \boldsymbol{\pi}' \in \mathbf{B}_\gamma$, $\boldsymbol{\pi} \neq \boldsymbol{\pi}'$, then $\boldsymbol{\pi} \cap \boldsymbol{\pi}'$ is empty. Put $\delta = N^{-1/2k}$.

Appealing to the previous lemma, we obtain $\mathcal{I}' \subseteq \mathcal{I}$ with $|\mathcal{I}'| \geq |\mathcal{I}|/2$ and $\mathbf{A}' \subset \mathbf{A}$ with $|\mathbf{A}'| \leq \delta|\mathbf{A}|$, having the property that for each $\gamma \in \mathcal{I}'$ there is some $\boldsymbol{\pi} \in \mathbf{B}_\gamma$ such that $\boldsymbol{\pi} \subseteq \mathbf{A}'$. Consequently, $\gamma \in Q_k(\mathbf{A}')$ for all $\gamma \in \mathcal{I}'$, in other words, $\mathcal{I}' \subseteq Q_k(\mathbf{A}')$.

Since $\mathcal{I}' \subseteq \mathbf{E}$, that ensures that $|Q_k(\mathbf{A}') \cap \mathbf{E}| \geq |\mathcal{I}'|$. Together with the hypothesis of the lemma, these observations yield the inequalities

$$\frac{1}{2}|\mathcal{I}| \leq |\mathcal{I}'| \leq |Q_k(\mathbf{A}') \cap \mathbf{E}| \leq C_k|\mathbf{A}'| \leq C_k\delta|\mathbf{A}|,$$

which give the desired result.   □

Lemma 6.4.5. *Suppose there is some constant* N *such that every finite subset of* $\mathbf{E}$ *is a union of* N *sets, each of which is* k-*independent. Then* $\mathbf{E}$ *is a union of* N *sets that are* k-*independent.*

Proof. This is a limit argument. By assumption, if $\mathbf{F}$ is a finite subset of $\mathbf{E}$, then $\mathbf{F} = \bigcup_{j=1}^{N} \mathbf{H}_j^{(\mathbf{F})}$, where the sets $\mathbf{H}_j^{(\mathbf{F})}$ are k-independent and disjoint (possibly empty). The collection of finite subsets of $\mathbf{E}$ is a partially ordered set, with the usual inclusion ordering. Consider the characteristic functions

$1_{\mathbf{F}}$ and $1_{\mathbf{H}_j^{(\mathbf{F})}}$, $j = 1, \ldots, N$, as nets with respect to this partial ordering. These characteristic functions belong to the compact space $\{0, 1\}^{\mathbf{\Gamma}}$ and so by passing to successive subnets we can assume $1_{\mathbf{H}_j^{(\mathbf{F}_\beta)}} \to f_j$, $j = 1, \ldots, N$, and $1_{\mathbf{F}_\beta} \to f$ pointwise.

If $\gamma \notin \mathbf{E}$, then $1_{\mathbf{F}}(\gamma) = 0$ for all $\mathbf{F} \subseteq \mathbf{E}$, and thus $f(\gamma) = 0$. If $\gamma \in \mathbf{E}$, then $\gamma \in \mathbf{F}_\beta$ eventually. Therefore, $f(\gamma) = 1$, proving $f = 1_{\mathbf{E}}$. Because each f_j is $0, 1$-valued, $f_j = 1_{\mathbf{H}_j}$ for suitable $\mathbf{H}_j \subseteq \mathbf{\Gamma}$. A limiting argument shows that $\sum_{j=1}^N 1_{\mathbf{H}_j} = 1_{\mathbf{E}}$, and hence $\bigcup_{j=1}^N \mathbf{H}_j = \mathbf{E}$.

To see that each set $\mathbf{H}_j$ is k-independent, suppose $\prod_{i=1}^k \gamma_i^{\varepsilon_i} = 1$ with each $\varepsilon_i = \pm 1$. Since $\gamma_1, \ldots, \gamma_k \in \mathbf{H}_j^{(\mathbf{F}_\beta)}$ eventually, this contradicts the fact that every $\mathbf{H}_j^{(\mathbf{F}_\beta)}$ is k-independent. $\square$

Proof (of Theorem 6.4.1). Lemma 6.4.5 shows we may assume $\mathbf{E}$ is finite.

Take $N = N(k, C_k) = (2C_k)^{2k} + 1$. We proceed by contradiction and assume $\mathbf{E}$ is not the union of N k-independent sets. Then there must be a subset $\mathbf{F} \subseteq \mathbf{E}$ which is not a union of N k-independent sets, but every proper subset of $\mathbf{F}$ is such a union. In particular, for each $\gamma \in \mathbf{F}$ the proper subset, $\mathbf{F} \setminus \{\gamma\}$, is a union of N sets, $\mathbf{F}_{\gamma,1}, \ldots, \mathbf{F}_{\gamma,N}$, each of which is k-independent.

For each j, it must be true that $\gamma \in Q_k(\mathbf{F}_{\gamma,j})$ because otherwise the set $\mathbf{F}_{\gamma,j} \cup \{\gamma\}$ would still be k-independent and then the collection of sets $\mathbf{F}_{\gamma,1}, \ldots, \mathbf{F}_{\gamma,j} \cup \{\gamma\}, \ldots, \mathbf{F}_{\gamma,N}$ would be a decomposition of $\mathbf{F}$ into N k-independent sets.

Define $\mathcal{I}$ as in Lemma 6.4.4, but with $\mathbf{A}$ in the lemma equal to $\mathbf{F}$. The previous remark implies that $\mathbf{F} \subseteq \mathcal{I}$, and hence $|\mathcal{I}| \geq |\mathbf{F}|$. But, Lemma 6.4.4 implies $|\mathcal{I}| \leq 2C_k N^{-1/2k} |\mathbf{F}|$ and this is strictly less than $|\mathbf{F}|$ by the choice of N. This contradiction completes the proof of the theorem. $\square$

Recall (Definition 2.7.6) that a set $\mathbf{E} \subseteq \mathbf{\Gamma}$ has "step length tending to infinity" if for every finite subset $\mathbf{F}' \subseteq \mathbf{\Gamma}$ there exists a finite subset $\mathbf{F} \subseteq \mathbf{E}$ such that if $\chi, \psi \in \mathbf{E} \setminus \mathbf{F}$ and $\chi \neq \psi$, then $\chi\psi^{-1} \notin \mathbf{F}'$. We can deduce from Bourgain's Theorem 6.4.1 that every Sidon set is a finite union of sets with step length tending to infinity.

Corollary 6.4.6. *If there is a constant C such that all finite subsets $\mathbf{F} \subseteq \mathbf{E}$ satisfy the condition $|\mathbf{E} \cap Q_4(\mathbf{F})| \leq C|\mathbf{F}|$, then $\mathbf{E}$ is a finite union of sets with step length tending to infinity.*

Proof. Theorem 6.4.1 implies such a set is a finite union of 4-independent sets, and thus it suffices to check that each 4-independent set $\mathbf{E}$ has step length tending to infinity.

Assume otherwise. Then there is a finite set $\mathbf{F}' \subseteq \mathbf{\Gamma}$ such that for every finite subset $\mathbf{F} \subseteq \mathbf{E}$ there exist $\chi, \psi \in \mathbf{E} \setminus \mathbf{F}$ with $\chi \neq \psi$ and yet $\chi\psi^{-1} \in \mathbf{F}'$. Inductively choose infinite sequences, $\{\chi_n\}_{n=1}^\infty$, $\{\psi_n\}_{n=1}^\infty$ from $\mathbf{E}$, such that $\chi_n\psi_n^{-1} \in \mathbf{F}' \setminus \{1\}$ and $\chi_n, \psi_n \notin \{\chi_1, \psi_1, \ldots, \chi_{n-1}, \psi_{n-1}\}$.

Since $\mathbf{F}'$ is finite, by passing to a further subsequence, if necessary, we can assume $\chi_n \psi_n^{-1} \chi_1^{-1} \psi_1 = 1$ for all n, contradicting the assumption that $\mathbf{E}$ is 4-independent.   □

Corollary 6.4.7 (Déchamps-Gondim–Bourgain theorem). *Every Sidon set is a finite union of Sidon sets whose step length tends to infinity.*

It is easy to see that each set $\mathbf{E}$ whose step length tends to infinity has the property that for each finite subgroup $\mathbf{X} \subseteq \Gamma$, $\mathbf{E}$ is the union of a finite set and a set which intersects each coset of $\mathbf{X}$ in at most one point. Thus, Theorem 6.3.6 implies that every Sidon set whose step length tends to infinity is $\mathrm{Sidon}(U)$ for all non-empty, open sets U. More generally,

Corollary 6.4.8. *Every Sidon set is a finite union of sets that are $\mathrm{Sidon}(U)$ for all non-empty, open U.*

In Corollary 7.4.3, we will show that a Sidon set whose step length tends to infinity is $\mathrm{Sidon}(U)$ in a uniform sense.

6.5 Remarks and Credits

Introduction. López and Ross [123] provide a detailed report of what was known about Sidon sets up to the early 1970s. Some results are also in [88, Sect. 37] and [167, Sect. 5.7], both published earlier. A modern treatment is in Li and Queffélec's [119]. Sidon's proof that Hadamard sets have the property of Corollary 6.2.5 is in [174].

Characterizations. Theorem 6.2.2 and Corollary 6.2.5 are standard. Other equivalences can be found in [88, Sect. 37.2], [123, Chap. 1] and [167, Sect. 5.7].

Corollary 6.2.5 shows that if $f : \mathbf{E} \to \mathbb{C}$ is bounded, then $\sum_{\gamma \in \Gamma} \hat{f}(\gamma) \gamma$ converges uniformly, and therefore f agrees almost everywhere with a continuous function. Thus, if $\mathbf{E}$ is Sidon, then $L_{\mathbf{E}}^\infty = C_{\mathbf{E}}$. Sets with the latter property are known as *Rosenthal sets*, after Rosenthal's construction of examples of Rosenthal sets that are not Sidon [164] or [123, pp. 161-2]. That corollary also implies that if $\mathbf{E}$ is Sidon, then $C_{\mathbf{E}}$ is isomorphic to ℓ^1. The converse of this is true and due to Varopoulos [190]. It requires functional analysis techniques that we will not develop. We refer the reader to [119, Theorem 5.IV.15].

Weak Independence Properties. The term "dissociate set" was first used by Hewitt and Zuckerman in [90, 91]. In [89] they had proved that dissociate sets have the Fatou–Zygmund property. A form of Exercise 6.6.4(2) can also be found there. The term, "Rider set" is used because Rider [160] showed that the sets now named after him were Sidon. That improved an earlier result of Stečkin [176], where some of the ideas developed further in [89] also appear. See [123, Chap. 2] for a more detailed discussion of Rider sets. Proposition 6.2.14 is due to Pisier [150, Proposition 2.13]; the proof here is due to Crevier [26].

Li and Queffélec [119, Chap. 12:I.1] show that quasi-independent sets have Sidon constant at most $6\sqrt{6}$, as per Exercise 6.6.5. Another proof, using Rider's randomly continuous function characterization of Sidonicity [162], can be found in [149, Lemma 3.2]. When the quasi-independent set is (equivalent to) a set of Rademacher functions, then the Sidon constant is exactly $\pi/2$ if the set is infinite [171] (or see [119, p. 297]) and $N\sin(\pi/2N)$ if the Rademacher set has N elements [2].

Li and Queffélec [119, pp. 209-10] give a charming proof that Hadamard sets are Sidon.

A set $\mathbf{E} \subseteq \Gamma$ is called a *Rajchman set* if whenever $\mu \in M(G)$ and $\widehat{\mu} \in c_0(\Gamma \setminus \mathbf{E})$, (i.e., $\widehat{\mu}$ vanishes at infinity off $\mathbf{E}$), then $\mu \in M_0(G)$. The terminology comes from a classical result of Rajchman that states that $-\mathbb{N}$ is a Rajchman set. It is known that the union of a Rajchman set and a Sidon set is a Rajchman set [145]. Obviously, if $\mathbf{E}$ contains a translate of the support of the Fourier transform of a Riesz product measure, then $\mathbf{E}$ is not Rajchman. Host and Parreau [93] showed that the converse holds, as well.

The Union Theorem, the Fatou–Zygmund Property and Sidon(U). The union theorem was first proved by Drury [32]. Many proofs of it are known. See Remark 7.2.3 for one due to Pisier [149] (or at least noted by the reviewer of Pisier's paper) and [196] for a proof for the non-abelian case. Edwards, Hewitt and Ross [36] introduced the Fatou–Zygmund property, but it was Drury [33] who showed that every Sidon set has the property.

That continuous measures can be used to do the interpolation for Sidon sets (Theorem 6.3.5) was discovered independently by Hartman and Wells [80, 195]. This result is sometimes called the "Hartman–Wells theorem". Méla earlier proved Theorem 6.3.5 for I_0 sets [128, Theorem 6.1]. In Hartman's proof, it was shown that the complement of a Sidon set is dense in $\overline{\Gamma}$. Unlike I_0 sets, it is unknown if Sidon sets can cluster at a continuous character or even if they can be dense in $\overline{\Gamma}$. This problem will be revisited in Chaps. 8 and 10.

The Sidon(U) property was established by Déchamps-Gondim in [27], who used the terminology "strictly associated with all open sets". The notion of Sidon(U) first appears in [85] under the terminology, "U is appropriate to $\mathbf{E}$".

Complete proofs of all of these results can be found in [123, Chaps. 3-4, 7-8].

$\Lambda(p)$ Sets and Arithmetic Properties. Zygmund [6] and Rudin [166] are responsible for many of the basic facts about $\Lambda(p)$ sets and their relationship with Sidon sets. See also [88, Chap. 37] and [123, Chap. 5] where many characterizations can be found. Rudin [166] proved Theorem 6.3.9 for Sidon sets in $\mathbb{Z}$ (the proof for general Γ involves nothing new) and showed that the $\Lambda(p)$ constant $O(\sqrt{p})$ is best possible for an infinite set. Thirty years earlier, Zygmund had established the result for Hadamard sets [198] and proved Proposition 6.3.11 for Hadamard and Rademacher sets. Proposition 6.3.10 is due to Bonami [16, p. 359].

Rudin [166] also discovered many arithmetic properties of $\Lambda(p)$ sets in $\mathbb{Z}$, including Propositions 6.3.11, 6.3.12(2), and 6.3.15. Proofs for general groups and further arithmetic properties, including Proposition 6.3.16, can be found in [123, Chap. 6]. See also [9, 99].

Another accomplishment of Rudin was to construct examples to show that for each integer $q \geq 2$ there were subsets of integers which were $\Lambda(2q)$ sets but not $\Lambda(2q + \varepsilon)$ for any $\varepsilon > 0$. Later, Bourgain [19] used probabilistic methods to show that for each $p > 2$ there are $\Lambda(p)$ sets that are not $\Lambda(p + \varepsilon)$ for any $\varepsilon > 0$. The contrasting fact that each $\Lambda(p)$ set for $1 < p < 2$ is also $\Lambda(q)$ for some $q > p$ can be found in [5, 74].

$\Lambda(p)$ sets were also defined by Rudin for $p \leq 1$. These are sets **E** with the property that there is some $0 < q < p$ and constant C such that $\|f\|_p \leq C\|f\|_q$ for all $f \in \mathrm{Trig}_{\mathbf{E}}(G)$. It is unknown if such sets are $\Lambda(2)$.

Rudin's paper [166] continues to influence activity. For example, [24] addresses a consequence of an affirmative answer to a question posed by Rudin in that paper: Is the set of squares a $\Lambda(p)$ set for some p?

A set $\mathbf{E} \subseteq \mathbf{\Gamma}$ is said to be a B_2 set if $r_2(\mathbf{E}, \gamma) \leq 2$ for all γ. These are the sets such that $\gamma_1 \gamma_2 = \gamma_3 \gamma_4$ for $\gamma_j \in \mathbf{E}$ if and only if $\{\gamma_3, \gamma_4\}$ is a permutation of $\{\gamma_1, \gamma_2\}$. In additive groups, such as $\mathbb{Z}$, these are the sets for which all non-zero differences from the set are distinct. B_2 sets are examples of $\Lambda(4)$ sets. They are often called "Sidon sets" in number theory because the terminology B_2 set was first coined by Sidon. See [138] for a detailed annotated bibliography. However, such sets need not be Sidon sets in our harmonic analysis sense; Rudin's construction of $\Lambda(4)$ sets that were not $\Lambda(4 + \varepsilon)$ for any $\varepsilon > 0$ gives B_2 sets, and such sets cannot be Sidon (in our sense).

Szemerdi's famous theorem [182] says that each subset of integers of positive density contains arbitrarily long arithmetic progressions. It is much easier to prove a set of positive density contains parallelepipeds of arbitrarily large dimension; see [11, 133, 134] for a proof for subsets of $\mathbb{Z}$ and [73] for the general case. The Gowers uniformity norm, [49], important in the study of length four arithmetic progressions, is a natural way to control parallelepipeds. A set contains no parallelepipeds of dimension 2 if and only if it is a B_2 set, and therefore such sets are $\Lambda(4)$.

Proofs that $\Lambda(p)$ sets do not contain parallelepipeds of arbitrarily large dimension can be found in [35, 39, 73, 133]. In particular, no $\Lambda(p)$ set can contain arbitrarily long arithmetic progressions. It follows easily from Proposition 6.3.12 that a $\Lambda(p)$ set in $\mathbb{Z}$ has upper density zero. On the other hand, since one can construct sets $\mathbf{E} \subset \mathbb{Z}$ which contain arbitrarily large parallelepipeds and satisfy $|\mathbf{E} \cap [a, a + N| \ll N$, no density condition is sufficient to guarantee a set is Sidon or even $\Lambda(p)$ [73].

Decompositions of Sidon Sets. Corollary 6.4.7 for countable Sidon sets is due to Déchamps-Gondim [27]; Bourgain gave the general case in [17], where he also proved Theorem 6.4.1.

Bourgain's paper also contains a proof that if $G = \mathbb{Z}_k^a$ is a product of groups of order k where k has no repeated prime factors, then every Sidon set in $\mathbf{\Gamma}$ is a finite union of quasi-independent sets. This improves on a result of the Malliavins': When $G = \mathbb{Z}_p^a$ is a product of groups of prime order, p, then every Sidon set in $\mathbf{\Gamma}$ is a finite union of independent sets (which are I_0) [126]. Of course, this implies an answer of yes to [**P 1**] and no to [**P 2**] for these groups.

p-Sidon Sets. Let $1 \leq p < 2$ and p' be the conjugate index. A subset $\mathbf{E} \subseteq \mathbf{\Gamma}$ is called a *p-Sidon* set if for every $\phi \in \ell^{p'}(\mathbf{E})$ there is a measure μ such that $\widehat{\mu} = \phi$ on $\mathbf{E}$. Equivalently, $\mathbf{E}$ is p-Sidon if there is a constant S_p such that whenever $f \in \mathrm{Trig}_{\mathbf{E}}(G)$, then $\|\widehat{f}\|_{\ell^p} \leq S_p \|f\|_\infty$. Of course, a 1-Sidon set is simply a Sidon set, and the nestedness of the ℓ^p spaces implies that if $\mathbf{E}$ is p-Sidon for some p, then it is q-Sidon for each $q > p$. The notion of p-Sidon set was introduced by Edwards and Ross in [35]. They established a number of properties similar to those for Sidon sets and showed that if $\mathbf{D_1}, \mathbf{D_2}$ are infinite disjoint subsets of $\mathbf{\Gamma}$ whose union is dissociate, then $\mathbf{E} = \mathbf{D_1} \cdot \mathbf{D_2}$ is p-Sidon if and only if $p \geq 4/3$. Johnson and Woodward [98] improved this to show that if $\mathbf{D_1}, \ldots, \mathbf{D_n}$ are disjoint infinite subsets of $\mathbf{\Gamma}$ whose union is dissociate, then $\mathbf{D_1} \cdots \mathbf{D_n}$ is p-Sidon if and only if $p \geq 2n/(n+1)$. Blei in [14] introduced the notion of fractional cartesian products of sets to construct examples of sets that were p-Sidon for each given $1 < p < 2$ but not q-Sidon for any $q < p$. His constructions also provide examples of sets that are $\Lambda(p)$ with constant $O(p^a)$, for each given $p > 2$ and $a > 1/2$.

Exercises. Exercise 6.6.12 is adapted from [94, Chap. V]. Proofs of many of the other exercises can be found in [123] or [166]. For more on the Rudin–Shapiro polynomials, see [56, pp. 33 ff.] and [108, p. 34]. Independent unions are called "*n*-fold joins" by Asmar and Montgomery-Smith, where an estimate of their Sidon constants is given [4, Theorem 3.5] (the exact value appears to be unknown).

6.6 Exercises

Exercise 6.6.1. 1. Show that $\sum_{\gamma \in \mathbf{E}} |\widehat{f}(\gamma)| \leq C \|f\|_\infty$ for every $\mathbf{E}$-polynomial f if and only if $\|\mu\|_{M_d(\mathbf{E})} \leq C \|\widehat{\mu}\|_\infty$ for every $\mu \in M_d(\mathbf{E})$.
  2. Show that $\mathbf{E}$ is Sidon if and only if $\ell^\infty(\mathbf{E}) = B(\mathbf{E})$ if and only if $c_0(\mathbf{E}) \subseteq B(\mathbf{E})$.

Exercise 6.6.2. 1. Show that a Fatou–Zygmund set is a Sidon set.
  2. Show that an independent set is a Fatou–Zygmund set.
  3. Suppose that for each Hermitian $\varphi : \mathbf{E} \to \Delta$, there is some positive measure μ such that

$$\sup\{|\widehat{\mu}(\gamma) - \varphi(\gamma)| : \gamma \in \mathbf{E}\} < 1.$$

Show that there is a finite set $\mathbf{F}$ such that $\mathbf{E} \setminus \mathbf{F}$ is a Fatou–Zygmund set.

Exercise 6.6.3. Show that the sets Ω_n of the proof of Lemma 6.2.4 are closed.

Exercise 6.6.4. 1. Is every Hadamard set Rider? If not, does this depend on the ratio?

2. Give an example of a quasi-independent set in $\mathbb{Z}$ that is not a finite union of Hadamard sets but is I_0. Hint: See Example 2.2.7.

3. Show that every Hadamard set is a finite union of quasi-independent sets.

Exercise 6.6.5. 1. Show that the Sidon constant of a Rider set is bounded by a function of the Rider constant.

2. Adapt the argument of Theorem 6.2.11 to show that a quasi-independent set has Sidon constant at most 16. (Hint: $\sum_{n=2}^{\infty} 2^{-n} R_n(\mathbf{E}, \chi) \leq 1/2$ for $\chi \in \mathbf{E}$.)

3. Use the difference of two Riesz products to show that a quasi-independent set has Sidon constant at most $6\sqrt{6}$. (One should equal $a\psi/2$ on $\mathbf{F}$ and the other $-a\psi/2$.)

4. Use the difference of two Riesz products to show that $S(\mathbf{E}) \leq 48C^{3/2}$ if $\mathbf{E}$ is Rider with constant C.

Exercise 6.6.6. 1. Show that $\{3^j + 3^k : j < k\}$ is not a Sidon set.

2. Let $d \in \mathbb{N}$. Show that a Sidon set $\mathbf{E}$ cannot contain more than $dC \log N$ elements of a d-perturbed arithmetic progression, where C depends only on the Sidon constant. See Exercise 2.9.11 for the definition of "d-perturbed arithmetic progression".

Exercise 6.6.7. 1. Use Corollary 6.3.2 to prove that the union of two Sidon sets is a Sidon set.

2. Suppose $\mathbf{E}$ is a Sidon set and $\gamma \in \Gamma$. Prove that $S(\mathbf{E} \cup \{\gamma\}) \leq 2S(\mathbf{E}) + 1$.

Exercise 6.6.8. Suppose that $\mathbf{E} \subseteq \mathbb{N}$ is $\Lambda(p)$ for some $p > 1$.

1. Show that $\sum_{n \in \mathbf{E}} 1/n < \infty$.

2. Show there is some $1 < q < \infty$ such that if $1/q + 1/q' = 1$, then

$$\sup_N \frac{1}{N^{1/q'}} \left(\sum_{n=1}^{N} r_2(\mathbf{E}, n)^q \right)^{1/q} < \infty.$$

3. Show that the set of perfect squares in $\mathbb{Z}$ is not $\Lambda(4)$.

Exercise 6.6.9. Give an example of a set $\mathbf{E}$ which is I_0, but $\{\gamma^2 : \gamma \in \mathbf{E}\}$ is not even Sidon.

Exercise 6.6.10. Define trigonometric polynomials P_n, Q_n on $\mathbb{T}$ inductively by $P_0 = Q_0 = 1$,

$$P_{n+1}(t) = P_n(t) + e^{i2^n t} Q_n(t),$$

$$Q_{n+1}(t) = P_n(t) - e^{i2^n t} Q_n(t).$$

1. Prove $\|P_n\|_\infty \leq 2^{(n+1)/2}$.
2. Show there is a sequence of signs $r_k = \pm 1$ such that $P_n(t) = \sum_{k=0}^{2^n-1} r_k e^{ikt}$ for each n.
3. Exhibit $f \in C(\mathbb{T})$ such that $\widehat{f} \notin \ell^1$.

These are known as the Rudin–Shapiro polynomials.

Exercise 6.6.11. 1. Show that $S(\mathbf{E}) \leq \sqrt{N}$ if $\mathbf{E}$ is any subset of $\mathbf{\Gamma}$ of cardinality N.
2. Prove that there is a constant $c > 0$ such that $S(\{1, 2, \ldots, N\}) \geq c\sqrt{N}$ for all N.

Exercise 6.6.12. Let $\mathbf{E}_j \subseteq \mathbf{\Gamma}$ be disjoint. The union (finite or not) $\bigcup \mathbf{E}_j$ is a *dissociate union* if $\gamma_j \in \mathbf{E}_j$, $\epsilon_j \in \{-2, -1, 0, 1, 2\}$, $J \in \mathbb{N}$ and $1 = \prod_{j=1}^{J} \gamma_j^{\epsilon_j}$ imply all $\epsilon_j = 0$. Let $\bigcup_j \mathbf{E}_j$ be a dissociate union and p_j real-valued trigonometric polynomials with $\mathrm{Supp}\,\widehat{p_j} \subseteq \mathbf{E}_j \cup \mathbf{E}_j^{-1}$ for $1 \leq j \leq J$. Suppose that $1 + p_j \geq 0$ for all j. Show that $\mu = \prod_1^J (1 + p_j)\, dm_G$ is a probability measure with

$$
\widehat{\mu}(\gamma) = \begin{cases} 1 & \text{if } \gamma = 1, \\ \widehat{p_j}(\gamma) & \text{if } \gamma \in \mathbf{E}_j \cup \mathbf{E}_j^{-1}, \\ \prod_k \widehat{p_{j_k}}(\gamma_{j_k}) & \text{if } \gamma = \prod \gamma_{j_k} \text{ with } \gamma_{j_k} \in \mathbf{E}_{j_k} \cup \mathbf{E}_{j_k}^{-1}, j_k \text{ distinct, and} \\ 0 & \text{otherwise.} \end{cases}
$$

The measure μ (and weak* limits of such μ) is called a *generalized Riesz product*; see [94, p.175].

Exercise 6.6.13. Let $\mathbf{E} \subseteq \mathbf{\Gamma}$ be infinite.

1. Suppose $\{\gamma^2 : \gamma \in \mathbf{E}\}$ is infinite. Prove $\mathbf{E}$ contains an infinite dissociate subset.
2. Suppose $\{\gamma \in \mathbf{E} : \gamma^2 = 1\}$ is infinite. Prove that $\mathbf{E}$ contains an infinite independent subset.
3. Deduce that every infinite subset of $\mathbf{\Gamma}$ contains an infinite Fatou–Zygmund set of the same cardinality.

Exercise 6.6.14. Let $H \subseteq G$ be dense. Let $\mathbf{\Lambda} = \overline{\mathbf{\Gamma}}/(H^\perp)$ and $i : \mathbf{\Gamma} \to \mathbf{\Lambda}$ be the composition of natural mappings, $\mathbf{\Gamma} \hookrightarrow \overline{\mathbf{\Gamma}} \to \mathbf{\Lambda}$.

1. Show that i maps $\mathbf{\Gamma}$ one-to-one onto a dense subgroup of $\mathbf{\Lambda}$.
2. Show that each finite $\mathbf{E} \subset \mathbf{\Gamma}$ has Sidon constant of $\mathbf{E}$ in $\mathbf{\Gamma}$ the same as the Sidon constant of $i(\mathbf{E})$.

Chapter 7
Sidon Sets Are Proportional Quasi-independent

The Pisier characterization of Sidon sets as proportional quasi-independent is proved. A Sidon set has step length tending to infinity if and only if it is Sidon(U) with bounded constants.

7.1 Introduction

In the previous chapter we surveyed what was known about Sidon sets prior to 1980. In the 1980s the understanding of Sidon sets increased vastly as a result of (particularly) Pisier's and Bourgain's contributions. They used an interesting blend of probabilistic, combinatorial and analytical arguments to deduce various arithmetic characterizations of Sidon. In Theorem 6.4.1, Bourgain's k-independent set theorem, we already saw one example of this: every Sidon set is a finite union of k-independent sets. In Sect. 7.2 we prove the Pisier arithmetic characterization of Sidon sets as those sets that are proportional quasi-independent (defined in the following section). Part of the proof also involves showing that Sidon sets are characterized by the property of being $\Lambda(p)$ for all $p < \infty$, with $\Lambda(p)$ constant $O(\sqrt{p})$.

Applications of this characterization are given. The union property for Sidon sets is one immediate application. Another application is to characterize the Sidon sets with step length tending to infinity as those which are Sidon(U) in a uniform sense. This theorem of Déchamps-Gondim and Bourgain is established in Sect. 7.4.

In Sect. 7.3, Sidon sets are shown to possess a separation property known as the "Pisier ε-net condition". This property will be further investigated in Chap. 9, where it will be proven to be another characterization of Sidon and used to establish that Sidon sets can also be characterized as proportional I_0 or ε-Kronecker (in the latter case, provided $\overline{\Gamma}$ does not have "too many" elements of order two). Other results which use the Pisier arithmetic characterization theorem will also be given in Chaps. 8 and 9.

C.C. Graham and K.E. Hare, *Interpolation and Sidon Sets for Compact Groups*,
CMS Books in Mathematics, DOI 10.1007/978-1-4614-5392-5_7,
© Springer Science+Business Media New York 2013

7.2 The Pisier Characterization of Sidon Sets

In this section we prove the theorem highlighted in this chapter's introduction, the most significant item of which is (3), the arithmetic characterization of Sidon sets as those sets which are proportional quasi-independent.

Theorem 7.2.1 (Pisier's characterization theorem). *The following are equivalent for* $\mathbf{E} \subseteq \mathbf{\Gamma} \smallsetminus \{1\}$:

1. *The set* $\mathbf{E}$ *is Sidon.*
2. *There is a constant,* C_2, *such that*

   $$\|f\|_p \leq C_2 \sqrt{p} \|f\|_2 \quad \text{for all } f \in \mathrm{Trig}_{\mathbf{E}}(G) \text{ and all } 2 \leq p < \infty.$$

3. *There is a constant,* C_3, *such that each finite subset,* $\mathbf{F} \subseteq \mathbf{E}$, *has a quasi-independent subset,* $\mathbf{F}_1$, *with* $|\mathbf{F}| \leq C_3 |\mathbf{F}_1|$.
4. *There is a constant,* C_4, *such that for all finite,*[1] *complex-valued sequences* $(a_\gamma)_{\gamma \in \mathbf{E}}$, *there exists a quasi-independent subset* $\mathbf{F} \subseteq \mathbf{E}$ *satisfying*

   $$\sum_{\gamma \in \mathbf{E}} |a_\gamma| \leq C_4 \sum_{\gamma \in \mathbf{F}} |a_\gamma|. \tag{7.2.1}$$

5. *There is a constant,* C_5, *such that for each finite set* $\mathbf{F} \subseteq \mathbf{E}$ *there is a subset* $\mathbf{F}_1 \subseteq \mathbf{F}$ *such that* $\mathbf{F}_1$ *has Sidon constant at most* C_5 *and* $|\mathbf{F}| \leq C_5 |\mathbf{F}_1|$.

Remarks 7.2.2. (i) The equivalence of (1) and (2) shows that Sidon sets are characterized by the property of being $\Lambda(p)$ for all $p < \infty$, with the $\Lambda(p)$ constant at most $O(\sqrt{p})$. This is the minimal $\Lambda(p)$ constant for an infinite set. See Sects. 6.3.2 and 6.5 for more details about $\Lambda(p)$ sets.

(ii) The reciprocal of the constant C_3 from (3) is called a *Pisier constant* for $\mathbf{E}$.

(iii) We will use "proportional" quasi-independent (resp., Sidon) to express the relationship in (3) (resp. (5)) and refer to the equivalence of (1) and (3) as *(Pisier's) proportional quasi-independent characterization of Sidonicity.*

Remark 7.2.3. Although this was not the original proof, each of (2)–(5) immediately yields the important union property of Sidon sets (Corollary 6.3.3). In fact, the Sidon constant of a union is bounded by a constant that depends only on the Sidon constants of the uniands. That can be seen from analysis of the proof of Theorem 7.2.1 (see Exercise 7.6.1).

We will prove Theorem 7.2.1 by establishing two overlapping loops of equivalences. First, that (1) $\Rightarrow$ (2) $\Rightarrow$ (3) $\Rightarrow$ (4) $\Rightarrow$ (1) and, second, that (3) $\Leftrightarrow$ (5).

We begin with the implications (2) $\Rightarrow$ (3) $\Rightarrow$ (4) since they are the most difficult.

---

[1] By "finite" we mean, $a_\gamma = 0$ for all but a finite number of γ.

7.2.1 Proof of Theorem 7.2.1 (2) ⇒ (3) ⇒ (4)

The major task in proving (2) ⇒ (3) ⇒ (4) is to find the quasi-independent sets. The first step of the proof will be to show that under assumption (2) it is possible to find subsets of $\mathbf{E}$ which admit no "long" quasi-relations, where by a *quasi-relation* in $\mathbf{F} \subseteq \mathbf{\Gamma}$, we mean a product $\prod_{\gamma \in \mathbf{F}} \gamma^{c_\gamma} = 1$ where $c_\gamma \in \{-1, 0, 1\}$ and only a finite number of the c_γ are non-zero, and by the *length of the relation* we mean the sum $\sum_\gamma |c_\gamma|$. Note that a subset of $\mathbf{E}$ is quasi-independent if and only if it has no non-trivial quasi-relations.

This is a probabilistic argument and is detailed in Lemma 7.2.4. Random sets are considered and, by studying Riesz product-like functions, the number of long quasi-relations which these random sets admit are shown to be typically bounded. In these estimates, the L^q norms of finite sums of characters arise and this is where assumption (2) is used. With these estimates, one can then deduce that large subsets of random sets admit no quasi-relations, whence (2) ⇒ (3) follows.

Lemma 7.2.4. *Assume $\mathbf{E}$ satisfies Theorem 7.2.1(2) (with constant $C = C_2$) and $\mathbf{F}$ is a finite subset of $\mathbf{E}$ which has $|\mathbf{F}| \geq 2^{14} C^2$. Let*

$$\tau = 2^{-10} C^{-2} \text{ and } \ell = \left\lfloor \frac{\tau |\mathbf{F}|}{4} \right\rfloor. \tag{7.2.2}$$

Then there exists $\widetilde{\mathbf{F}} \subseteq \mathbf{F}$ such that $\tau |\mathbf{F}|/2 \leq |\widetilde{\mathbf{F}}|$ and $\widetilde{\mathbf{F}}$ has no quasi-relations of length more than ℓ.

Here, $\lfloor x \rfloor$ is the greatest integer $\leq x$.

Proof (of Lemma 7.2.4). We note that $C \geq 1/\sqrt{2}$, so $\tau \leq 2^{-9}$ and $\ell \geq 4$.

For $\gamma \in \mathbf{F}$, let $\xi_\gamma(\omega)$ be independent, Bernoulli random variables defined on the probability space $(\mathbf{\Omega}, \mathbf{\Sigma}, \mathbb{P})$, with $\mathbb{P}(\xi_\gamma = 1) = \tau$ and $\mathbb{P}(\xi_\gamma = 0) = 1 - \tau$. We will show that one of the random sets $\widetilde{\mathbf{F}} = \mathbf{F}(\omega) = \{\gamma \in \mathbf{F} : \xi_\gamma(\omega) = 1\}$ will do.

For each $\omega \in \mathbf{\Omega}$ define the function F_ω on G by

$$F_\omega(x) = \sum_{n=\ell+1}^{|\mathbf{F}|} \sum_{\substack{\mathbf{H} \subseteq \mathbf{F} \\ |\mathbf{H}|=n}} \prod_{\gamma \in \mathbf{H}} \xi_\gamma(\omega) 2 \Re \gamma(x).$$

By the independence of the ξ_γ,

$$\iint_{\mathbf{\Omega} \times G} F_\omega(x) dm_G(x) d\mathbb{P}(\omega) = \sum_{n=\ell+1}^{|\mathbf{F}|} \tau^n \left[\sum_{\substack{\mathbf{H} \subseteq \mathbf{F} \\ |\mathbf{H}|=n}} \int_G \prod_{\gamma \in \mathbf{H}} 2 \Re \gamma(x) dm_G(x) \right].$$

Let $N(\omega) = \int_G F_\omega(x) dm_G(x)$. Since ξ_γ is $0, 1$-valued, $N(\omega)$ is integer-valued, and $N(\omega)$ is at least the number of quasi-relations of length $\ell + 1$ or more (Exercise 7.6.4). (If $\mathbf{H}$ has no elements of order two, then $N(\omega)$ is equal to the number of quasi-relations of length $\ell + 1$ or more, though we will not use this fact.) Consequently, if $N(\omega) = 0$, then $\mathbf{F}(\omega)$ has no quasi-relations of length at least $\ell + 1$. Thus, we are interested in estimating $\mathbb{P}(N(\omega) \neq 0) = \mathbb{P}(N(\omega) \geq 1)$.

For each set $\mathbf{H}$ of cardinality n, the expression $\int_G \prod_{\gamma \in \mathbf{H}} 2\Re e\gamma(x) dm_G$ appears $n!$ times in the expansion of

$$\int_G \Big[\sum_{\gamma \in \mathbf{F}} 2\Re e\gamma(x) \Big]^n dm_G. \tag{7.2.3}$$

Moreover, the additional terms that occur in (7.2.3) are non-negative.

Since Theorem 7.2.1 (2) is assumed, that implies

$$0 \leq \int_\Omega N(\omega) \, d\mathbb{P}(\omega) \leq \sum_{n=\ell+1}^{|\mathbf{F}|} \frac{\tau^n}{n!} \int_G \Big[\sum_{\gamma \in \mathbf{F}} 2\Re e\gamma \Big]^n dm_G$$

$$\leq \sum_{n=\ell+1}^{|\mathbf{F}|} \frac{2^n \tau^n}{n!} \Big\| \sum_{\gamma \in \mathbf{F}} \gamma \Big\|_n^n \leq \sum_{n=\ell+1}^{|\mathbf{F}|} \frac{2^n \tau^n}{n!} \Big(C\sqrt{n} \, \Big\| \sum_{\gamma \in \mathbf{F}} \gamma \Big\|_2 \Big)^n$$

$$\leq \sum_{n=\ell+1}^{|\mathbf{F}|} \Big(\frac{6\tau C \sqrt{|\mathbf{F}|}}{\sqrt{n}} \Big)^n \leq \sum_{n=\ell+1}^{\infty} \Big(\frac{6\tau C \sqrt{|\mathbf{F}|}}{\sqrt{\ell+1}} \Big)^n,$$

where we have used the inequality $(n/3)^n \leq n!$ on the third line. It is a routine exercise to check that $6\tau C \sqrt{|\mathbf{F}|}/\sqrt{\ell+1} \leq 3/8$, and therefore

$$\int_\Omega N(\omega) \, d\mathbb{P}(\omega) \leq \sum_{n=\ell+1}^{\infty} \Big(\frac{3}{8} \Big)^n < 2^{-\ell}.$$

Since $\ell \geq 4$ and $N(\omega)$ is either 0 or at least 1,

$$\mathbb{P}(N(\omega) \geq 1) \leq 2^{-\ell} \leq \frac{1}{16}. \tag{7.2.4}$$

We will also want $|\mathbf{F}(\omega)| \geq \tau |\mathbf{F}|/2$. For this, we will apply Lemma C.2.2, a consequence of Markov's inequality, with the independent random variables $X_\gamma = \xi_\gamma - \tau$ (for $\gamma \in \mathbf{F}$), $c^2 = \tau |\mathbf{F}|$ and $a = \tau |\mathbf{F}|/2$. Note that

$$\sum_{\gamma \in \mathbf{F}} \mathbb{E}(X_\gamma^2) = |\mathbf{F}|(\tau - \tau^2) \leq c^2,$$

so $c^2 = \tau |\mathbf{F}|$ is a valid choice for c^2 in that lemma. Thus, by Lemma C.2.2,

$$\mathbb{P}\Big(\Big| \sum_{\gamma \in \mathbf{F}} X_\gamma \Big| \geq \frac{\tau |\mathbf{F}|}{2} \Big) \leq 2 \exp\Big(-\frac{\tau |\mathbf{F}|}{12} \Big). \tag{7.2.5}$$

Because $\sum_{\gamma \in \mathbf{F}} X_\gamma = \sum (\xi_\gamma - \tau) = |\mathbf{F}(\omega)| - \tau |\mathbf{F}|$,

$$\mathbb{P}\left(\left|\sum_{\gamma\in F}X_\gamma\right|\geq\frac{\tau|\mathbf{F}|}{2}\right)=\mathbb{P}\left(\left||\mathbf{F}(\omega)|-\tau|\mathbf{F}|\right|\geq\frac{\tau|\mathbf{F}|}{2}\right)$$

$$\geq\mathbb{P}\left(|\mathbf{F}(\omega)|\leq\frac{\tau|\mathbf{F}|}{2}\right).\qquad(7.2.6)$$

But $\tau|\mathbf{F}|\geq 16$, so by (7.2.5) and (7.2.6),

$$\mathbb{P}\left(|\mathbf{F}(\omega)|\leq\frac{\tau|\mathbf{F}|}{2}\right)\leq 2\exp\left(\frac{-\tau|\mathbf{F}|}{12}\right)\leq 2e^{-4/3}\leq\frac{9}{16}.$$

Since (7.2.4) states $\mathbb{P}(N(\omega)\geq 1)\leq 1/16$, the probability that $N(\omega)=0$ and $|\mathbf{F}(\omega)|\geq\tau|\mathbf{F}|/2$ simultaneously occur is at least $3/8$. Thus, there is an ω such that $\widetilde{\mathbf{F}}=\mathbf{F}(\omega)$ has no quasi-relations of length greater than ℓ and $|\widetilde{\mathbf{F}}|\geq\tau|\mathbf{F}|/2$.   □

Proof (of Theorem 7.2.1 (2) ⇒ (3)). Let $\mathbf{F}$ be any non-empty, finite subset of $\mathbf{E}$. If $|\mathbf{F}|<2^{14}(C_2)^2$, then any singleton set, $\mathbf{F}'$, will suffice with $C_3=2^{14}(C_2)^2$.

Otherwise, let the set $\widetilde{\mathbf{F}}$ be given by Lemma 7.2.4. Let $\prod_{\gamma\in\widetilde{\mathbf{F}}}\gamma^{c_\gamma}$ be any quasi-relation of maximal length (at most ℓ) and let $\mathbf{F}'=\{\gamma:c_\gamma=0\}$. We claim $\mathbf{F}'$ is quasi-independent. Indeed, if $\prod_{\gamma\in\mathbf{F}'}\gamma^{d_\gamma}=1$ is a quasi-relation, then $(\prod_{\gamma\in\mathbf{F}'}\gamma^{d_\gamma})\cdot(\prod_{\gamma\in\widetilde{\mathbf{F}}}\gamma^{c_\gamma})$ is also a quasi-relation of greater length, unless all the $d_\gamma=0$, contradicting the maximality of the choice of $\{\gamma^{c_\gamma}:\gamma\in\widetilde{\mathbf{F}}\}$.

Finally,

$$|\mathbf{F}'|\geq|\widetilde{\mathbf{F}}|-\ell\geq\frac{\tau}{2}|\mathbf{F}|-\ell\geq\frac{\tau}{4}|\mathbf{F}|\geq 2^{-12}(C_2)^{-2}|\mathbf{F}|.$$

Thus, (3) is satisfied with $C_3=2^{14}(C_2)^2$.   □

To prove (3) ⇒ (4), it will be helpful to develop a method for piecing together finite quasi-independent sets to obtain a large quasi-independent set. A complicated combinatorial argument is involved. To make the flow of the argument clearer, we have put the details in the appendix. Here is the combinatorial result we need.

Proposition B.1.2. *There is a constant $A>10$ such that whenever the finite sets $\mathbf{E}_1,\ldots,\mathbf{E}_J$ are quasi-independent, pairwise disjoint and satisfy $|\mathbf{E}_{j+1}|/|\mathbf{E}_j|\geq A$ for $1\leq j<J$, then we can find subsets $\mathbf{E}'_j\subseteq\mathbf{E}_j$ such that $\bigcup_{j=1}^J\mathbf{E}'_j$ is quasi-independent and $|\mathbf{E}'_j|\geq|\mathbf{E}_j|/10$ for $1\leq j\leq J$.*

Proof (of Theorem 7.2.1 (3) ⇒ (4)). Let $\{a_\gamma:\gamma\in\mathbf{E}\}$ be a finite sequence and let $\mathbf{F}_0=\{\gamma:a_\gamma\neq 0\}$. Without loss of generality, $\sum_{\gamma\in\mathbf{F}_0}|a_\gamma|=1$. Let A be given by Proposition B.1.2. For each $k=0,1,2,\ldots$, define

$$\mathbf{E}_k=\left\{\gamma\in\mathbf{F}_0:\frac{1}{A^{k+1}}<|a_\gamma|\leq\frac{1}{A^k}\right\}.$$

By the hypothesis of (3), we can find quasi-independent sets $\mathbf{E}'_k\subseteq\mathbf{E}_k$ such that $|\mathbf{E}_k|\leq C_3|\mathbf{E}'_k|$.

Now,

$$1 = \sum_{k \geq 0} \sum_{\gamma \in \mathbf{E}_k} |a_\gamma| \leq \sum_{k \geq 0} \frac{1}{A^k} |\mathbf{E}_k| \leq AC_3 \sum_{k \geq 0} \frac{1}{A^{k+1}} |\mathbf{E}'_k| \leq AC_3 \sum_{k \geq 0} \sum_{\gamma \in \mathbf{E}'_k} |a_\gamma|.$$

Define $0 = k_1 < k_2 < \cdots$ inductively by

$$k_{j+1} = \min\{k > k_j : |\mathbf{E}'_k| > A |\mathbf{E}'_{k_j}|\}$$

and $k_{j+1} = \infty$ if there is no such finite k. Since $\mathbf{E}'_k \subseteq \mathbf{E}_k$ and $|\mathbf{E}'_k| \leq A |\mathbf{E}'_{k_j}|$ if $k_j < k < k_{j+1}$, we obtain[2]

$$\sum_{j \geq 1} \sum_{k=k_j+1}^{k_{j+1}-1} \sum_{\gamma \in \mathbf{E}'_k} |a_\gamma| \leq \sum_{j \geq 1} \sum_{k=k_j+1}^{k_{j+1}-1} \frac{|\mathbf{E}'_k|}{A^k} \leq \sum_{j \geq 1} \sum_{k_j < k} \frac{1}{A^k} A |\mathbf{E}'_{k_j}|$$

$$= \frac{A^2}{A-1} \sum_{j \geq 1} \frac{1}{A^{k_j+1}} |\mathbf{E}'_{k_j}| \leq \frac{A^2}{A-1} \sum_{j \geq 1} \sum_{\gamma \in \mathbf{E}'_{k_j}} |a_\gamma|$$

$$\leq 2A \sum_{j \geq 1} \sum_{\gamma \in \mathbf{E}'_{k_j}} |a_\gamma|.$$

This implies that

$$\frac{1}{AC_3} \leq \sum_{k \geq 1} \sum_{\gamma \in \mathbf{E}'_k} |a_\gamma| = \sum_{j \geq 1} \sum_{k=k_j+1}^{k_{j+1}-1} \sum_{\gamma \in \mathbf{E}'_k} |a_\gamma| + \sum_{j \geq 1} \sum_{\gamma \in \mathbf{E}'_{k_j}} |a_\gamma|$$

$$\leq (1 + 2A) \sum_{j \geq 1} \sum_{\gamma \in \mathbf{E}'_{k_j}} |a_\gamma|. \tag{7.2.7}$$

Since $|\mathbf{E}'_{k_{j+1}}| \geq A |\mathbf{E}'_{k_j}|$, by Proposition B.1.2 there are subsets, $\mathbf{F}_j \subseteq \mathbf{E}'_{k_j}$, with $|\mathbf{F}_j| \geq |\mathbf{E}'_{k_j}|/10$, such that $\mathbf{F} = \bigcup_{j=1}^J \mathbf{F}_j$ is quasi-independent. Because $\mathbf{F}_j \subseteq \mathbf{E}'_{k_j}$, $\sum_{\gamma \in \mathbf{F}_j} |a_\gamma| \geq |\mathbf{F}_j|/A^{k_j+1}$, and thus

$$\sum_{\gamma \in \mathbf{F}} |a_\gamma| \geq \sum_{j \geq 1} \frac{1}{A^{k_j+1}} |\mathbf{F}_j| \geq \frac{1}{10A} \sum_{j \geq 1} \frac{1}{A^{k_j}} |\mathbf{E}'_{k_j}|$$

$$\geq \frac{1}{10A} \sum_{j \geq 1} \sum_{\gamma \in \mathbf{E}'_{k_j}} |a_\gamma| \geq \frac{1}{10A^2 C_3(1 + 2A)},$$

where the final inequality comes from (7.2.7). Therefore, $C_4 = 10A^2 C_3(1 + 2A)$ will do.   □

---

[2] Since $\mathbf{F}_0$ is finite, the sums here are all finite sums.

7.2.2 Proof of Theorem 7.2.1: $(1) \Rightarrow (2)$, $(4) \Rightarrow (1)$

We now complete the first loop of implications.

Proof (of $(1) \Rightarrow (2)$). This is Theorem 6.3.9. $\qquad\square$

Proof (of $(4) \Rightarrow (1)$). This is a variation of the proof of Theorem 6.2.11. We will show that there is a constant S such that $\|\widehat{f}\|_1 \leq S\|f\|_\infty$ for all $f \in \text{Trig}_{\mathbf{E}}(G)$. By Theorem 6.2.2(4), that will prove that $\mathbf{E}$ is Sidon. It remains to find S.

There will be no loss of generality in assuming that $\mathbf{E}$ is symmetric. Let C_4 be as in (4) and $f \in \text{Trig}_{\mathbf{E}}(G)$. There will also be no loss of generality in assuming that f and $\widehat{f}$ are both real-valued. Then $\widehat{f}(\gamma) = \widehat{f}(\gamma^{-1})$ for $\gamma \in \mathbf{E}$. Choose the quasi-independent set $\mathbf{F} \subseteq \mathbf{E}$ such that (7.2.1) holds for the finite sequence $(a_\gamma)_{\gamma \in \mathbf{E}} = (\widehat{f}(\gamma))$.

For $\gamma \in \mathbf{E}$, let $\psi(\gamma) = \text{sgn}\,\widehat{f}(\gamma)$. Since $\widehat{f}(\gamma) = \widehat{f}(\gamma^{-1})$, ψ is Hermitian. Let $0 < a < 1/(4C_4)$ and let $\mu = \mu_{\mathbf{F},a}$ be the Riesz product as in (6.2.11). As shown there, $\|\mu\| \leq 2$. We define

$$\beta(\gamma) = \frac{a}{2C_4}\psi(\gamma) - \widehat{\mu}(\gamma).$$

Then

$$|\beta(\gamma)| \leq 2a^2 \text{ for } \gamma \in \mathbf{F} \cup \mathbf{F}^{-1} \text{ and } |\widehat{\mu}(\chi)| \leq 2a^2 \text{ for } \chi \notin \mathbf{F} \cup \mathbf{F}^{-1} \cup \{1\}. \tag{7.2.8}$$

We now estimate $\sum_{\gamma \in \Gamma}|\widehat{f}(\gamma)| = \sum_{\gamma \in \mathbf{E}}|\widehat{f}(\gamma)|$. First,

$$\frac{1}{C_4}\sum_{\gamma \in \Gamma}|\widehat{f}(\gamma)| \leq \sum_{\gamma \in \mathbf{F}}|\widehat{f}(\gamma)| \leq \sum_{\gamma \in (\mathbf{F} \cup \mathbf{F}^{-1}) \cap \mathbf{E}}|\widehat{f}(\gamma)| = \sum_{\gamma \in (\mathbf{F} \cup \mathbf{F}^{-1}) \cap \mathbf{E}}\widehat{f}(\gamma)\psi.$$

Using that, (7.2.8) and Parseval's formula, we have

$$\frac{a}{2(C_4)^2}\sum_{\gamma \in \Gamma}|\widehat{f}(\gamma)| \leq \frac{a}{2C_4}\sum_{\gamma \in (\mathbf{F} \cup \mathbf{F}^{-1}) \cap \mathbf{E}}|\widehat{f}(\gamma)|$$

$$= \sum_{\gamma \in (\mathbf{F} \cup \mathbf{F}^{-1}) \cap \mathbf{E}}\widehat{f}(\gamma)\widehat{\mu}(\gamma) + \sum_{\gamma \in (\mathbf{F} \cup \mathbf{F}^{-1}) \cap \mathbf{E}}\widehat{f}(\gamma)\beta(\gamma)$$

$$= \sum_{\gamma \in \Gamma}\widehat{f}(\gamma)\widehat{\mu}(\gamma) - \sum_{\gamma \in \mathbf{E} \setminus (\mathbf{F} \cup \mathbf{F}^{-1})}\widehat{f}(\gamma)\widehat{\mu}(\gamma) + \sum_{\gamma \in (\mathbf{F} \cup \mathbf{F}^{-1}) \cap \mathbf{E}}\widehat{f}(\gamma)\beta(\gamma)$$

$$\leq 2\|f\|_\infty + 4a^2\sum_{\gamma \in \mathbf{E}}|\widehat{f}(\gamma)|.$$

Then the choice $a = 1/(16C_4^2)$ gives the estimate, $S(\mathbf{E}) \leq 128C_4^4$. $\qquad\square$

7.2.3 Proof of Theorem 7.2.1: (3) ⇒ (5)
and (5) ⇒ (3)

Here we complete the second loop of implications and thus complete the proof
of Theorem 7.2.1.

Proof (of (3) ⇒ (5)). Since a quasi-independent set is Sidon with Sidon
constant independent of the set (Corollary 6.2.12), (5) follows from (3).   □

Proof (of (5) ⇒ (3)). Examination of the proofs of (1) ⇒ (2) and (2) ⇒
(3) shows that *finite* Sidon sets having a common bound to their Sidon constant are proportional quasi-independent, with the constant of proportionality depending only on that common bound. The hypotheses of (5) give a
common bound to the Sidon constants of the finite subsets, $\mathbf{F} \subset \mathbf{E}$, and so
(5) ⇒ (3).   □

7.3 Pisier's ε-Net Condition

The next proposition, known as the Pisier ε-net condition, is an important application of the Pisier characterization of Sidon sets. It will be used in Chap. 9
to prove that Sidon sets can also be characterized as those which are proportional I_0. Before stating the proposition, we give a lemma.

Lemma 7.3.1. *Let* $\{\gamma_1, \ldots, \gamma_N\} \subseteq \mathbf{E} \setminus \{1\} \subseteq \Gamma$ *be a quasi-independent set.*
Suppose $c_n \in \Delta$ *for* $1 \leq n \leq N$ *and* $0 < \tau < 1$. *Let*

$$X = \{x \in G : \inf_n |\Re(c_n \gamma_n(x))| > \tau\}.$$

Then $m_G(X) \leq 2(1 + \tau^2)^{-N}$.

Proof. Let $X_+ = \{x \in G : \inf_n \Re(c_n \gamma_n(x)) > \tau\}$ and let $p_n(x) = 1 + \tau \Re(c_n \gamma_n(x))$ for $1 \leq n \leq N$. We note that $\prod_{n=1}^{N} p_n(x) \geq 0$ for all x and
that $(1 + \tau^2)^N \leq \prod_{n=1}^{N} p_n(x)$ for $x \in X_+$. Also,

$$1 = \left(\prod_{n=1}^{N} p_n\right)^\frown(1) = \int \prod_{n=1}^{N} p_n(x) dx \geq (1 + \tau^2)^N m_G(X_+),$$

where the first equality comes from the quasi-independence. So $m_G(X_+) \leq (1 + \tau^2)^{-N}$. Redefining $p_n = 1 - \tau \Re(c_n \gamma_n)$, we see that (the obviously
defined) X_- also satisfies $m_G(X_-) \leq (1 + \tau^2)^{-N}$, from which the conclusion
follows.   □

Proposition 7.3.2. *Suppose that* $\mathbf{E} \subseteq \boldsymbol{\Gamma} \smallsetminus \{1\}$ *is Sidon. Then there exists* $\varepsilon > 0$ *such that for every finite* $\mathbf{F} \subseteq \mathbf{E}$, *there is a set* $Y = \{y_1, \ldots, y_N\} \subseteq G$ *such that* $N \geq 2^{\varepsilon |\mathbf{F}|}$ *and*

$$\varepsilon \leq \sup_{\gamma \in \mathbf{F}} |\gamma(y_n) - \gamma(y_m)| \text{ for } 1 \leq n \neq m \leq N. \tag{7.3.1}$$

We will say that $\mathbf{E}$ *satisfies a Pisier ε-net condition* if the conclusion of Proposition 7.3.2 holds for some $0 < \varepsilon < 1$.

Proof (of Proposition 7.3.2). If $\mathbf{F}' \subseteq \mathbf{F}$, then

$$\sup_{\gamma \in \mathbf{F}'} |\gamma(x) - \gamma(y)| \leq \sup_{\gamma \in \mathbf{F}} |\gamma(x) - \gamma(y)|,$$

and so by Theorem 7.2.1(3), there is no loss of generality in assuming $\mathbf{F} \subseteq \mathbf{E}$ is quasi-independent.

Fix $0 < \tau < 1$ and pick M_0 so large that $(1 + \tau^2)^{M_0} > 8$. If $|\mathbf{F}| < M_0$, then take any $\gamma \in \mathbf{F}$ and choose any $y_1, y_2 \in G$ such that $|\gamma(y_1) - \gamma(y_2)| \geq |1 - e^{2\pi i/3}|$. Then $Y = \{y_1, y_2\}$ suffices with the choice of suitably small ε.

Thus, there is also no loss of generality in assuming $|\mathbf{F}| \geq M_0$. Let

$$X = \{x \in G : \inf_{\gamma \in \mathbf{F}} |\mathfrak{Re}\gamma(x)| > \tau\}$$

and choose the largest integer $K \geq 2$ such that if $|\mathbf{F}| = M$, then

$$2^{K+1} < (1 + \tau^2)^M.$$

(This is possible by the choice of M_0.) We will find $x_1, \ldots, x_K \in G$ such that for each pair of distinct products $x = x_{k(1)} \cdots x_{k(j)}$ and $y = x_{\ell(1)} \cdots x_{\ell(m)}$, it is the case that $\sup_{\gamma \in \mathbf{F}} |\gamma(x) - \gamma(y)| > 1 - \tau$. The set Y will then be $\{\prod_{k=1}^{K} x_j^{m_k} : m_k = 0, 1\} \smallsetminus \{e\}$. The choice of K ensures there is a constant $\varepsilon_1 > 0$, which does not depend on $|\mathbf{F}|$, so that $|Y| \geq 2^K - 1 \geq 2^{\varepsilon_1 |\mathbf{F}|}$.

We may assume that x and y have no common factors. Our interest is in differences of the form $|\gamma(xy^{-1}) - 1|$. This motivates the next paragraph.

For each non-trivial choice of $m = (m_1, \ldots, m_K) \in \{0, 1\}^K$, let

$$X_m = \left\{(x_1, \ldots, x_K) \in GK : \inf_{\gamma \in \mathbf{F}} \left| \mathfrak{Re}\left(\gamma\left(\prod_1 Kx_k m_k\right)\right) \right| > \tau \right\}.$$

If we let $s = \sum_j m_j$, then it is easy to see that

$$X_m \subseteq \prod_{j:m_j \neq 0} \{x_j : \inf_{\gamma \in \mathbf{F}} |\mathfrak{Re}\gamma(x_j)| > \tau\} \times G^{K-s}.$$

It follows from Lemma 7.3.1 that $m_{G^K}(X_m) \leq 2(1+\tau^2)^{-M}$. The choice of K ensures that $m_{G^K}(\bigcup_{m \neq 0} X_m) < 1$ and this proves there exists $(x_1, \ldots, x_K) \in G^K$ such that for all non-trivial $(m_1, \ldots m_K) \in \{0,1\}^K$ there exists $\gamma \in \mathbf{F}$ with $\left| \Re \left(\gamma(\prod_k x_k^{m_k}) \right) \right| \leq \tau$ and so $\left| \gamma(\prod x_k^{\pm m_k}) - 1 \right| \geq 1 - \tau$. It is now straightforward to check that there is an $0 < \varepsilon \leq 1 - \tau$ satisfying the conclusions of the proposition.   □

The following corollary will also be used later.

Corollary 7.3.3. *Let $V \subseteq G$ and suppose $\mathbf{E} \subseteq \Gamma$ is a Sidon set. Assume that finitely many translates of the subset V cover G. Then there are constants $\varepsilon = \varepsilon(\mathbf{E}) > 0$ and $\alpha = \alpha(\mathbf{E}, V)$ with the property that for any finite subset $\mathbf{F} \subseteq \mathbf{E}$ having cardinality at least α, there is a set $V_0 \subseteq V$ having cardinality at least $2^{\varepsilon|\mathbf{F}|/2}$ and satisfying*

$$\sup_{\gamma \in \mathbf{F}} |\gamma(x) - \gamma(y)| \geq \varepsilon \text{ for } x \neq y \in V_0. \tag{7.3.2}$$

Proof. Assume $G = \bigcup_{k=1}^K a_k V$, fix a finite $\mathbf{F} \subseteq \mathbf{E} \smallsetminus \{1\}$ and obtain the set Y from Proposition 7.3.2 satisfying (7.3.1). By the pigeon hole principle, at least $N/K \geq 2^{\varepsilon|\mathbf{F}|}/K$ of the points $\{y_1, \ldots, y_N\} \subset G$ satisfying (7.3.1) belong to one of the sets $a_k V$. For any two such points, w, z, we have $w = a_k x, z = a_k y$ for some $x, y \in V$. Of course, $|\gamma(x) - \gamma(y)| = |\gamma(z) - \gamma(w)|$ for all γ. Thus, there is a subset V_0 of V with at least $2^{\varepsilon|\mathbf{F}|}/K$ elements, such that for $x \neq y \in V_0$, $\sup_{\gamma \in \mathbf{F}} |\gamma(x) - \gamma(y)| \geq \varepsilon$.

Pick α such that $K \leq 2^{\alpha\varepsilon/2}$. It follows that if $|\mathbf{F}| \geq \alpha$, then there will be at least $2^{\varepsilon|\mathbf{F}|/2}$ points in V satisfying (7.3.2).   □

7.4 Characterizing Sidon Sets with Step Length Tending to Infinity

Recall (Definition 2.7.6) that $\mathbf{E}$ has step length tending to infinity if for every finite set $\mathbf{F}' \subseteq \Gamma$, there exists a finite subset $\mathbf{F} \subseteq \mathbf{E}$ such that if $\chi, \psi \in \mathbf{E} \smallsetminus \mathbf{F}$ and $\chi \neq \psi$, then $\chi\psi^{-1} \notin \mathbf{F}'$. In this section we complete a line of investigation, begun with Corollary 6.4.7, which stated that every Sidon set is a finite union of sets with step length tending to infinity.

Definition 7.4.1. A set $\mathbf{E} \subseteq \Gamma$ is *Sidon(U) with bounded constants* if there exists a constant K such that for every non-empty, open $U \subseteq G$ there exists a finite $\mathbf{F} \subseteq \mathbf{E}$ such that for every $\varphi \in \ell^\infty(\mathbf{E})$ there exists $\mu \in M(U)$ such that $\hat{\mu} = \varphi$ on $\mathbf{E} \smallsetminus \mathbf{F}$ and $\|\mu\| \leq K$.

An $I_0(U)$ set with bounded constants is clearly Sidon(U) with bounded constants but not conversely; see Exercise 7.6.2.

Theorem 7.2.1 (4) has the following consequence. Recall (see Example 5.2.11) that the analogous statement for I_0 sets is not true.

Theorem 7.4.2 (Déchamps-Gondim–Bourgain-bounded constants). *If $\mathbf{E} \subseteq \Gamma$ is Sidon, then the following are equivalent:*

1. $\mathbf{E}$ *is Sidon(U) with bounded constants.*
2. $\mathbf{E}$ *has step length tending to infinity.*

Combined with Corollary 6.4.7, this yields the following:

Corollary 7.4.3. *If $\mathbf{E} \subseteq \Gamma$ is a Sidon set, then $\mathbf{E}$ is a finite union of Sidon sets that are Sidon(U) with bounded constants.*

Proof (of Theorem 7.4.2 (1) $\Rightarrow$ (2)). This is similar to the proof of Proposition 5.2.9: just replace "$M_d(V)$" with "$M(V)$" and "$I_0(U)$" with "Sidon(U)". $\qquad\square$

We now turn to the other direction of the proof of Theorem 7.4.2. We will find a $\delta > 0$ (depending only on $\mathbf{E}$) with the property that for each non-empty, open set $U \subseteq G$ there is a finite set $\mathbf{E}'$ such that for each $q \in \mathrm{Trig}_{\mathbf{E} \smallsetminus \mathbf{E}'}(G)$ there is a measure $\mu \in M(U)$ with

$$\|\mu\| \leq 2 \quad \text{and} \quad \left| \int q \mathrm{d}\mu \right| \geq \delta \sum |\widehat{q}| \qquad (7.4.1)$$

and that will give the uniform bound on the Sidon(U) constants.

The interpolating measure μ will again be built from a suitable Riesz product but in this case scaled by the restriction to U of a non-negative polynomial p which will "almost" be supported on U. Theorem 7.2.1 (by way of Lemma 7.4.5) will be used to extract a suitable quasi-independent set on which to build the Riesz product. The assumption that $\mathbf{E}$ has step length tending to infinity will be used to estimate the integral $|\int q \mathrm{d}\mu|$. The main difficulty is in finding the finite subset $\mathbf{E}'$. We do that in the second of the two following lemmas, the first lemma being a preliminary technical result. The proof proper of Theorem 7.4.2 (2) $\Rightarrow$ (1) is given at the end of this section.

Lemma 7.4.4. *Suppose $\mathbf{F}$ is a finite quasi-independent set and $a_\gamma \in \Delta$ for $\gamma \in \mathbf{F}$. Then*

$$1 = \left\| \prod_{\gamma \in \mathbf{F}} (1 + \Re(a_\gamma \gamma)) \right\|_1 = \left\| \prod_{\gamma \in \mathbf{F}} (1 + \Re\gamma) \right\|_1. \qquad (7.4.2)$$

Suppose, in addition, that p is a non-negative, trigonometric polynomial with $\widehat{p} \geq 0$ and $\mathbf{F}$ has the property that whenever $\gamma \neq \chi \in \mathbf{F}$, then $\gamma\chi^{-1} \notin \mathrm{Supp}\ \widehat{p}$. Then

$$\left\| p \prod_{\gamma \in \mathbf{F}} (1 + \Re(a_\gamma \gamma)) \right\|_1 \leq \left\| p \prod_{\gamma \in \mathbf{F}} (1 + \Re\gamma) \right\|_1. \qquad (7.4.3)$$

Proof (of Lemma 7.4.4). For (7.4.2), we just calculate the size of the Fourier coefficient at **1**, which is 1 in both cases.

For (7.4.3), as both the polynomial p and the two Riesz products are non-negative functions, it is again the case that the L^1-norms are the Fourier coefficients at **1**.

On the right hand side, all Fourier coefficients of all the factors (including p) are non-negative, so their contributions to the coefficient at **1** do not cancel. On the left side, some Fourier coefficients may be negative, leading to cancellation when computing the coefficient at **1**. On the left, furthermore, the Fourier coefficients of the Riesz product factor's second term have absolute values at most $\frac{1}{2}$ (or 1 if γ has order 2), while those on the right are exactly $\frac{1}{2}$ (or 1). Thus, the Fourier coefficients on the left can only have smaller absolute values than those on the right, and therefore (7.4.3) follows.   □

Lemma 7.4.5. *Suppose $\mathbf{E} \subset \mathbf{\Gamma} \setminus \{1\}$ is Sidon, $\mathbf{F}' \subset \mathbf{\Gamma}$ is finite and $C = C_4$ is given by Theorem 7.2.1 (4) for $\mathbf{E}$. Assume that $p \geq 0$ is a trigonometric polynomial with non-negative Fourier coefficients and having $\widehat{p}(1) = 1$. Let Supp $\widehat{p} = \mathbf{E}_0$. Then there exists a second finite subset $\mathbf{E}' \subseteq \mathbf{\Gamma}$, with $\mathbf{E}' \supseteq \mathbf{E}_0 \cup \mathbf{F}'$, such that if $q = \sum c_\gamma \gamma$ is an $(\mathbf{E} \setminus \mathbf{E}')$-polynomial, then there exists a quasi-independent set $\mathbf{F} \subseteq \mathbf{E} \setminus \mathbf{E}'$ such that*

$$\sum_{\gamma \in \mathbf{F}} |c_\gamma| > \frac{1}{2C} \sum_{\gamma \in \mathbf{E} \setminus \mathbf{E}'} |c_\gamma| \text{ and} \tag{7.4.4}$$

$$\int p(x) \prod_{\gamma \in \mathbf{F}} (1 + \Re\gamma(x)) \, dx = \left\| p(x) \prod_{\gamma \in \mathbf{F}} (1 + \Re\gamma) \right\|_1 < 2. \tag{7.4.5}$$

Proof (of Lemma 7.4.5). Suppose that the conclusion were false. Then for each positive integer, R, there are disjoint finite sets $\mathbf{E}_1, \ldots, \mathbf{E}_R \subseteq \mathbf{E}$ and scalars c_γ such that $\sum_{\gamma \in \mathbf{E}_r} |c_\gamma| = 1$ and for which a quasi-independent set satisfying both (7.4.4) and (7.4.5) does not exist. Obtain such sets for $R > 2C|\mathbf{E}_0|$. We apply Theorem 7.2.1 (4), to the subset $\bigcup_{r=1}^R \mathbf{E}_r$ and scalars (c_γ), to obtain a quasi-independent set $\mathbf{F}$ such that

$$\sum_{\gamma \in \bigcup_{r=1}^R \mathbf{E}_r \cap \mathbf{F}} |c_\gamma| > \frac{1}{C} \sum_{\gamma \in \bigcup_{r=1}^R \mathbf{E}_r} |c_\gamma| = \frac{R}{C}. \tag{7.4.6}$$

The Riesz product, $\prod_{\gamma \in \mathbf{F}}(1 + \Re\gamma(x))$, can be expressed as

$$\prod_{\gamma \in \mathbf{F}} (1 + \Re\gamma(x)) = \sum_{r=1}^R \left(-1 + \prod_{\gamma \in \mathbf{E}_r \cap \mathbf{F}} (1 + \Re\gamma(x)) \right) + T(x),$$

where $T(x)$ is a polynomial with non-negative Fourier coefficients. Since also $\widehat{p} \geq 0$, it follows that

$$\int p(x) \sum_{r=1}^{R} \left(-1 + \prod_{\gamma \in \mathbf{E}_r \cap \mathbf{F}} (1 + \mathfrak{Re}\gamma(x)) \right) dx \qquad (7.4.7)$$

$$\leq \int p(x) \prod_{\gamma \in \mathbf{F}} (1 + \mathfrak{Re}\gamma(x)) dx \leq |\mathbf{E}_0|,$$

with the final inequality coming from the facts that $|\widehat{p}| \leq 1$ and $\mathbf{E}_0$ is the support of $\widehat{p}$.

We assert that (7.4.6) and (7.4.7) imply there is r such that

$$\sum_{\gamma \in \mathbf{E}_r \cap \mathbf{F}} |c_\gamma| > \frac{1}{2C} \quad \text{and} \quad \int p(x) \prod_{\gamma \in \mathbf{E}_r \cap \mathbf{F}} (1 + \mathfrak{Re}\gamma(x)) dx < 2.$$

Indeed, to have (7.4.7) holding, we must have

$$\int p(x) \left(-1 + \prod_{\gamma \in \mathbf{E}_r \cap \mathbf{F}} (1 + \mathfrak{Re}\gamma(x)) \right) dx \leq 1$$

for at least $R - |\mathbf{E}_0|$ choices of r. But by (7.4.6), if we had only $|\mathbf{E}_0|$ choices of r for which $\sum_{\gamma \in \mathbf{E}_r \cap \mathbf{F}} |c_\gamma| > 1/2C$, then, since $\sum_{\gamma \in \mathbf{E}_r} |c_\gamma| = 1$, those choices would contribute at most $|\mathbf{E}_0|$ to (7.4.6), and the remaining $R - |\mathbf{E}_0|$ choices would contribute $(R - |\mathbf{E}_0|)/2C$, which is insufficient. That gives r.

Since $\mathbf{E}_r \cap \mathbf{F}$ is quasi-independent, that gives a contradiction and establishes the existence of a subset $\mathbf{E}'$ with properties (7.4.4)–(7.4.5). Enlarging $\mathbf{E}'$ if necessary, we may assume $\mathbf{E}_0 \cup \mathbf{F}' \subseteq \mathbf{E}'$. □

Proof (of Theorem 7.4.2 (2) ⇒ (1)). Without loss of generality $1 \notin \mathbf{E}$. Fix the non-empty, open set U. To begin, choose $C = C_4$ satisfying Theorem 7.2.1 (4) for $\mathbf{E}$. Use Exercise C.4.9 to find a trigonometric polynomial p on G such that $p \geq 0$, $\widehat{p} \geq 0$, $\int p \, dx = 1$ and $|p(x)| < \varepsilon$ on $G \setminus U$, where $\varepsilon > 0$ will be chosen later. Let $\mathbf{E}_0$ be the support of $\widehat{p}$ (which is symmetric) and use the assumption that $\mathbf{E}$ has step length tending to infinity to find a finite set $\mathbf{F}' \subseteq \mathbf{E}$ such that $\gamma \chi^{-1} \notin \mathbf{E}_0$ for all distinct $\gamma, \chi \in \mathbf{E} \setminus \mathbf{F}'$.

Let $\mathbf{E}'$ be given by Lemma 7.4.5 for $\mathbf{E}, \mathbf{E}_0, \mathbf{F}'$ and p. Now fix $\sum_{\gamma \in \mathbf{E} \setminus \mathbf{E}'} c_\gamma \gamma$ and obtain the associated quasi-independent set $\mathbf{F}$.

For any set of $a_\gamma \in \Delta$, Lemma 7.4.4 and (7.4.5) imply

$$\left\| p \prod_{\gamma \in \mathbf{F}} (1 + \mathfrak{Re}(a_\gamma \gamma)) \right\|_1 < 2. \qquad (7.4.8)$$

In particular, this is true when $a_\gamma = sb_\gamma$, where $b_\gamma = \overline{c_\gamma}/|c_\gamma|$ (and $b_\gamma = 0$ if $c_\gamma = 0$) and $s \in (0,1]$ is to be determined. It remains to show that the measure $\mu \in M(U)$ given by

$$\mu = p(x) 1_U \prod_{\gamma \in \mathbf{F}} (1 + \mathfrak{Re}(a_\gamma \gamma^{-1})),$$

which has norm at most 2 by (7.4.8), has a sufficiently large integral against $q = \sum_{\gamma \in \mathbf{E} \smallsetminus \mathbf{E}'} c_\gamma \gamma$.

Let

$$w = \left| \int_G p(x) \prod_{\gamma \in \mathbf{F}} \left(1 + \mathfrak{Re}(a_\gamma \gamma^{-1}) \right) \sum_{\gamma \in \mathbf{E} \smallsetminus \mathbf{E}'} c_\gamma \gamma \, dx \right|.$$

Note that

$$\prod_{\gamma \in \mathbf{F}} \left(1 + \mathfrak{Re}(a_\gamma \gamma^{-1}(x)) \right) = 1 + s \sum_{\gamma \in \mathbf{F}} \mathfrak{Re}(b_\gamma \gamma^{-1}(x)) + \sum_{d \geq 2} s^d P_d, \qquad (7.4.9)$$

where

$$P_d(x) = \sum_{\substack{\mathbf{H} \subseteq \mathbf{F} \\ |\mathbf{H}| = d}} \prod_{\gamma \in \mathbf{H}} \mathfrak{Re}(b_\gamma \gamma^{-1}(x)).$$

Since $\mathbf{F} \subseteq \mathbf{E} \smallsetminus \mathbf{E}_0$, $\int p(x)(\sum_{\gamma \in \mathbf{F}} c_\gamma \gamma(x)) \, dx = 0$, and thus a lower bound for w is

$$w \geq s \left| \int p(x) \sum_{\gamma \in \mathbf{F}} \mathfrak{Re}(b_\gamma \gamma^{-1}(x)) \sum_{\gamma \in \mathbf{E} \smallsetminus \mathbf{E}'} c_\gamma \gamma(x) \, dx \right|$$
$$- \sum_{d \geq 2} s^d \left| \int P_d(x) p(x) \sum_{\gamma \in \mathbf{E} \smallsetminus \mathbf{E}'} c_\gamma \gamma(x) \, dx \right|. \qquad (7.4.10)$$

We want to bound the last term in (7.4.10). For all d,

$$\left| \int p(x) \, P_d(x) \sum_{\gamma \in \mathbf{F}} c_\gamma \gamma(x) dx \right| \leq \| (p \, P_d)^\frown \|_\infty \sum_{\gamma \in \mathbf{F}} |c_\gamma|.$$

Lemma 7.4.4 implies that

$$\| (P_d \, p)^\frown \|_\infty = \left\| \left[\sum_{\substack{\mathbf{H} \subseteq \mathbf{F} \\ |\mathbf{H}| = d}} \prod_{\gamma \in \mathbf{H}} \mathfrak{Re}(b_\gamma \gamma^{-1}(x)) p \right]^\frown \right\|_\infty$$
$$\leq \left\| \left[\prod_{\gamma \in \mathbf{F}} (1 + \mathfrak{Re}\gamma) p \right]^\frown \right\|_\infty \leq \left\| \prod_{\gamma \in \mathbf{F}} (1 + \mathfrak{Re}\gamma) p \right\|_1 < 2.$$

We now apply the last estimate to bounding w:

$$w \geq s \left| \int p \sum_{\gamma \in \mathbf{F}} \mathfrak{Re}(b_\gamma \gamma^{-1}) \sum_{\gamma \in \mathbf{E} \smallsetminus \mathbf{E}'} c_\gamma \gamma \, dx \right| - \frac{2s^2}{1-s} \sum_{\mathbf{E} \smallsetminus \mathbf{E}'} |c_\gamma|. \qquad (7.4.11)$$

In (7.4.11) we may replace "$\mathfrak{Re}(b_\gamma \gamma^{-1})$" with "$\mathfrak{Im}(b_\gamma \gamma^{-1})$" and obtain the same estimate. Hence,

$$w \geq \frac{s}{2} \left| \int p \sum_{\gamma \in \mathbf{F}} (b_\gamma \gamma^{-1}) \sum_{\gamma \in \mathbf{E} \setminus \mathbf{E}'} c_\gamma \gamma \, dx \right| - \frac{2s^2}{1-s} \sum_{\mathbf{E} \setminus \mathbf{E}'} |c_\gamma|. \qquad (7.4.12)$$

Recall that $\mathbf{F}'$ was chosen from the step length tending to infinity property to ensure that if $\gamma \neq \psi \in \mathbf{E} \setminus \mathbf{F}'$, then $\gamma^{-1}\psi \notin \mathbf{E}_0$ (the support of $\hat{p}$). In particular, this is true for all $\gamma \neq \psi \in \mathbf{E} \setminus \mathbf{E}'$, and thus the choice of b_γ gives

$$\int p \left(\sum_{\gamma \in \mathbf{F}} b_\gamma \gamma^{-1} \right) \left(\sum_{\lambda \in \mathbf{E} \setminus \mathbf{E}'} c_\lambda \lambda \right) dx = \int p \left(\sum_{\gamma \in \mathbf{F}} b_\gamma c_\gamma \right) dx$$

$$= \sum_{\gamma \in \mathbf{F}} |c_\gamma| \int p \, dx = \sum_{\gamma \in \mathbf{F}} |c_\gamma|.$$

This fact, together with (7.4.4) and (7.4.12), implies

$$w \geq \frac{s}{2} \sum_{\gamma \in \mathbf{F}} |c_\gamma| - \frac{2s^2}{1-s} \sum_{\gamma \in \mathbf{E} \setminus \mathbf{E}'} |c_\gamma| \geq \left(\frac{s}{4C} - \frac{2s^2}{1-s} \right) \sum_{\gamma \in \mathbf{E} \setminus \mathbf{E}'} |c_\gamma|.$$

Since $p \leq \varepsilon$ on U^c,

$$\left| \int_{U^c} p(x) \prod_{\gamma \in \mathbf{F}} (1 + \mathfrak{Re}(a_\gamma \gamma(x))) \sum_{\gamma \in \mathbf{E} \setminus \mathbf{E}'} c_\gamma \gamma(x) dx \right|$$

$$\leq \varepsilon \left\| \prod_{\gamma \in \mathbf{F}} (1 + \mathfrak{Re}(a_\gamma \gamma(x))) \right\|_1 \sum_{\gamma \in \mathbf{E} \setminus \mathbf{E}'} |c_\gamma| \leq \varepsilon \sum_{\gamma \in \mathbf{E} \setminus \mathbf{E}'} |c_\gamma|.$$

Putting these bounds together yields

$$\left| \int \sum_{\gamma \in \mathbf{E} \setminus \mathbf{E}'} c_\gamma \gamma(x) d\mu \right| = \left| \int_U p(x) \prod_{\gamma \in \mathbf{F}} (1 + \mathfrak{Re}(a_\gamma \gamma(x))) \sum_{\gamma \in \mathbf{E} \setminus \mathbf{E}'} c_\gamma \gamma(x) dx \right|$$

$$\geq w - \left| \int_{U^c} p \prod_{\gamma \in \mathbf{F}} (1 + \mathfrak{Re}(a_\gamma \gamma(x))) \sum_{\gamma \in \mathbf{E} \setminus \mathbf{E}'} c_\gamma \gamma dx \right|$$

$$\geq \left(\frac{s}{4C} - \frac{2s^2}{1-s} - \varepsilon \right) \sum_{\gamma \in \mathbf{E} \setminus \mathbf{E}'} |c_\gamma|.$$

Substituting $s = 1/18C$ and $\varepsilon = 1/288C^2$ gives the desired result,

$$\sup_{u \in U} \left| \sum_{\gamma \in \mathbf{E} \smallsetminus \mathbf{E}'} c_\gamma \gamma(u) \right| \|\mu\| \geq \left| \int \sum_{\gamma \in \mathbf{E} \smallsetminus \mathbf{E}'} c_\gamma \gamma(x) \mathrm{d}\mu \right| \geq \frac{1}{288C^2} \sum_{\gamma \in \mathbf{E} \smallsetminus \mathbf{E}'} |c_\gamma|.$$

Since $\|\mu\|_{M(G)} \leq 2$, that proves $\mathbf{E} \smallsetminus \mathbf{E}'$ is Sidon(U) with constant $\delta = 1/576C^2$. $\square$

7.5 Remarks and Credits

The Pisier Characterization. Theorem 7.2.1 has a complicated history. (1) $\Rightarrow$ (2) is due to Zygmund [199] and Rudin [166]; the other direction is due to Pisier [147, 148]. The equivalence of Theorem 7.2.1 (1) and (3) is also due to Pisier [150, 151]. The equivalence of Theorem 7.2.1 (1) and (5) can be found in Pisier's paper [149]. Related characterizations may be found in Pisier's [151].

Bourgain in [18, 19] later provided a different proof of the equivalence of (1)–(3) by including the then new (4) in the circle of equivalences. It is from this later proof that the argument given in this chapter is derived, following the excellent French exposition of Li and Queffélec [119, pp. 482–499].

We give an indirect proof that the Pisier ε-net condition implies Sidonicity in Theorem 9.2.1. The characterization of Sidonicity by the ϵ-net condition is in [150].

Sidon Sets with Step Length Tending to Infinity. Theorem 7.4.2 is due to Déchamps–Gondim [27, Theorem 5.2] for the case that $G = Z_p^{\mathbb{N}}$, where p is prime; the general case is due to Bourgain [19]. Its proof is adapted from [19, Corollary 2]. Corollary 7.4.3 is due to Déchamps(-Gondim) (see [123, Theorem 9.1]) for countable $\mathbf{E}$ and Bourgain for the general case. Sets that are Sidon(U) with bounded constants are called "Sidon sets of the first type" in [27, 130].

Other Characterizations of Sidonicity.

1. Random continuity. Let $\{\pi_n\}$ be the Rademacher functions on $\mathbb{D}$. Given a trigonometric polynomial $P(t) = \sum_{n=1}^N a_n \gamma_n(t)$ defined on G and $\omega \in \mathbb{D}$, let $P^\omega(t) = \sum_{n=1}^N \pi_n(\omega) a_n \gamma_n(t)$ and put

$$\|P\|_R = \mathbb{E}\left(\|P^\omega\|_\infty \right) = \int_{\mathbb{D}} \sup_{t \in G} |P^\omega(t)| \mathrm{d}m(\omega).$$

If $\mathbf{E} \subseteq \mathbf{\Gamma}$ is Sidon, then for all $\mathbf{E}$-polynomials P and each $\omega \in \mathbb{D}$,

$$\|P\|_\infty \leq \sum |\widehat{P}(\gamma_n)| = \sum |\pi_n(\omega)\widehat{P}(\gamma_n)| = \sum |\widehat{P^\omega}(\gamma_n)| \leq S\left(\mathbf{E}\right) \|P^\omega\|_\infty.$$

Hence, if $\mathbf{E}$ is Sidon, then $\|P\|_\infty \leq S(\mathbf{E})\|P\|_R$ for all $\mathbf{E}$-polynomials P. Rider in [162] showed that the converse is true, as well: $\mathbf{E}$ is Sidon if and only if there is a constant C such that $\sum \left|\widehat{P}(\gamma_n)\right| \leq C\|P\|_R$ for all $\mathbf{E}$-polynomials P. This characterization was used by Marcus and Pisier [127, p. 118ff] to give another proof of the equivalence of Theorem 7.2.1 (1) and (2). An exposition of Rider's result can be found in [119, p. 220ff, pp. 499ff.].

2. Orlicz spaces. Pisier [149] gave a characterization using the Orlicz space $L^{\psi_q}(G)$, where $\psi_q(x) = \exp(|x|^q) - 1$: $\mathbf{E}$ is Sidon if and only if there is some $q \geq 2$ and constant C such that for every finite $\mathbf{F} \subseteq \mathbf{E}$ we have $\|\sum_{\gamma \in \mathbf{F}} \gamma\|_{\psi_q} \leq C|\mathbf{F}|^{1/p}$. Here $1/p + 1/q = 1$.

3. Arithmetic diameter. The arithmetic diameter (or "diametre banachique") of a finite dimensional Banach space X is the least integer $N = N(X)$ such that there is a subspace $Y \subseteq \ell_N^\infty$ and an isomorphism $T : X \to Y$ with $\|T\|\,\|T^{-1}\| \leq 2$. Bourgain [18] proved that $\mathbf{E} \subseteq \Gamma$ is Sidon if there exist $\delta > 0$ such that $N(C_{\mathbf{F}}(G)) \geq 2^{\delta|\mathbf{F}|}$ for all finite subsets $\mathbf{F} \subseteq \mathbf{E}$. See also [119, pp. 545–546]. A simple proof that this condition implies that $\mathbf{E}$ satisfies Pisier's ε-net condition can be found at [113, p. 892].

4. Cotype. A Banach space X has $cotype$ $q \in [2,\infty)$ if for some $C > 0$ and all finite sequences $x_j \in X$, $\left(\sum \|x_j\|^q\right)^{1/q} \leq C\left(\int \|\sum \epsilon_j x_j\|^2 d\mathbb{P}\right)^{1/2}$, where the ϵ_j are independent Bernoulli random variables and $d\mathbb{P}$ is the related probability measure. Pisier [146] and Kwapień and Pełczyński [117] proved (independently) that $\mathbf{E}$ is Sidon if $C_{\mathbf{E}}$ has cotype 2. A proof is in [119, p. 226]. This was improved by Bourgain and Milman [21] or [119, pp. 226, 544]: either $C_{\mathbf{E}}(G)$ has cotype q for some $2 \leq q < \infty$ and $\mathbf{E}$ is Sidon or $C_{\mathbf{E}}(G)$ has no finite cotype.

5. Unconditional structure. Hare and Tomczak–Jaegermann [77] characterize Sidon sets in terms of an "unconditional structure" of related invariant subspaces. Sidon sets are also characterized as $\Lambda(2)$ sets such that for some $p > 2$ the dual of $L_{\mathbf{E}}^p(G)$ is a complemented subspace of a Banach lattice, that is, has Gordon–Lewis unconditional structure. See [113, p. 886ff] for a discussion.

Exercises. Exercise 7.6.7 is [158, Theorems 16-17]. The proof asked for here is simpler than the original.

7.6 Exercises

Exercise 7.6.1. Show that the Sidon constant of a union of two sets is bounded by a function that depends only on the Sidon constants of the uniands.

Exercise 7.6.2. Can a set be Sidon(U) with bounded constants and $I_0(U)$ for all non-empty, open U but not $I_0(U)$ with bounded constants? Hint: Consider a union of two Hadamard sets.

Exercise 7.6.3. Suppose there are constants C and $\delta > 0$ with the property that for each finite set $\mathbf{F} \subseteq \mathbf{E}$ there is a subset $\mathbf{F}_1 \subseteq \mathbf{F}$ such that $\mathbf{F}_1$ has Sidon constant at most C and $\delta|\mathbf{F}| \leq |\mathbf{F}_1|$. Determine a value of $\delta_0 > 0$ such that every finite subset $\mathbf{F} \subseteq \mathbf{E}$ contains a quasi-independent subset $\mathbf{F}_1$ with $\delta_0|\mathbf{F}| \leq |\mathbf{F}_1|$.

Exercise 7.6.4. Show that $N(\omega)$ (defined on page 120) is at least the number of quasi-relations of length $\ell+1$ or greater, with equality if Γ has no elements of order two.

Exercise 7.6.5. Let $\mathbf{E}_j \subseteq \Gamma$ be disjoint. The union (finite or not) $\bigcup \mathbf{E}_j$ is an *independent union* if $\gamma_j \in \mathbf{E}_j$, $\epsilon_j \in \mathbb{Z}$, $J \in \mathbb{N}$ and $1 = \prod_{j=1}^{J} \gamma_j^{\epsilon_j}$ imply all $\epsilon_j = 0$. Suppose that $\bigcup \mathbf{E}_j$ is an independent union and that the $\mathbf{E}_j$ are Sidon sets with $C = \sup_j S(E_j) < \infty$. Show that $\bigcup \mathbf{E}_j$ is Sidon with $S(\mathbf{E}) \leq 6C$.
We do not know if the independent union of $I_0(N, 1/2)$ sets is I_0 [**P 6**].

Exercise 7.6.6. 1. Suppose $\mathbf{F}_n$ are finite, quasi-independent subsets of $\mathbb{Z} \setminus \{0\}$. Let $k_1 = 1$ and $k_n > \sum\{|\ell| : \ell \in \bigcup_{j=1}^{n-1} k_j \mathbf{F}_j\}$ for $n \geq 2$. Show that $\bigcup_{n=1}^{\infty} k_n \mathbf{F}_n$ is quasi-independent.
2. Suppose $\mathbf{F}_n$ are finite Sidon sets in $\mathbb{Z}$ with uniformly bounded Sidon constants for $n = 1, 2, \ldots$. Show that there exist $k_n \in \mathbb{N}$ such that $\bigcup_{n=1}^{\infty} k_n \mathbf{F}_n$ is Sidon.
3. Suppose $\mathbf{F}_n \subset \Gamma_n$ are quasi-independent subsets of the discrete groups Γ_n for $n = 1, 2, \ldots$. Show that $\bigcup_{n=1}^{\infty} \mathbf{F}_n \subset \bigoplus \Gamma_n$ is quasi-independent.

Exercise 7.6.7. 1. Given $\mathbf{E} \subseteq \Gamma$, let $\mathcal{N}(\mathbf{E})$ be the least N such that $\mathbf{E}$ is $I_0(N, 1/2)$. Show that $\mathcal{N}(\mathbf{E}) = \sup\{\mathcal{N}(\mathbf{F}) : \mathbf{F} \subseteq \mathbf{E} \text{ is finite}\}$.
2. Put $\eta(\mathbf{E}, N) = \infty$ if $\mathbf{E}$ is not a finite union of $I_0(N, 1/2)$ sets and otherwise let $\eta(\mathbf{E}, N)$ be the least k such that $\mathbf{E}$ is the union of k $I_0(N, 1/2)$ sets. Prove that

$$\eta(\mathbf{E}, N) = \sup\{\eta(\mathbf{F}, N) : \mathbf{F} \text{ is a finite subset of } \mathbf{E}\}. \qquad (7.6.1)$$

Hint: See Bourgain's technique in Lemma 6.4.5.
3. Show that if every Sidon subset of $\mathbb{Z}$ is a finite union of I_0 sets, then there is a function $f : \mathbb{R}^+ \to \mathbb{Z}^+$ such that for all Sidon sets $\mathbf{E} \subseteq \mathbb{Z}$ with Sidon constant $S = S(\mathbf{E})$,

$$\eta(\mathbf{E}, f(S)) \leq f(S).$$

Hint: Use Exercise 7.6.6, as well as the previous parts of this exercise.
4. Show that if every Sidon subset of every discrete Γ is a finite union of I_0 sets, then there is a function $f : \mathbb{R}^+ \to \mathbb{Z}^+$ such that for all Sidon sets $\mathbf{E}$ (in any Γ) with Sidon constant $S = S(\mathbf{E})$,

$$\eta(\mathbf{E}, f(S)) \leq f(S).$$

Chapter 8
How Thin Are Sidon Sets in the Bohr Compactification?

If a Sidon set clusters at $n \in \mathbb{Z}$ then there is a Sidon set that is dense in $\overline{\mathbb{Z}}$. Statistical evidence for non-density of Sidon sets in $\mathbb{Z}$ is given.

8.1 Introduction

Throughout the book we have focused on the theme of thinness from two perspectives, the "size" of a given set $\mathbf{E}$ when viewed as a subset of Γ, or as a subset of $\overline{\Gamma}$. The definitive statement on the size of Sidon sets in Γ is Pisier's characterization of Sidon as those sets which are proportional quasi-independent, Theorem 7.2.1 (3).

In this chapter, we explore what is known about the thinness of Sidon sets as subsets of $\overline{\mathbb{Z}}$. It is a long-standing problem whether Sidon subsets can be dense in the Bohr compactification of Γ [**P 2**]. Although there are no definitive answers in general, there is an answer for one class of groups: When $G = \mathbb{Z}_p^a$ where p is prime, then [126] every Sidon set in the dual of G is a finite union of independent sets. Since independent sets are I_0 (Exercise 3.7.11), this gives an answer of no to [**P 2**] (and yes to [**P 1**]) in that case.

For $\mathbb{Z}$ there are some partial answers. We first prove that the existence of a Sidon set in $\mathbb{Z}$ that clusters at one continuous character is equivalent to the existence of a Sidon set that is dense in $\overline{\mathbb{Z}}$. Then, we show that, in a probabilistic sense, Sidon sets in $\mathbb{Z}$ are not dense in $\overline{\mathbb{Z}}$. The first of these results relies upon Pisier's proportional characterization. The second uses deep probabilistic techniques.

C.C. Graham and K.E. Hare, *Interpolation and Sidon Sets for Compact Groups*, 135
CMS Books in Mathematics, DOI 10.1007/978-1-4614-5392-5_8,
© Springer Science+Business Media New York 2013

8.2 If a Sidon Set Clusters at a Continuous Character

Theorem 8.2.1 (Ramsey cluster). *Suppose there is a Sidon set in $\mathbb{Z}$ which clusters at some integer. Then there is a Sidon set which is dense in $\overline{\mathbb{Z}}$.*

The plan of the proof of Theorem 8.2.1 is this: Let $(m_\ell, N_\ell, p_\ell)_{\ell=1}^\infty$ be a *fixed*, exhaustive enumeration of $\mathbb{N} \times \mathbb{N} \times \mathbb{Z}$. We recall that an integer, q, is a cluster point of $\mathbf{E} \subseteq \mathbb{Z}$ in the Bohr topology if and only if for every $\varepsilon > 0$, $J \in \mathbb{N}$ and $(t_1, \ldots, t_J) \in \mathbb{T}^J$, there exists $m \in \mathbf{E}$ such that

$$\max_{1 \le j \le J} \left| e^{imt_j} - e^{iqt_j} \right| < \varepsilon. \tag{8.2.1}$$

We will say $\mathbf{E}$ *approximates q to within ε on $\mathbb{T}^J$* if for each $(t_1, \ldots, t_J) \in \mathbb{T}^J$ there exists $m \in \mathbf{E}$ such that (8.2.1) holds.

Assume $\mathbf{E}$ clusters at q. A compactness argument will be given to prove there are finite sets $\mathbf{E}_\ell \subseteq \mathbf{E}$ which approximate q to within $1/m_\ell$ on $\mathbb{T}^{N_\ell}$. Let $\mathbf{S} = \bigcup_\ell (p_\ell + k_\ell(\mathbf{E}_\ell - q))$. For any set of dilation factors, k_ℓ, the set $\mathbf{S}$ will be seen to be dense in $\overline{\mathbb{Z}}$. We will use Theorem 7.2.1(3) to show that for a suitably, rapidly growing sequence, $\{k_\ell\}$, the subset $\mathbf{S}$ is also Sidon.

We begin the proof with preliminary lemmas.

Lemma 8.2.2. *Suppose $\mathbf{E} \subseteq \mathbb{Z}$ clusters at $q \in \mathbb{Z}$. For each $\varepsilon > 0$ and $J \in \mathbb{N}$ there is a finite set $\mathbf{E}' \subseteq \mathbf{E}$ which approximates q to within ε on $\mathbb{T}^J$.*

Proof. Since $\mathbf{E}$ clusters at q, for each $(t_1, \ldots, t_J) \in \mathbb{T}^J$ there is some $m \in \mathbf{E}$ (depending on $(t_1, \ldots, t_J)$) such that (8.2.1) holds. By continuity of the functions e^{imx} and e^{iqx} on $\mathbb{T}$, there exists a neighbourhood, $U = U_{t_1, \ldots, t_J} \subseteq \mathbb{T}^J$ of $(t_1, \ldots, t_J)$, for which (8.2.1) holds with $(t_1, \ldots, t_J)$ replaced by each and every $(u_1, \ldots, u_J) \in U$. Since $\mathbb{T}^J$ is compact, finitely many of the sets $U_{t_1, \ldots, t_J}$ cover it. Let $\mathbf{E}'$ be the finite set of $m's$ which correspond to the finitely many J-tuples, $(t_1, \ldots, t_J)$. $\qquad\square$

Lemma 8.2.3. *Suppose $k, q, p \in \mathbb{Z}$ and $J \ge 1$. If $\mathbf{S} \subseteq \mathbb{Z}$ approximates q to within $\varepsilon > 0$ on $\mathbb{T}^J$, then $p + k(\mathbf{S} - q)$ approximates p to within ε on $\mathbb{T}^J$.*

Proof. Let $(t_1, \ldots, t_J) \in \mathbb{T}^J$ and pick $m \in \mathbf{S}$ which approximates q at $(kt_1, \ldots, kt_J)$ to within ε, that is, $\max_{1 \le j \le J} \left| e^{imkt_j} - e^{iqkt_j} \right| < \varepsilon$. Simple calculation shows that

$$\max_{1 \le j \le J} \left| e^{i(p+k(m-q)t_j)} - e^{ipt_j} \right| = \max_{1 \le j \le J} \left| e^{imkt_j} - e^{iqkt_j} \right| < \varepsilon. \qquad\square$$

Lemma 8.2.4. *Suppose $\mathbf{E} \subseteq \mathbb{Z}$ clusters at integer q. For each ℓ there is a finite subset $\mathbf{E}_\ell \subseteq \mathbf{E}$ such that for all $k_\ell \in \mathbb{N}$, the set $\mathbf{S} = \bigcup_\ell (p_\ell + k_\ell(\mathbf{E}_\ell - q))$ is dense in $\overline{\mathbb{Z}}$.*

Proof. It will be enough to prove that for every $p \in \mathbb{Z}$, $\varepsilon > 0$ and $J \in \mathbb{N}$, the set $\mathbf{S}$ approximates p to within ε on $\mathbb{T}^J$. Fix such p, ε and J, and choose a positive integer m such that $1/m < \varepsilon$. Take ℓ such that $(m, J, p) = (m_\ell, N_\ell, p_\ell)$.

We appeal to Lemma 8.2.2 to obtain a finite set $\mathbf{E}_\ell \subseteq \mathbf{E}$ which approximates q to within $1/m_\ell$ on $\mathbb{T}^{J_\ell}$ and then apply Lemma 8.2.3. $\quad\square$

It remains to see that one can choose integers, k_ℓ, so that $\bigcup_\ell (p_\ell + k_\ell(\mathbf{E}_\ell - q))$ is still Sidon. For this, it is convenient to adapt the notation for the set of all quasi-words from a given (finite) set $\mathbf{F} \subseteq \mathbb{Z}$ (see Definition 6.2.9):

$$Q(\mathbf{F}) = \left\{ \sum_{x \in \mathbf{F}} \epsilon_x x : \epsilon_x = 0, \pm 1 \right\}. \tag{8.2.2}$$

Lemma 8.2.5. *Suppose finite quasi-independent sets $\mathbf{E}_\ell$ are given and that integers $k_1, k_2, \ldots$ satisfy the inequalities*

$k_1 > |p_1||\mathbf{E}_1|$ *and*

$$k_\ell > \max \left\{ |x| : x \in Q\left(\bigcup_{j=1}^{\ell-1} (p_j + k_j \mathbf{E}_j) \right) \right\} + |p_\ell||\mathbf{E}_\ell| \text{ for } 2 \leq \ell < \infty. \tag{8.2.3}$$

Then the sets $p_\ell + k_\ell \mathbf{E}_\ell$ are pairwise disjoint and $\bigcup_{\ell=1}^{\infty}(p_\ell + k_\ell \mathbf{E}_\ell)$ is quasi-independent.

Proof. To simplify notation, put $\mathbf{F}_\ell = p_\ell + k_\ell \mathbf{E}_\ell$. The disjointness follows easily from the choice of k_ℓ, so it suffices to show $\bigcup_{\ell=1}^{L} \mathbf{F}_\ell$ is quasi-independent for each L. The argument proceeds by induction on L.

$L = 1$: Suppose $\sum \epsilon_x x = 0$ for $x \in \mathbf{F}_1$, $\epsilon_x = 0, \pm 1$. Each $x \in \mathbf{F}_1$ can be written in a unique way as $p_1 + k_1 u_x$ for some $u_x \in \mathbf{E}_1$, and thus we have $p_1 \sum_{x \in \mathbf{F}_1} \epsilon_x = -k_1 \sum \epsilon_x u_x$. We note that if not all $\epsilon_x = 0$, then, since $\mathbf{E}_1$ is quasi-independent, $\sum \epsilon_x u_x$ is a non-zero integer. Therefore, $k_1 \leq |k_1 \sum \epsilon_x u_x| = |p_1 \sum_{x \in \mathbf{F}_1} \epsilon_x| \leq |p_1||\mathbf{E}_1|$. That contradicts the definition of k_1 and shows $\mathbf{F}_1$ is quasi-independent.

Now assume that $L \geq 2$, $\bigcup_{\ell=1}^{L-1} \mathbf{F}_\ell$ is quasi-independent and

$$\sum_{x \in \mathbf{F}_L} \epsilon_x x + \sum_{y \in \bigcup_{\ell=1}^{L-1} \mathbf{F}_\ell} \epsilon_y y = 0. \tag{8.2.4}$$

Writing $x \in \mathbf{F}_L$ as $x = p_L + k_L u_x$ for some $u_x \in \mathbf{E}_L$ gives

$$k_L \sum_{x \in \mathbf{F}_L} \epsilon_x u_x = -p_L \sum_{x \in \mathbf{F}_L} \epsilon_x - \sum_{y \in \bigcup_{\ell=1}^{L-1} \mathbf{F}_\ell} \epsilon_y y.$$

If $\sum_{x \in \mathbf{F}_L} \epsilon_x u_x \neq 0$, then

$$k_L \leq \left| k_L \sum_{x \in \mathbf{F}_L} \epsilon_x u_x \right| \leq \left| p_L \sum_{x \in \mathbf{F}_L} \epsilon_x \right| + \left| \sum_{y \in \bigcup_{\ell=1}^{L-1} \mathbf{F}_\ell} \epsilon_y y \right|$$

$$\leq p_L |\mathbf{F}_L| + \max \left\{ |z| : z \in Q \left(\bigcup_{\ell=1}^{L-1} \mathbf{F}_\ell \right) \right\}. \qquad (8.2.5)$$

But inequality (8.2.5) contradicts the definition of k_L. That shows that $\sum_{x \in \mathbf{F}_L} \epsilon_x u_x = 0$. Since the u_x belong to the quasi-independent set $\mathbf{E}_L$, it follows that $\epsilon_x = 0$ for all x. But then (8.2.4) can hold only if $\sum \{\epsilon_y y : y \in \bigcup_{j=1}^{L-1} \mathbf{F}_\ell\} = 0$, and then, by the induction assumption, we must have all $\epsilon_y = 0$. Hence, $\bigcup_{\ell=1}^{L} (p_\ell + k_\ell \mathbf{E}_\ell)$ is quasi-independent.   □

We now complete the proof of the theorem.

Proof (of Theorem 8.2.1). Since translates of Sidon sets are Sidon, there is no loss of generality in assuming 0 is a cluster point of the Sidon set $\mathbf{E}$. Apply Lemma 8.2.4 to choose finite subsets $\mathbf{E}_\ell \subseteq \mathbf{E}$ such that for all choices of integers, k_ℓ, the set $\bigcup_{\ell=1}^{\infty} (p_\ell + k_\ell \mathbf{E}_\ell)$ is dense in $\overline{\mathbb{Z}}$. Now pick integers k_ℓ satisfying (8.2.3) and let $\mathbf{S} = \bigcup_{\ell=1}^{\infty} (p_\ell + k_\ell \mathbf{E}_\ell)$.

Let $\lambda > 0$ be a Pisier constant for $\mathbf{E}$ (Remarks 7.2.2), meaning that each finite subset $\mathbf{H}$ of $\mathbf{E}$ contains a further finite subset $\mathbf{F} \subseteq \mathbf{H}$ that is quasi-independent, with $|\mathbf{F}| \geq \lambda |\mathbf{H}|$. By Pisier's proportional quasi-independent characterization of Sidonicity, it will suffice to show λ is also a Pisier constant for $\mathbf{S}$.

To see that λ has that property, let $\mathbf{H} \subseteq \mathbf{S}$ be a finite subset and put $\mathbf{H}_\ell = \mathbf{H} \cap (p_\ell + k_\ell \mathbf{E}_\ell)$. Using the fact that λ is a Pisier constant for $\mathbf{E}$ and that the map $n \mapsto p_\ell + k_\ell n$ is one-to-one from $\mathbf{E}_\ell \to \mathbf{S}$, we can find a quasi-independent subset $\mathbf{F}_\ell \subseteq \mathbf{E}_\ell$ with $|\mathbf{F}_\ell| \geq \lambda |\mathbf{H}_\ell|$. Put $\mathbf{F} = \bigcup_{\ell=1}^{\infty} (p_\ell + k_\ell \mathbf{F}_\ell)$. Since the integers k_ℓ satisfy condition (8.2.3) with the sets $\mathbf{E}_\ell$ in (8.2.3) replaced by $\mathbf{F}_\ell$, Lemma 8.2.5 implies $\mathbf{F}$ is quasi-independent. Furthermore, because the sets $p_\ell + k_\ell \mathbf{F}_\ell$ are disjoint,

$$|\mathbf{F}| = \sum |p_\ell + k_\ell \mathbf{F}_\ell| = \sum |\mathbf{F}_\ell| \geq \lambda \sum |\mathbf{H}_\ell| = \lambda |\mathbf{H}|.$$

This proves $\mathbf{S}$ is proportional quasi-independent and hence Sidon.   □

Essentially the same argument proves the analogous result for quasi-independent sets.

Proposition 8.2.6. *If there is a quasi-independent set which clusters at some integer, then there is a quasi-independent set which is dense in $\overline{\mathbb{Z}}$.*

8.3 Probabilistic Evidence for a Sidon Set Having No Bohr Cluster Points in Γ

In this section we will prove that random sets of integers, of the appropriate size, are almost surely Sidon and do not cluster at any integers, while sets that are too "big" to be Sidon are almost surely dense in $\overline{\mathbb{Z}}$.

8.3.1 Random Sets of Integers

Let Ω be a probability space and let X_n be independent, Poisson random variables defined on Ω, with Poisson parameters p_n.[1] By a "random set of integers" we mean a set $\mathbf{E} = \mathbf{E}(\omega)$ given by

$$\mathbf{E}(\omega) = \{n \in \mathbb{N} : X_n(\omega) > 0\} = \{n \in \mathbb{N} : X_n(\omega) \geq 1\}.$$

Here are the precise statements of the theorems proved in this section.

Theorem 8.3.1 (Kahane–Katznelson). *If there is a constant α such that $p_n \leq \alpha/n$ for all n, then almost surely $\mathbf{E}(\omega)$ is Sidon and has no integers as cluster points in $\overline{\mathbb{Z}}$.*

Theorem 8.3.2 (Kahane–Katznelson). *If $\lim_n np_n = \infty$, then almost surely $\mathbf{E}(\omega)$ is not Sidon and is dense in $\overline{\mathbb{Z}}$.*

8.3.2 Proof of Theorem 8.3.1

First, we reduce α. Choose an integer $q > 9\alpha$. For each $i = 0, 1, \ldots, q-1$, let $Y_n^{(i)} = X_{qn+i}$ and consider the random sets $\mathbf{E}^{(i)}(\omega) = \{n \in \mathbb{N} : Y_n^{(i)}(\omega) > 0\}$. Then $\mathbf{E}(\omega) = \bigcup_{i=0}^{q-1} \mathbf{E}^{(i)}(\omega)$. Since a finite union of Sidon sets is Sidon, and the closure of a finite union of sets is the union of their closures, it will be enough to show that almost surely each set $\mathbf{E}^{(i)}(\omega)$ is Sidon and has no integers as cluster points in $\overline{\mathbb{Z}}$.

The functions $Y_n^{(i)}$ are independent, Poisson random variables with parameters $p_{qn+i} \leq \alpha/(qn+i) < 1/(9n)$. Hence, there is no loss of generality in assuming $\alpha < 1/9 < 1/\ln 64$ (it will be convenient to have both bounds).

---

[1] Appendix C.2 summarizes basic background material in probability theory.

Almost Surely E Is Sidon When $p_n = O(1/n)$

This part of the theorem will be shown by establishing that, under these assumptions, $\mathbf{E}(\omega)$ is a finite union of quasi-independent sets for almost every ω. Since finite unions of quasi-independent sets are Sidon (Cors. 6.2.12 and 6.3.3) that will prove that $\mathbf{E}$ is a.s. Sidon. We continue to use the notation $Q(\mathbf{F})$ for the set of all quasi-words from $\mathbf{F}$.

Lemma 8.3.3. *For almost every $\omega \in \Omega$, there exists $M(\omega)$ such that for all $J \geq M(\omega)$, the set $\mathbf{E}(\omega) \cap [4^{J-1}, \infty) \cap Q\big(\mathbf{E}(\omega) \cap [1, 4^J]\big)$ is empty.*

First, we will use Lemma 8.3.3 to prove that $\mathbf{E}(\omega)$ is a.s. Sidon, and then we will prove Lemma 8.3.3. Indeed, we claim that Lemma 8.3.3 implies that for all $\omega \in \Omega_0$, a set of full measure[2], the set $\mathbf{E}(\omega) \smallsetminus [1, 4^{M(\omega)}]$ is quasi-independent. Of course, that would ensure that $\mathbf{E}(\omega)$ is Sidon for all $\omega \in \Omega_0$, being a union of a quasi-independent set and a finite set.

To prove the claim, suppose otherwise, say $\sum_{\ell=1}^{L} \epsilon_\ell n_\ell = 0$ for some $\epsilon_\ell = 0, \pm 1$, $n_\ell \in \mathbf{E}(\omega) \smallsetminus [1, 4^{M(\omega)}]$ and $\omega \in \Omega_0$. We may assume $n_L = \max(n_1, \ldots, n_L)$ and that $\epsilon_L \neq 0$. Thus, $n_L \in Q(\{n_1, \ldots, n_{L-1}\})$. Pick the integer k such that $n_L \in \mathbf{E}(\omega) \cap (4^{k-1}, 4^k]$. Then also $\{n_1, \ldots, n_{L-1}\} \subseteq [1, 4^k]$. Since $n_\ell \notin [1, 4^{M(\omega)}]$, we have $k > M(\omega)$. Hence,

$$n_L \in \mathbf{E}(\omega) \cap Q\big(\mathbf{E}(\omega) \cap [1, 4^k]\big) \cap [4^{k-1}, \infty).$$

But this is an impossibility since that last intersection is empty by Lemma 8.3.3.

Remark 8.3.4. Exercise 8.5.3 shows that, if $\sum_n p_n < \infty$, then for each $q > 1$ the set $\mathbf{E}(\omega)$ is a.s. a union of a finite set with a Hadamard set of ratio q.

Proof (of Lemma 8.3.3). For $m \geq 1$, let $\mathbf{L}_m(\omega) = \mathbf{E}(\omega) \cap [1, 4^m]$. For $\omega \in \Omega$, define $K(\omega)$ to be the least positive integer k for which $|\mathbf{L}_m| \leq m$ for all $m \geq k$, if such k exist, and otherwise, $K(\omega) = \infty$.

We will use the conditional probability formula, $\mathbb{P}(A \mid B)\,\mathbb{P}(B) = \mathbb{P}(A \cap B)$, for events

$$A = A_k = \{\omega : \mathbf{E}(\omega) \cap [4^{k-1}, \infty) \cap Q(\mathbf{L}_k) \neq \emptyset\} \ \text{ and}$$
$$B = B_k = \{\omega : K(\omega) \leq k\}.$$

There are several steps to the proof.

Step I: $K(\omega) < \infty$ almost surely. Equivalently, we show $\mathbb{P}(B_k) \to 1$ as $k \to \infty$. To prove the assertion of Step I, let M_k denote the random variable $M_k = M_k(\omega) = |\mathbf{L}_k|$ for $1 \leq k < \infty$. Of course, $M_k(\omega) \geq k$ if and only if $X_n(\omega)$ is non-zero for at least k choices of $n \in [1, 4^k]$. Thus,

---

$$\{\omega : M_k(\omega) \geq k\} \subseteq \left\{\omega : \sum_{n=1}^{4^k} X_n(\omega) \geq k\right\} \text{ and}$$

$$\mathbb{P}(M_k \geq k) \leq \mathbb{P}\left(\omega : \sum_{n=1}^{4^k} X_n \geq k\right).$$

But $\sum_{n=1}^{4^k} X_n$ is a Poisson distribution with parameter

$$m_k = \sum_{n=1}^{4^k} p_n \leq \sum_{n=1}^{4^k} \frac{\alpha}{n} \leq \frac{k}{3}.$$

(Here we used the assumption that $\alpha < 1/(\ln 64)$.) Thus, a geometric series argument (Exercise 8.5.6) gives

$$\mathbb{P}\left(\sum_{n=1}^{4^k} X_n \geq k\right) \leq e^{-m_k} \sum_{j=k}^{\infty} \frac{m_k^j}{j!} \leq C\left(\frac{e}{3}\right)^k. \qquad (8.3.1)$$

Hence, $\sum_{k=1}^{\infty} \mathbb{P}(M_k \geq k) < \infty$. The Borel–Cantelli Lemma C.2.3 tells us that $\mathbb{P}(M_k \geq k \text{ i.o.}) = 0$. Therefore, for a.e. ω, there exists an integer $K(\omega)$ such that $M_k(\omega) \leq k$ for all $k \geq K(\omega)$. That completes Step I.

Step II: For every finite set $\mathbf{F}$,

$$\mathbb{P}\big(Q(\mathbf{F}) \cap \mathbf{E}(\omega) \cap [4^{k-1}, \infty) \neq \emptyset\big) \leq \frac{\alpha 3^{|\mathbf{F}|}}{4^{k-1}}.$$

The argument will be similar to Step I. Let b_k be the Poisson parameter of $\sum_{n \in Q(\mathbf{F}), n \geq 4^{k-1}} X_n$. Then the probability that $Q(\mathbf{F}) \cap \mathbf{E}(\omega) \cap [4^{k-1}, \infty)$ is not empty is bounded above by

$$\mathbb{P}\left(\sum_{\substack{n \in Q(\mathbf{F}) \\ n \geq 4^{k-1}}} X_n \geq 1\right) = 1 - e^{-b_k} \leq b_k \leq \sum_{\substack{n \in Q(\mathbf{F}) \\ n \geq 4^{k-1}}} \frac{\alpha}{n} \leq \frac{\alpha |Q(\mathbf{F})|}{4^{k-1}} \leq \frac{\alpha 3^{|\mathbf{F}|}}{4^{k-1}},$$

since $|Q(\mathbf{F})| \leq 3^{|\mathbf{F}|}$. That completes Step II.

Step III: Completing the proof of Lemma 8.3.3. Step II gives the estimate $\mathbb{P}(A_k | B_k) \leq \alpha 3^k / 4^{k-1}$ and thus $\mathbb{P}(A_k \cap B_k) \leq 3^k \alpha / 4^{k-1}$. Using the Borel–Cantelli lemma we see that $\mathbb{P}(A_k \cap B_k \text{ i.o. }) = 0$. Step I tells us that $\mathbb{P}(B_k) \to 1$ therefore, also, $\mathbb{P}(A_k \text{ i.o.}) = \mathbb{P}(A_k \cap B_k \text{ i.o.}) = 0$. $\qquad \square$

Almost Surely E(ω) Has No Integers as Cluster Points When $p_n = O(1/n)$

We introduce new notation for this part of the theorem. It is convenient to take $\mathbb{T} = [-1/2, 1/2]$ here. For $\omega \in \Omega$, let

$$D_m^\omega = \{t \in \mathbb{T} : t\,([m, \infty) \cap \mathbf{E}(\omega)) \subseteq [-1/4, 1/4]\} \text{ and } D_\infty^\omega = \bigcup_{m=1}^\infty D_m^\omega.$$

The key step is to prove:

Lemma 8.3.5. *Almost surely D_∞^ω is dense in $\mathbb{T}$.*

Once this lemma is proved, the argument can be completed quickly. Indeed, suppose $q \in \mathbb{Z}$ is non-zero. Since D_∞^ω is almost surely dense in $\mathbb{T}$, there is a subset $\Omega_1 \subseteq \Omega$, of full measure, such that for each $\omega \in \Omega_1$ we can choose $t \in D_\infty^\omega$ such that $|qt - 1/2| < 1/16$. Let $\mathbf{U} = \{n \in \mathbb{Z} : |nt - qt| < 1/16\}$, a basic neighbourhood of q in $\mathbb{Z}$ with respect to the relative topology from $\overline{\mathbb{Z}}$.

Since $t \in D_\infty^\omega$, there is some m such that $t \in D_m^\omega$. This means that for all $N \in \mathbf{E}(\omega) \cap [m, \infty)$, $|Nt| < 1/4$. But then $N \notin \mathbf{U}$. That shows $\mathbf{E}(\omega) \cap \mathbf{U}$ is a finite set and therefore q cannot be a limit point of $\mathbf{E}(\omega)$ for any $\omega \in \Omega_1$.

A translation argument (Exercise 8.5.4) shows that $0 \notin \overline{\mathbf{E}(\omega)}$ a.s.

Proof (of Lemma 8.3.5). We note that if $\mathbf{E} \subseteq \mathbf{E}'$, then $t\,([m, \infty) \cap \mathbf{E}) \subseteq t\,([m, \infty) \cap \mathbf{E}')$, so we may assume $p_n = \alpha/n$ for all n.

We proceed by contradiction and suppose D_∞^ω is not dense for all ω in a subset of Ω of positive measure. For each such ω there is some closed (non-trivial) interval, I^ω, with rational endpoints, such that $D_\infty^\omega \cap I^\omega$ is empty. Because there are only countably many such intervals, there is a single interval, J_0, such that $D_\infty^\omega \cap J_0 = \emptyset$ for all ω in a (further) subset, $\Omega_1 \subseteq \Omega$, of positive measure. Thus, for every $t \in J_0$ and $\omega \in \Omega_1$, $t \notin D_m^\omega$ for every positive integer m.

That implies that if we put $I_N = \{t : Nt \notin [-1/4, 1/4]\}$, then for every $\omega \in \Omega_1$ and integer m, $J_0 \subseteq \bigcup_{N \in \mathbf{E}(\omega)} I_N$ where the union is over $N \geq m$. But the sets I_N are open and J_0 is compact, so for each $\omega \in \Omega_1$ and m, there is a finite cover, say $J_0 \subseteq I_{N_1} \cup \cdots \cup I_{N_k}$, with the $N_i \in \mathbf{E}(\omega) \cap [m, \infty)$. Set $N_m^\omega = \max N_i$.

If f is a function supported on $[-1/4, 1/4]$, then $f(Nt) = 0$ for each $t \in I_N$. In particular, this is true when f is the even, positive-definite, piecewise linear function supported on $[-1/4, 1/4]$ given by

$$f(t) = 16(1/4 - t) \text{ for } t \in [0, 1/4].$$

Let $c_n(t) = \mathbb{E}(f(nt)^{X_n(\omega)})$, where we define $f(nt)^{X_n(\omega)} = 1$ if $X_n(\omega) = 0$, even if $f(nt) = 0$. If A is any positive constant and X is a Poisson random variable with parameter β, then

$$\mathbb{E}(A^{X(\omega)}) = \sum_{k=0}^{\infty} \mathbb{P}(X(\omega) = k)A^k = \sum_k e^{-\beta}\beta^k A^k / k! = e^{\beta(A-1)}. \qquad (8.3.2)$$

Thus,

$$c_n(t) = \exp\left(\frac{\alpha}{n}(f(nt) - 1)\right) > 0. \qquad (8.3.3)$$

Define independent random variables, $F_n^\omega(t)$, and random measures, $_m\mu_n$, by

$$F_n^\omega(t) = \frac{1}{c_n}f(nt)^{X_n(\omega)} \text{ and } _m\mu_n^\omega = \prod_{j=m}^{n} F_j^\omega(t)dt \text{ for } 1 \leq m \leq n.$$

Notice $F_N^\omega(t) = 0$ if $t \in I_N$ and $N \in \mathbf{E}(\omega)$. Hence, if $\omega \in \Omega_1$ and $n \geq N_m^\omega$, then $\prod_{j=m}^{n} F_j^\omega(t) = 0$ for all $t \in J_0$ and

$$_m\mu_n^\omega(J_0) = \int_{J_0} \prod_{j=m}^{n} F_j^\omega(t)dt = 0 \text{ for all } n \geq N_m^\omega \text{ and } \omega \in \Omega_1. \qquad (8.3.4)$$

Since the functions F_n^ω are non-negative, independent, random variables with expectation 1, $\{\prod_{j=m}^{n} F_j^\omega(t)\}_{n \geq m}$ is a martingale for each $m \in \mathbb{N}$ and $t \in \mathbb{T}$. It follows from the independence of the functions F_j^ω and properties of conditional expectation that $\{_m\mu_n^\omega(I)\}_{n \geq m}$ is also a martingale for each interval I and positive integer m. Moreover, for each interval I, $_m\mu_n^\omega(I) \geq 0$ and, by the definition of the $c_n(t)$,

$$\mathbb{E}(_m\mu_n^\omega(I)) = \mathbb{E}\left(\int_I \prod_{j=m}^{n} F_j^\omega dt\right) = \int_I \prod_{j=m}^{n} \mathbb{E}(F_j^\omega)dt = |I|. \qquad (8.3.5)$$

The martingale convergence Theorem C.2.7, says that the martingales $\{_m\mu_n^\omega(I)\}_n$ converge almost surely, with limit denoted $_m\mu^\omega(I)$. Let Ω_2 be the set of $\omega \in \Omega$ such that for every $m \in \mathbb{N}$ and for every interval I with rational endpoints, $\{_m\mu_n^\omega(I)\}$ converges. Then Ω_2^c is a set of measure zero and $\Omega_1 \cap \Omega_2$ has positive measure. Moreover, (8.3.4) implies that $_m\mu^\omega(J_0) = 0$ on $\Omega_1 \cap \Omega_2$, for all m.

We need more than almost sure convergence, however.

Claim. For each I and m, the martingale $\{_m\mu_n^\omega(I)\}_n$ is L^2 bounded.

It is in proving this claim that the hypothesis on $p_n = \alpha/n$ is used.

Assuming the claim, the martingale convergence Theorem C.2.8 for L^2 bounded martingales and (8.3.5) imply that $\mathbb{E}(_m\mu(I)) = \lim_n \mathbb{E}(_m\mu_n(I)) = |I|$ for each integer m and interval I.

The set

$$\mathcal{T} = \{\omega : \forall j \ \exists m \geq j \text{ with } _m\mu^\omega(J_0) > 0\}$$

is a tail event and hence by the zero-one law has probability 0 or 1. But, for every I, $_m\mu(I) = 0$ implies $_j\mu(I) = 0$ for all $j \leq m$. Therefore,

$$\mathcal{T} = \{\omega : \exists m \text{ with } _m\mu(J_0) > 0\}.$$

Since $\mathbb{E}(_m\mu(I)) > 0$ for all m, $\mathcal{T}$ cannot have probability zero. But $\mathcal{T}$ is disjoint from $\Omega_1 \cap \Omega_2$, so it cannot have probability one. This contradiction will complete the proof of Lemma 8.3.5 once we establish the claim that the martingales $\{_m\mu^\omega(I)\}_n$ are L^2 bounded.

Proof (of the claim). It will be enough to prove the martingales, $\{_m\mu_n^\omega(\mathbb{T})\}_n$, are L^2 bounded. Linearity, independence and Fubini's theorem give

$$\mathbb{E}\left(\left(_m\mu_n(\mathbb{T})\right)^2\right) = \mathbb{E}\left(\int_\mathbb{T}\int_\mathbb{T} \prod_{j=m}^{n} \prod_{k=m}^{n} F_j^\omega(s)F_k^\omega(t)\,ds\,dt\right)$$

$$= \int_\mathbb{T}\int_\mathbb{T}\left(\prod_{j=m}^{n} \mathbb{E}\left(F_j^\omega(s)F_j^\omega(t)\right)\right)ds\,dt.$$

As $F_n^\omega(t) = f(nt)^{X_n(\omega)}\exp\left(\frac{-\alpha}{n}(f(nt)-1)\right)$, (8.3.2) applied with $A = f(jt)f(js)$ gives

$$\mathbb{E}\left(\left(_m\mu_n(\mathbb{T})\right)^2\right)$$

$$= \int_\mathbb{T}\int_\mathbb{T}\prod_{j=m}^{n} \mathbb{E}\left([f(jt)f(js)]^{X_j}\right)\exp\left(\frac{-\alpha}{j}[(f(jt)-1)+(f(js)-1)]\right)ds\,dt$$

$$= \int_\mathbb{T}\int_\mathbb{T}\prod_{j=m}^{n}\exp\left(\frac{\alpha}{j}[f(jt)f(js)-1-(f(jt)-1)-(f(js)-1)]\right)ds\,dt$$

$$\leq C_m\int_\mathbb{T}\int_\mathbb{T}\exp\left(\sum_{j=1}^{n}\frac{\alpha}{j}[f(jt)f(js)-1-(f(jt)-1)-(f(js)-1)]\right)ds\,dt.$$

The C_m comes from inserting the terms from 1 to $m-1$.

In the (absolutely summable) Fourier series $f(s) = \sum_n \hat{f}(n)e^{ins}$, we note that $\hat{f}(n) \geq 0$ for all n, $\hat{f}(0) = 1$ and f is even. Thus, by symmetry,

$$f(s)f(t) - 1 - (f(s)-1) - (f(t)-1)$$
$$= \sum_{k,\ell\neq 0}\hat{f}(k)\hat{f}(\ell)e^{iks}e^{i\ell t} = 2\sum_{k,\ell\geq 1}\hat{f}(k)\hat{f}(\ell)\cos 2\pi(ks + \ell t).$$

For $n \geq 1$ we put

$$R_n(x) = \sum_{j=1}^{n}\frac{\cos 2\pi jx}{j}. \tag{8.3.6}$$

We will use below an inequality of Zygmund, [199, V.2.28], which says there is a constant C, independent of n, such that

$$|R_n(x)| \le C - \log|\sin \pi x|. \tag{8.3.7}$$

In terms of R_n, we have

$$\mathbb{E}\big((_m\mu_n(\mathbb{T}))^2\big) \le C_m \int_{\mathbb{T}} \int_{\mathbb{T}} \prod_{k,\ell \ge 1} \exp\big(\alpha \widehat{f}(k)\widehat{f}(\ell)R_n(ks + \ell t)\big) \mathrm{d}s\,\mathrm{d}t.$$

Hölder's inequality gives the bound

$$\mathbb{E}\big((_m\mu_n(\mathbb{T}))^2\big) \le C_m \prod_{k,l \ge 1} \left(\int_{\mathbb{T}} \int_{\mathbb{T}} \exp\big(r_{k,\ell}\alpha \widehat{f}(k)\widehat{f}(\ell)R_n(ks + \ell t)\big) \mathrm{d}s\,\mathrm{d}t \right)^{1/r_{k,\ell}}$$

whenever $r_{k,\ell}$ are non-negative numbers whose reciprocals sum to 1. We will make the choice

$$r_{k,\ell}^{-1} = \left(\sum_{i,j \ne 0} \widehat{f}(i)\widehat{f}(j) \right)^{-1} \widehat{f}(k)\widehat{f}(\ell),$$

so that for all k, ℓ,

$$r_{k,\ell}\widehat{f}(k)\widehat{f}(\ell) = \sum_{i,j \ne 0} \widehat{f}(i)\widehat{f}(j) = \left(\sum_{j \ne 0} \widehat{f}(j) \right)^2 = (f(0) - 1)^2 = 9.$$

Upon making a change of variable and using (8.3.7), we have

$$\int_{\mathbb{T}} \exp\big(r_{k,\ell}\alpha \widehat{f}(k)\widehat{f}(\ell)R_n(ks + \ell t)\big) \mathrm{d}s$$

$$= \int_{\mathbb{T}} \exp\big(r_{k,\ell}\alpha \widehat{f}(k)\widehat{f}(\ell)R_n(ks)\big) \mathrm{d}s$$

$$\le \int_{\mathbb{T}} \exp\big(9\alpha(-\log|\sin \pi ks| + C)\big) \mathrm{d}s$$

$$\le C' \int_{\mathbb{T}} |\sin \pi ks|^{-9\alpha} \mathrm{d}s.$$

The final integral is bounded independent of k because α was chosen less than $1/9$. Since $\sum r_{k,\ell}^{-1} = 1$ it follows that $\mathbb{E}\big((_m\mu_n(\mathbb{T}))^2\big)$ is bounded by a constant independent of n.

That proves the claim and completes the proof of Lemma 8.3.5.   □

8.3.3 Proof of Theorem 8.3.2

Almost Surely E Is Not Sidon When $np_n \to \infty$

If $\mathbf{E}(\omega)$ is Sidon, then there exists (Corollary 6.3.13) a constant $C = C_\omega$ such that

$$| \mathbf{E}(\omega) \cap [1, N] | \leq C \log N \text{ for all } N. \tag{8.3.8}$$

We will show this a.s. fails to be true.

Let $S_k(\omega) = | \mathbf{E}(\omega) \cap [2^{2^k}, 2^{2^{k+1}}] |$. Define the indicator random variables, $Z_n(\omega) = 1$ if $X_n(\omega) \neq 0$ (equivalently, $n \in \mathbf{E}(\omega)$) and $Z_n(\omega) = 0$ otherwise. Then also

$$S_k(\omega) = \sum_{n=2^{2^k}}^{2^{2^{k+1}}} Z_n.$$

Put $p_n = a_n/n$. We may assume $a_n/n \to 0$ and $\{a_n\}$ is increasing to infinity. Then $\mathbb{E}(Z_n) = 1 - e^{-a_n/n}$ and

$$\mathbb{E}(S_k) = \sum_{n=2^{2^k}}^{2^{2^{k+1}}} 1 - e^{-a_n/n} \geq \sum_{n=2^{2^k}}^{2^{2^{k+1}}} \frac{a_n}{3n} \geq C_0 a_{2^{2^k}} 2^k,$$

where the constant C_0 is independent of k.

By Markov's inequality (C.2.1) applied to $X = (S_k - \mathbb{E}(S_k))^2$,

$$\mathbb{P}\left(|S_k - \mathbb{E}(S_k)| > \frac{\mathbb{E}(S_k)}{2} \right) \leq \frac{\mathbb{E}(|S_k - \mathbb{E}(S_k)|^2)}{(\mathbb{E}(S_k)/2)^2}. \tag{8.3.9}$$

A computation using $Z_n^2 = Z_n$ and the independence of the Z_n shows that the right-hand side of (8.3.9) is bounded by

$$\frac{4 \sum \mathbb{E}(Z_n)(1 - \mathbb{E}(Z_n))}{(\sum \mathbb{E}(Z_n))^2} \leq \frac{4}{\mathbb{E}(S_k)} \leq \frac{C_1}{2^k}.$$

That implies the probability that $S_k \leq \mathbb{E}(S_k)/2$ is at most $C_1/2^k$. By the Borel–Cantelli lemma, $\mathbb{P}(S_k \leq \mathbb{E}(S_k)/2 \text{ i.o.}) = 0$. Thus, for almost every ω and for all sufficiently large k,

$$\frac{S_k(\omega)}{2^{k+1}} = \frac{| \mathbf{E}(\omega) \cap [2^{2^k}, 2^{2^{k+1}}] |}{2^{k+1}} \geq \frac{\mathbb{E}(S_k)}{2^{k+2}} \geq \frac{C_0}{4} a_{2^{2^k}} \to \infty,$$

which contradicts (8.3.8). $\qquad\square$

Almost Surely $\mathbf{E}(\omega)$ Is Dense When $np_n \to \infty$

To show $\mathbf{E} = \mathbf{E}(\omega)$ is dense in $\overline{\mathbb{Z}}$ it is enough to prove that the closures of $\mathbf{E}t$ and $\mathbb{Z}t$ in $\mathbb{T}^M$ coincide for all positive integers M and for all $t \in \mathbb{T}^M$ (Exercise 8.5.5). In fact, since homomorphisms are uniquely determined by their values on a generating set, it is enough to show this for the subset $\{(t_1, \ldots, t_M) : t_m \in S\} \subseteq \mathbb{T}^M$, where the set S generates $\mathbb{T}$. So our strategy will be to prove that, with probability one, $\mathbf{E}(\omega)t \cap O \neq \emptyset$ for every non-empty, open set $O \subseteq \mathbb{T}^M$ and every $t \in \{(t_1, \ldots, t_M) : t_m \in S\}$, where S is the set of irrationals in $\mathbb{T}$.

Let $M = 1$. Then $\overline{\mathbb{Z}}t = \mathbb{T}$ for all $t \in S$, so it suffices to show $\overline{\mathbf{E}}t = \mathbb{T}$. There is no loss of generality in assuming the open set O is an interval I. Choose α such that $\alpha |I| > 1$ and choose N_0 such that for all $n \geq N_0$, $np_n \geq \alpha$. Let I' be a proper subinterval of I, with $d(I', I^c) = d > 0$. To show that $\mathbf{E}t \cap I$ is non-empty, it will be enough to prove that there exists $t' \in \mathbb{T}$ and

$$n \in \mathbf{E}_N^\omega := \mathbf{E}(\omega) \cap [N_0, N]$$

such that $nt' \in I'$ and $|t - t'| < d/N$ because these assumptions ensure $nt \in I$.

Pick the subinterval I' such that $\alpha |I'| > 1$. Choose a function $f \in C^2(\mathbb{T})$ that is a suitably good approximation to $1_{I'}$ so that $0 \leq f \leq 1_{I'}$ and $\alpha \widehat{f}(0) > 1$. Then pick $\varepsilon > 0$ such that

$$\alpha \widehat{f}(0)(1 - \varepsilon) > 1.$$

For each t, the set $(\mathbf{E}_N^\omega t) \cap I'$ is empty if and only if $\sum_{n \in \mathbf{E}_N^\omega} 1_{I'}(nt) = 0$ if and only if $\sum_{N_0 \leq n \leq N} X_n(\omega) 1_{I'}(nt) = 0$. Since $\sum_{N_0 \leq n \leq N} X_n 1_{I'}(nt)$ is a random variable with Poisson parameter $\sum_{N_0 \leq n \leq N} p_n 1_{I'}(nt)$, we can calculate the probability that $\mathbf{E}_N t \cap I'$ is empty:

$$\mathbb{P}\left(\mathbf{E}_N t \cap I' = \emptyset\right) = \mathbb{P}\left(\sum_{N_0 \leq n \leq N} X_n 1_{I'}(nt) = 0\right) = \exp\left(\sum_{N_0 \leq n \leq N} -p_n 1_{I'}(nt)\right)$$

$$\leq \exp\left(\sum_{N_0 \leq n \leq N} \frac{-\alpha}{n} f(nt)\right). \qquad (8.3.10)$$

Since $f \in C^2(\mathbb{T})$, the Fourier series of f is absolutely convergent, so we may write

$$f(t) = \widehat{f}(0) + \sum_{1 \leq |j| \leq J} \widehat{f}(j)e^{ijt} + \sum_{|j| > J} \widehat{f}(j)e^{ijt} = \widehat{f}(0) + g(t) + h(t),$$

where J is chosen such that $\|h\|_\infty < \varepsilon \widehat{f}(0)$. Then

$$\exp\left(\sum_{N_0 \le n \le N} \frac{-\alpha}{n} f(nt)\right)$$

$$= \exp\left(\sum_{N_0 \le n \le N} \frac{-\alpha \widehat{f}(0)}{n}\right) \exp\left(\sum_{N_0 \le n \le N} \frac{-\alpha g(nt)}{n}\right) \exp\left(\sum_{N_0 \le n \le N} \frac{-\alpha h(nt)}{n}\right).$$

$$(8.3.11)$$

Clearly, $\exp\left(\sum_{N_0 \le n \le N} \frac{-\alpha \widehat{f}(0)}{n}\right) \le C_0 \exp(-\alpha \widehat{f}(0) \log N) = C_0 N^{-\alpha \widehat{f}(0)}$ and

$$\exp\left(\sum_{N_0 \le n \le N} \frac{-\alpha h(nt)}{n}\right) \le \exp\left(\alpha \|h\|_\infty \log N\right) \le N^{\varepsilon \alpha \widehat{f}(0)}. \qquad (8.3.12)$$

Since $\sum_{n=N_0}^{N} (\sin 2\pi njt)/n$ is uniformly bounded (see Exercise 8.5.8),

$$\sum_{N_0 \le n \le N} \frac{g(nt)}{n} = \sum_{1 \le j \le J}\left(a_j \sum_{N_0 \le n \le N} \frac{\cos 2\pi njt}{n} + ib_j \sum_{N_0 \le n \le N} \frac{2\sin \pi njt}{n}\right)$$

$$= \sum_{1 \le j \le J} a_j (R_N(jt) - R_{N_0-1}(jt)) + C_N(t), \qquad (8.3.13)$$

where the $C_N(t)$ are uniformly bounded over all N and t and the a_j, b_j are the real Fourier coefficients of g. Here, R_n is given by (8.3.6).

Fix $\delta > 0$ and put

$$G = G(\delta) = \{t : |\sin \pi jt| \ge \delta \text{ for } 1 \le j \le J\}.$$

As noted in (8.3.7), there is a constant C such that $\sup_n |R_n(t)| \le C - \log|\sin \pi t|$. Thus, for every $t \in G$ and $1 \le j \le J$,

$$|R_N(jt) - R_{N_0-1}(jt)| \le 2(C - \log|\sin \pi jt|) \le 2(C + |\log \delta|) = C_1,$$

where C_1 depends only on δ (but not on N, N_0, j or t). Because $C_N(t)$ is uniformly bounded,

$$\exp\left(\sum_{n=N_0}^{N} \frac{-\alpha g(nt)}{n}\right) \le e^{C_N(t)} \exp\left(\alpha \sum_{j=1}^{J} |a_j| |R_N(jt) - R_{N_0-1}(jt)|\right)$$

$$\le C_2 \qquad (8.3.14)$$

for every $t \in G$, where C_2 depends only on δ. Using (8.3.10)–(8.3.14), we conclude that for every $t \in G$,

$$\mathbb{P}\left(\mathbf{E}_N t \cap I' = \emptyset\right) \leq \exp\left(-\sum_{N_0 \leq n \leq N} \frac{\alpha}{n} f(nt)\right) \leq C_0 C_2 N^{-\alpha \widehat{f}(0)(1-\varepsilon)}.$$

Integrating over all $t \in G$ gives

$$\int_G \mathbb{P}\left(\mathbf{E}_N t \cap I' = \emptyset\right) dt \leq C_0 C_2 N^{-\alpha \widehat{f}(0)(1-\varepsilon)}.$$

Choose $M_N = \lfloor \frac{N}{d} \rfloor$. Since each $t \in \mathbb{T}$ can be written as $x + m/M_N$ where $x \in [0, 1/M_N]$ and $m = 0, 1, \ldots, M_N - 1$,

$$\int_G \mathbb{P}\left(\mathbf{E}_N t \cap I' = \emptyset\right) dt = \sum_{m=1}^{M_N} \int_{G \cap [\frac{m-1}{M_N}, \frac{m}{M_N}]} \mathbb{P}\left(\mathbf{E}_N t \cap I' = \emptyset\right) dt$$

$$= \int_0^{1/M_N} \sum_{\{m : x + m/M_N \in G\}} \mathbb{P}\left(\mathbf{E}_N(x + \frac{m}{M_N}) \cap I' = \emptyset\right) dx.$$

Since $M_N/N \to 1/d$ as $N \to \infty$, there must be an $x \in [0, 1/M_N]$ such that

$$\sum_{\{m : x + m/M_N \in G\}} \mathbb{P}\left(\mathbf{E}_N(x + m/M_N) \cap I' = \emptyset\right) \leq C_0 C_2 M_N N^{-\alpha \widehat{f}(0)(1-\varepsilon)}$$

$$\leq C_3 N^{1 - \alpha \widehat{f}(0)(1-\varepsilon)}.$$

Fix such an x. The probability that there is some m with $x + m/M_N \in G$ and $\mathbf{E}_N(x + m/M_N) \cap I' = \emptyset$ is at most $C_3 N^{1 - \alpha \widehat{f}(0)(1-\varepsilon)}$.

Put $G_N = \{t \in G : \exists t' = x + m/M_N \in G, |t - t'| < 1/M_N\}$. If there exists $t \in G_N$ such that $\mathbf{E}_N t \cap I = \emptyset$, then there exists $t' = x + m/M_N$ with $|t - t'| < d/N$ and hence $\mathbf{E}_N t' \cap I'$ is empty. Since $\alpha \widehat{f}(0)(1 - \varepsilon) > 1$,

$$\mathbb{P}\left(\exists t \in G_N : \mathbf{E}_N t \cap I = \emptyset\right) \leq C_3 N^{1 - \alpha \widehat{f}(0)(1-\varepsilon)} \to 0 \text{ as } N \to \infty.$$

Using this, we will now deduce that $\mathbb{P}\left(\exists t \in G : \mathbf{E}t \cap I = \emptyset\right) = 0$. Take $\eta > 0$ and choose $N_k \to \infty$ such that

$$\sum_k \mathbb{P}\left(\exists t \in G_{N_k} : \mathbf{E}_{N_k} t \cap I = \emptyset\right) < \eta.$$

Let $t \in G$. Since G is open, there exists n_t such that if $n \geq n_t$ and $|t - t'| < 1/M_n$, then $t' \in G$. Thus, $t \in G_n$ for all $n \geq n_t$ and so $G \subseteq \bigcup G_{N_k}$. Furthermore, $\mathbf{E}t \cap I = \emptyset$ only if $\mathbf{E}_{N_k} t \cap I = \emptyset$ for all k, hence

$$\mathbb{P}\left(\exists t \in G : \mathbf{E}t \cap I = \emptyset\right) \leq \sum_k \mathbb{P}\left(\exists t \in G_{N_k} : \mathbf{E}t \cap I = \emptyset\right)$$

$$\leq \sum_k \mathbb{P}\left(\exists t \in G_{N_k} : \mathbf{E}_{N_k} t \cap I = \emptyset\right) < \eta.$$

Since $\eta > 0$ was arbitrary, this proves that for every fixed δ,

$$\mathbb{P}\left(\exists t \in G = G(\delta) : \mathbf{E}t \cap I = \emptyset\right) = 0.$$

Recall that S is the set of irrationals in $\mathbb{T}$. If $t \in S$, then $\sin \pi jt \neq 0$ for every integer j. Thus, if δ_n is chosen tending to 0, then $S \subseteq \bigcup_{n=1}^{\infty} G(\delta_n)$. Hence, $\mathbb{P}\left(\exists t \in S : \mathbf{E}t \cap I = \emptyset\right) = 0$, as well, and therefore $\mathbf{E}t \cap I \neq \emptyset$ for all $t \in S$ with probability one.

This completes the proof for the case $N = 1$. The general case follows by similar reasoning.   □

8.4 Remarks and Credits

Theorem 8.2.1 is due to Ramsey [158]. The probabilistic results, Theorems 8.3.1 and 8.3.2, are due to Kahane and Katznelson [104], where more is proven: (i) Under the assumptions of Theorem 8.3.1, the closures of the random sets, $\mathbf{E}(\omega)$, are also shown to have $m_{\overline{\mathbb{Z}}}$-measure zero almost surely. (ii) Under the assumptions of 8.3.2, the $\mathbf{E}(\omega)$ are a.s. "analytic" (see [56]).

In [105], the random sets are shown to be a.s. uniformly distributed in $\overline{\mathbb{Z}}$ under slightly more restrictive hypotheses than that of Theorem 8.3.2. These results improve earlier results of Katznelson and Malliavin [109].

Exercise 8.5.8 can be found in [108, p. 22] and [7, p. 90].

8.5 Exercises

Exercise 8.5.1. Construct p_n such that $\limsup np_n = \infty$ and $E(\omega)$ is a.s. Sidon.

Exercise 8.5.2. Prove that if there is a k-independent set which clusters at an integer, then there is a k-independent set which is dense in $\overline{\mathbb{Z}}$.

Exercise 8.5.3. Suppose X_n are independent, Poisson random variables, with Poisson parameters p_n. Let $\mathbf{E}(\omega) = \{n : X_n(\omega) \geq 1\}$. Prove that if $\sum_{n=1}^{\infty} p_n < \infty$, then for each $q > 1$, $\mathbf{E}(\omega)$ is a.s. the union of a finite set with a Hadamard set of ratio q.

Exercise 8.5.4. Finish the explanation of how the proof of Theorem 8.3.1 can be completed once Lemma 8.3.5 is proven by showing that $0 \notin \overline{\mathbf{E}(\omega)}$ a.s. (see p. 142).

Exercise 8.5.5. Show that $\mathbf{E} \subseteq \mathbb{Z}$ is dense in $\overline{\mathbb{Z}}$ if and only if for all positive integers N and for all $t \in \mathbb{T}^N$ the closures of $\mathbf{E}t$ and $\mathbb{Z}t$ in $\mathbb{T}^N$ coincide.

Exercise 8.5.6. Prove the inequality stated in (8.3.1).

Exercise 8.5.7. Prove the generalized Hölder's inequality for $1 \leq p_j < \infty$: If $\sum_{j=1}^{J} 1/p_j = 1$, then

$$\left| \int \prod_{j=1}^{J} f_j \right| \leq \prod_{j=1}^{J} \left(\int |f_j|^{p_j} \right)^{1/p_j}.$$

Exercise 8.5.8. Show that $\sum_{n=1}^{N} (\sin 2\pi njt)/n$ is uniformly bounded over N and t.

Chapter 9
The Relationship Between Sidon and I_0

Sidon sets are proportional I_0. When Γ has few elements of order a power of two, Sidon sets are also proportional ε-Kronecker. A set satisfying a Pisier ε-net condition is Sidon. The Ramsey–Wells–Bourgain $B_d(\mathbf{E}) = B(\mathbf{E})$ characterization of I_0 is proved.

9.1 Introduction

In this chapter the relationship between Sidon and I_0 will be further explored. We begin with several results that provide evidence to support a positive answer to the question, "Is every Sidon set a finite union of I_0 sets?" [**P 1**], which is unknown even for $\mathbb{Z}$.

First, a combinatorial argument, due to Pajor and stated in the next section, is used in Sect. 9.2 to show that if a set satisfies a Pisier ε-net condition, then it is proportional I_0. Of course, proportional I_0 implies proportional Sidon, and Theorem 7.2.1 shows this implies Sidonicity. Since Proposition 7.3.2 says Sidon sets satisfy a Pisier ε-net condition, that gives a second set of equivalences to Sidonicity, in addition to those of Theorem 7.2.1.

Similar arguments are used in Sect. 9.3 to show that if Γ does not contain too many elements of order 2, then every Sidon set in Γ is characterized by the property that it is proportional ε-Kronecker for an $\varepsilon < \sqrt{2}$ depending only on the Sidon constant. An example is given to show that not all Sidon sets are proportional ε-Kronecker.

Lastly, in Sect. 9.4, we prove the Ramsey–Wells–Bourgain characterization of I_0 as those sets satisfying $B(\mathbf{E}) = B_d(\mathbf{E})$. In the proof of that result, the Pisier ε-net characterization of Sidon is used; that proof also introduces and uses sup-norm partitions.

C.C. Graham and K.E. Hare, *Interpolation and Sidon Sets for Compact Groups*,   153
CMS Books in Mathematics, DOI 10.1007/978-1-4614-5392-5_9,
© Springer Science+Business Media New York 2013

9.2 Sidon Sets Are Proportional I_0

Recall the notation $AP(\mathbf{E}, N, \varepsilon)$, introduced in Notation 3.2.7: $AP(\mathbf{E}, N, \varepsilon)$ is the set of all $\varphi \in \mathrm{Ball}(\ell^\infty(\mathbf{E}))$ for which there is a discrete measure μ, of length N, such that $|\varphi(\gamma) - \widehat{\mu}(\gamma)| \leq \varepsilon$ for all $\gamma \in \mathbf{E}$. Corollary 3.2.17 characterizes I_0 sets as those sets $\mathbf{E}$ having the property that for some N and $\varepsilon < 1$, $AP(\mathbf{E}, N, \varepsilon)$ contains $\mathbb{Z}_2^{\mathbf{E}}$.

Theorem 9.2.1. *Let $\mathbf{E} \subseteq \Gamma$ with $1 \notin \mathbf{E}$. The following are equivalent:*

1. *The set $\mathbf{E} \subseteq \Gamma$ is Sidon.*
2. *There exists $\varepsilon > 0$ such that $\mathbf{E}$ satisfies a Pisier ε-net condition: For every finite $\mathbf{F} \subseteq \mathbf{E}$, there is a set $Y \subseteq G$ such that $|Y| \geq 2^{\varepsilon|\mathbf{F}|}$ and*

$$\sup_{\gamma \in \mathbf{F}} |\gamma(x) - \gamma(y)| \geq \varepsilon \text{ for all } x \neq y \in Y. \tag{9.2.1}$$

3. *There exist constants $\delta > 0$ and $\varepsilon < 1$ such that each finite subset $\mathbf{F} \subseteq \mathbf{E}$ contains a finite subset $\mathbf{F}' \subseteq \mathbf{F}$ with $|\mathbf{F}'| \geq \delta|\mathbf{F}|$ and $AP(\mathbf{F}', 2, \varepsilon) \supseteq \mathbb{Z}_2^{\mathbf{F}'}$.*
4. *There exist constants C and $\delta > 0$ such that each finite subset $\mathbf{F} \subseteq \mathbf{E}$ contains a finite subset $\mathbf{F}' \subseteq \mathbf{F}$ whose I_0 constant is at most C and has $|\mathbf{F}'| \geq \delta|\mathbf{F}|$.*

Remark 9.2.2. Only $(2) \Rightarrow (3)$ is new to this theorem. The fact that Sidon sets satisfy (2) is the content of Proposition 7.3.2. That $(3) \Rightarrow (4)$ is a consequence of Corollary 3.2.17. The implication $(4) \Rightarrow (1)$ follows from the Pisier characterization Theorem 7.2.1 (5) since proportional I_0 sets are proportional Sidon and the Sidon constant is bounded by the I_0 constant.

In the proof of $(2) \Rightarrow (3)$ and for the characterization of Sidon sets as proportional ε-Kronecker, established in the next section, we will make use of the following combinatorial results. To state these, we introduce the notation $P_I : X^N \to X^I$ for the natural projection map, where X is any set and I is a subset of $\{1, \ldots, N\}$.

Proposition 9.2.3 (Pajor). *Let X be a finite set of integers. For each $\beta > 0$ there exists $\delta > 0$, depending only on β and the cardinality of X, such that for all $N \geq 1$ and $S \subseteq X^N$, with $|S| > 2^{\beta N}$, there are distinct integers $a, b \in X$ and $I \subseteq \{1, \ldots, N\}$ having the property that $|I| \geq \delta N$ and $\{a, b\}^I \subseteq P_I(S)$.*

Proposition 9.2.4 (Pajor). *Suppose $X = X^+ \cup X^-$ where $|X^+| = p \geq 1$ and $|X^-| = q \geq 1$. For $N \geq 1$ define a mapping $\pi : X^N \to \mathbb{Z}_2^N$ by*

$$\pi(x)_n = \begin{cases} +1 & \text{if } x_n \in X^+ \text{ and} \\ -1 & \text{if } x_n \in X^-, \end{cases}$$

for $x \in X^N$. There exists $\tau > 0$ (dependent on p and q, but not N) such that if $S \subseteq X^N$ has $\pi(S) = \mathbb{Z}_2^N$, then there exist $t \in X^+$, $u \in X^-$ and $I \subseteq \{1, \ldots, N\}$, with $|I| \geq \tau N$ and $P_I(S) \supseteq \{t, u\}^I$.

We also introduce further terminology used throughout this section and the next. We will say that two arcs, I_1, I_2, on the unit circle are *separated by a gap of length at least* β when the shorter of the other two arcs of the circle, complementary to the original pair, has length at least β. We call this shorter arc the *gap* between I_1, I_2 and write $\operatorname{gap}(I_1, I_2) = \beta$ for the length of that shorter arc. (If the two complementary arcs have the same length, either can be taken as the gap.)

Proof (of Theorem 9.2.1 (2) $\Rightarrow$ (3)). We begin by sketching the idea of the proof. So assume the existence of a Pisier ε-net, as in (9.2.1).

Let $\mathbf{F} = \{\gamma_j\}_1^J \subseteq \mathbf{E}$ be finite. Let $M > 4\pi/\varepsilon$ and partition $\mathbb{T}$ into M pairwise disjoint half-open arcs, $I_1, \ldots, I_M$, having a common length, $\lambda < \varepsilon$. Define $\mathrm{i} : \mathbb{T} \to \{1, \ldots, M\}$ by $\mathrm{i}(z) = m$ if $z \in I_m$, $1 \le m \le M$. Then each $y \in Y$ defines an element, $f(y)$, of $\{1, \ldots, M\}^{\mathbf{F}}$ by $f(y) = \big(\mathrm{i}(\gamma_1(y)), \ldots, \mathrm{i}(\gamma_J(y))\big)$. Because of (9.2.1), for every pair $x \ne y \in Y$, there exists a j such that $\mathrm{i}(\gamma_j(x)) \ne \mathrm{i}(\gamma_j(y))$, and therefore $|f(Y)| \ge 2^{\varepsilon|\mathbf{F}|}$. Applying Proposition 9.2.3, we find $\delta = \delta(\varepsilon)$ and $\mathbf{F}' \subseteq \mathbf{F}$ with $|\mathbf{F}'| \ge \delta|\mathbf{F}|$ and $1 \le a < b \le M$ with $\{a, b\}^{\mathbf{F}'} \subseteq P_{\mathbf{F}'}(f(Y))$.

Thus, we see that for every subset $\mathbf{A} \subseteq \mathbf{F}'$, there exists $g \in Y$ such that

$$\gamma(g) \in \begin{cases} I_a & \text{if } \gamma \in \mathbf{A} \text{ and} \\ I_b & \text{if } \gamma \in \mathbf{F}' \setminus \mathbf{A}. \end{cases}$$

Suppose these intervals, I_a, I_b, have a gap of at least β. We will see that then there is an angle t and constant $\sigma > 0$, depending only on β, such that $\nu = \frac{1}{2i}(e^{it}\delta_{g^{-1}} - e^{-it}\delta_g)$ has $\hat{\nu}(\gamma) > \sigma$ if $\gamma \in \mathbf{F}'$ and $\hat{\nu}(\gamma) < -\sigma$ if $\gamma \in \mathbf{F}' \setminus \mathbf{A}$. Thus, $\mathbf{F}'$ would satisfy Theorem 9.2.1(3).

Having given the general idea, we will next show how to complete the argument once given the existence of such intervals. Thus, we suppose that there are positive constants $0 < \beta, \delta, \lambda \le \pi/4$ (all depending only on ε) such that for each finite $\mathbf{F} \subseteq \mathbf{E}$ there are a set $\mathbf{F}' \subseteq \mathbf{F}$ and two arcs $I_1, I_2 \subseteq \mathbb{T}$ with

$$|\mathbf{F}'| \ge \delta|\mathbf{F}|, \quad \text{length } I_1 = \text{length } I_2 \le \lambda, \quad \operatorname{gap}(I_1, I_2) \ge \beta \tag{9.2.2}$$

and

$$\forall \, \mathbf{A} \subseteq \mathbf{F}' \; \exists g \in Y \text{ with } \; \gamma(g) \in \begin{cases} I_1 & \text{if } \gamma \in \mathbf{A} \text{ and} \\ I_2 & \text{if } \gamma \in \mathbf{F}' \setminus \mathbf{A}. \end{cases} \tag{9.2.3}$$

Lemma 9.2.5 will prove that Theorem 9.2.1(2) implies this set-up.

We now apply (9.2.2)–(9.2.3), assuming Lemma 9.2.5. Given $\varphi \in \mathbb{Z}_2^{\mathbf{F}'}$, put $\mathbf{A} = \{\gamma \in \mathbf{F}' : \varphi(\gamma) = 1\}$ and obtain $g \in G$ such that if μ is the point mass measure at g^{-1}, then $\hat{\mu}(\gamma) = \gamma(g) \in I_1$ if $\gamma \in \mathbf{A}$ and $\hat{\mu}(\gamma) = \gamma(g) \in I_2$ if $\gamma \in \mathbf{F}' \setminus \mathbf{A}$. By multiplying these arcs by a suitable choice e^{it} and replacing μ by $e^{it}\mu$, we can assume the gap separating I_1 and I_2 is centred at 1 and that there is a discrete, length one measure $\mu_{\mathbf{A}}$ with

$$\widehat{\mu_{\mathbf{A}}}(\gamma) \in \begin{cases} I_1 & \text{if } \gamma \in \mathbf{A} \text{ and} \\ I_2 & \text{if } \gamma \in \mathbf{F}' \setminus \mathbf{A}. \end{cases}$$

Since the gap centred at 1 is the smaller of the two separating arcs, without loss of generality (replacing μ by $-\mu$, if necessary), I_1 intersects quadrant 1 and I_2 quadrant 4. Because length I_1 = length $I_2 \leq \pi/4$ and gap$(I_1, I_2) \geq \beta$, whenever $e^{i\theta} \in I_1$ we have $\Im me^{i\theta} \geq \sin(\beta/2) =: \sigma > 0$, while if $e^{i\theta} \in I_2$, then $\Im me^{i\theta} \leq -\sin(\beta/2) = -\sigma < 0$. Hence,

$$\Im\widehat{\mu_{\mathbf{A}}}(\gamma) \in \begin{cases} [\sigma, 1] & \text{if } \gamma \in \mathbf{A} \text{ and} \\ [-1, -\sigma] & \text{if } \gamma \in \mathbf{F}' \smallsetminus \mathbf{A}. \end{cases}$$

Let $\nu = (\mu_{\mathbf{A}} - \widehat{\mu_A})/2i$. Then ν is a discrete measure of length two with $\widehat{\nu}(\gamma) = \Im\widehat{\mu_A}(\gamma)$, and

$$|\varphi(\gamma) - \widehat{\nu}(\gamma)| \leq 1 - \sigma \text{ for all } \gamma \in \mathbf{F}'.$$

This shows $\varphi \in AP(\mathbf{F}', 2, 1-\sigma)$ for all $\varphi \in \mathbb{Z}_2^{\mathbf{F}'}$. Because both σ and δ depend only on ε, the proof now needs only the following lemma. $\qquad\square$

Lemma 9.2.5. *Let* $\mathbf{F} \subseteq \Gamma$ *be a finite set. Assume* $Y \subseteq G$ *has cardinality at least* $2^{\varepsilon|\mathbf{F}|}$ *and* $\sup_{\gamma \in \mathbf{F}} |\gamma(x) - \gamma(y)| \geq \varepsilon$ *for all* $x \neq y \in Y$. *Then there exist constants* $0 < \beta, \delta, \lambda \leq \pi/4$, *depending only on* ε, *a subset* $\mathbf{F}'$ *of* $\mathbf{F}$ *and two arcs,* $I_1, I_2 \subseteq \mathbb{T}$ *having the same length and satisfying* (9.2.2)–(9.2.3).

Proof. Without loss of generality $\varepsilon \leq \pi/8$. Choose any positive integer L such that $2\pi/L < \varepsilon/2$. Put $\lambda' = 2\pi/L$ and $\beta' \leq \lambda'/M$ where M is an integer satisfying $M^{-1} \leq 1 - 2^{-\varepsilon/2}$. The quantities β, λ and δ will be defined later.

Partition $\mathbb{T}$ into L disjoint arcs:

$$T_\ell = \{e^{i\theta} : \ell\lambda' \leq \theta < (\ell+1)\lambda'\}, \quad 0 \leq \ell < L,$$

and further partition each T_ℓ into M disjoint arcs:

$$U_{\ell,m} = \{e^{i\theta} : \ell\lambda' + m\beta' \leq \theta < \ell\lambda' + (m+1)\beta'\}, 0 \leq m < M.$$

Let $\mathbf{F} = \{\gamma_j\}_{j=1}^J$ and put $Y_0 = Y$. For $j = 1, \ldots, J$, define Y_j inductively, as follows. For $0 \leq \ell < L$ and $0 \leq m < M$, let

$$Y_\ell^j = \{g \in Y_{j-1} : \gamma_j(g) \in T_\ell\} \text{ and } Y_{\ell,m}^j = \{g \in Y_{j-1} : \gamma_j(g) \in U_{\ell,m}\}, \text{ so}$$

$$Y_{j-1} = \bigcup_{\ell=0}^{L-1} Y_\ell^j \text{ and } Y_\ell^j = \bigcup_{m=0}^{M-1} Y_{\ell,m}^j.$$

For each pair j, ℓ, pick the index $m = m(j, \ell)$ for which the cardinality of $Y_{\ell,m}^j$ is minimal. Clearly, $|Y_{\ell,m}^j| \leq M^{-1}|Y_\ell^j|$, and therefore

$$\left|\bigcup_{\ell=0}^{L-1} Y_{\ell,m(j,\ell)}^j\right| \leq M^{-1}|Y_{j-1}|.$$

Finally, put $Y_j = Y_{j-1} \setminus \bigcup_{\ell=0}^{L-1} Y_{\ell,m(j,\ell)}^j$. Then $|Y_j| \geq (1 - M^{-1})|Y_{j-1}|$ for all j, and therefore $|Y_J| \geq (1 - M^{-1})^J |Y_0|$. The choice of M implies that $|Y_J| \geq 2^{-J\varepsilon/2}|Y_0| \geq 2^{J\varepsilon/2}$.

For each pair (j, ℓ), let $I_{j,\ell}$ be the arc between $U_{\ell-1,m(j,\ell-1)}$ and $U_{\ell,m(j,\ell)}$. Since $U_{\ell,m} \subseteq T_\ell$, length $I_{j,\ell} \leq 2\lambda' - 2\beta' < \varepsilon$. Some of the arcs $I_{j,\ell}$ may be empty, but that will not matter.

For $g \in Y_J$, define $h_g \in \ell^\infty(\mathbf{F})$ by $h_g(\gamma_j) = \ell_j$ where $\gamma_j(g) \in I_{j,\ell_j}$. The construction of Y_J guarantees that these functions are all distinct, and hence there are at least $2^{J\varepsilon/2}$ such functions, a lower bound on $|Y_J|$.

We now appeal to the combinatorial Proposition 9.2.3, with $X = \{0, 1, \ldots, L - 1\}$ and $S = \{h_g : \mathbf{F} \to X : g \in Y_J\} \subseteq X^{\mathbf{F}}$. There are a constant $\delta > 0$ depending only on ε (and L, but L depends on ε), indices a, b with $0 \leq a < b < L$ and a subset $\mathbf{F}' \subseteq \mathbf{F}$ with $|\mathbf{F}'| \geq \delta|\mathbf{F}|$ such that $\{a, b\}^{\mathbf{F}'} \subseteq \{h_g|_{\mathbf{F}'} : g \in Y_J\}$. That identifies two arcs, $I_{j,a}$ and $I_{j,b}$, for each index $j = 1, \ldots, J$, such that

$$\mathbf{A} \subseteq \mathbf{F}' \Rightarrow \exists g \in Y_J \text{ with } \gamma_j(g) \in \begin{cases} I_{j,a} & \text{if } \gamma_j \in \mathbf{A} \text{ and} \\ I_{j,b} & \text{if } \gamma_j \in \mathbf{F}' \setminus \mathbf{A}. \end{cases}$$

Case I: Suppose first that $|b - a| \geq 2 \bmod L$. Note that

$$I_1 := \{e^{i\theta} : (a - 1)\lambda' + \beta' \leq \theta < (a + 1)\lambda' - \beta'\} \supseteq I_{j,a} \text{ and}$$
$$I_2 := \{e^{i\theta} : (b - 1)\lambda' + \beta' \leq \theta < (b + 1)\lambda' - \beta'\} \supseteq I_{j,b}.$$

The two arcs I_1, I_2 are separated by a gap of size at least $(b - a - 2)\lambda' + 2\beta' \geq 2\beta'$ and have length $2\lambda' - 2\beta'$. Moreover, for some $g \in Y_0$, $\gamma(g)$ belongs to I_1 if $\gamma \in \mathbf{A}$ and belongs to I_2 if $\gamma \in \mathbf{F}' \setminus \mathbf{A}$.
Case II: If $|b - a| = 1 \bmod L$, then for suitable y_j,

$$I_{j,a} \subseteq [(a - 1)\lambda' + \beta', y_j] \text{ and } I_{j,b} \subseteq [y_j + \beta', (b + 1)\lambda' - \beta'].$$

In Case II, put

$$I_1 := [-3\lambda' + 2\beta', -\beta'] \supseteq I_{j,a} - I_{j,b} \text{ and } I_2 := [\beta', 3\lambda' - 2\beta'] \supseteq I_{j,b} - I_{j,a}.$$

The arcs I_1, I_2 are of length $3\lambda' - 3\beta'$ and are separated by at least $2\beta'$. By choosing appropriate $g_1, g_2 \in Y_0$ and putting $g = g_1 g_2^{-1} \in Y_0 Y_0^{-1}$, we have $\gamma(g) \in I_1$ if $\gamma \in \mathbf{A}$ and $\gamma(g) \in I_2$ if $\gamma \in \mathbf{F}' \setminus \mathbf{A}$.

In either case, the construction produces two arcs, I_1, I_2, having the required properties, with common length at most $\lambda = 3\lambda' \leq 3\pi/16 \leq \pi/4$ and separated by a gap of length at least $\beta = 2\beta'$. $\qquad \square$

Corollary 9.2.6. *Let $\mathbf{E} \subseteq \Gamma$ with $1 \notin \mathbf{E}$. The following are equivalent:*

1. The set $\mathbf{E}$ is Sidon.

2. *There exists $\delta > 0$ such that each finite subset $\mathbf{F} \subset \mathbf{E}$ has a subset $\mathbf{F}' = \{\gamma_1, \ldots, \gamma_N\}$ with $N \geq \delta |\mathbf{F}|$ and the property that whenever $c_n \in \Delta$ for $1 \leq n \leq N$ and $0 < \tau < 1$, then*

$$m_G\{x \in G : \inf_n |\mathfrak{Re}(c_n\gamma_n(x))| > \tau\} \leq 2(1+\tau^2)^{-N}.$$

Proof. (1) $\Rightarrow$ (2) is immediate from Theorem 7.2.1 (1) $\Rightarrow$ (3) and Lemma 7.3.1.

(2) $\Rightarrow$ (1). The proof of Proposition 7.3.2 shows that $\mathbf{E}$ satisfies a Pisier ε-net condition. An application of Theorem 9.2.1 completes the proof. $\quad\square$

9.3 Sidon Sets Are Proportional ε-Kronecker

Given that Sidon sets are characterized as both proportional quasi-independent and proportional I_0, it is natural to ask whether they may also be characterized as proportional ε-Kronecker, meaning that there exists $\tau > 0$ and $\varepsilon < \sqrt{2}$ such that for every finite $\mathbf{F} \subseteq \mathbf{E}$ there is some ε-Kronecker set $\mathbf{F}' \subseteq \mathbf{F}$ with $|\mathbf{F}'| \geq \tau|\mathbf{F}|$.

Of course, if $\mathbf{E}$ is the Rademacher set in $\widehat{\mathbb{D}}$, then $\mathbf{E}$ is I_0 but not proportional ε-Kronecker for any $\varepsilon < \sqrt{2}$. The example below shows that a set $\mathbf{E}$ may have no elements of order 2, be I_0, but not be proportional ε-Kronecker. Theorem 9.3.2 gives conditions (involving elements of order 2 and powers thereof) under which that characterization does hold.

Example 9.3.1. An I_0 set that is not proportional ε-Kronecker for any $\varepsilon < \sqrt{2}$: We use the I_0 set of Example 4.3.4, $\mathbf{E} = \{(j, \pi_j) : j \in \mathbb{N}\} \subseteq \mathbb{Z} \oplus \widehat{\mathbb{D}}$ where $\{\pi_j\}$ is the Rademacher set in $\widehat{\mathbb{D}}$.

Suppose $\mathbf{E}$ were proportional angular $(\pi/2 - \varepsilon)$-Kronecker, for some $\varepsilon > 0$, with proportionality constant τ. For M even and sufficiently large, we select a set, $\{w_1, \ldots, w_M\}$, of Mth roots of unity that is ε-dense in $\mathbb{T}$. For each m, let p_m be any Mth root of w_m. Put $N = M^3$. Szemerédi's theorem [49, 182] says there exists $L = L(\tau, N)$ such that any subset of $\{1, \ldots, L\}$ with density at least τ contains an arithmetic progression of length N. By the proportionality assumption, the set $\{(\ell, \pi_\ell) : \ell = 1, \ldots, L\}$ contains an angular $(\pi/2 - \varepsilon)$-Kronecker subset $\{(\ell, \pi_\ell) : \ell \in \mathbf{F}_L\}$, where $\mathbf{F}_L$ is a subset of $\{1, \ldots, L\}$ of density τ, and hence contains a further subset $\mathbf{F}' = \{(jk_0 + d_0, \pi_{jk_0+d_0}) : j = 1, \ldots, N\}$.

Define numbers $t_i^{(m)} \in \mathbb{T}$ recursively, as follows: Let $t_1^{(1)} = 1$. Define

$$t_{i+1}^{(m)} = t_i^{(m)}/(p_m w_n) \text{ for } i = (n-1)M + 1, \ldots, nM \text{ and } m, n = 1, \ldots, M.$$

Then set $t_1^{(m+1)} = t_{M^2+1}^{(m)}$. We think of these as ordered by fixing m and ordering in index i and then letting m increase. Denote this ordered set by $\{z_r\}$.

Suppose $\varphi : \mathbf{F}' \to \mathbb{T}$ satisfies $\varphi(rk_0 + d_0, \pi_{rk_0+d_0}) = z_r$. Since $\mathbf{F}'$ is angular $(\pi/2 - \varepsilon)$-Kronecker, there exist a point (x, y) and error term ε_r such that $\varepsilon_r \le \pi/2 - \varepsilon$ so that

$$z_r = e^{i(rk_0+d_0)x} \pi_{rk_0+d_0}(y) e^{i\varepsilon_r}.$$

Because $\pi_m(y) = \pm 1$ for all m, y, there are $s_r = \pm 1$ with

$$\frac{z_r}{z_{r+1}} = s_r e^{-ik_0 x} e^{i(\varepsilon_r - \varepsilon_{r+1})}.$$

Let I be the union of the two subintervals of $\mathbb{T}$ of length 2ε, which are angular distance at least $\pi/2 - \varepsilon$ away from both $\pm e^{-ik_0 x}$. For each m, the set $\{p_m w_n : n = 1, \ldots, M\}$ is ε-dense, and hence there is some $n = n(m)$ such that $p_m w_n \in I$. That compels all the pairs $(t_i^{(m)}, t_{i+1}^{(m)})$ with ratio $p_m w_n$ (and there are M such pairs, those with $i = (n-1)M + 1, \ldots, nM$) to have error terms opposite in sign. Thus, $\varepsilon_i^{(m)}$ alternates in sign as i varies. But

$$\frac{t_{(n-1)M+1}^{(m)}}{t_{nM+1}^{(m)}} = (p_m w_n)^M = w_m = s'_{m,n} e^{-iMk_0 x} e^{i(\varepsilon'_{m,n})}, \tag{9.3.1}$$

where $s'_{m,n} = \pm 1$ and $\varepsilon'_{m,n} = \varepsilon_{(n-1)M+1}^{(m)} - \varepsilon_{nM+1}^{(m)}$. Since M is even, $\varepsilon_{(n-1)M+1}^{(m)}$ and $\varepsilon_{nM+1}^{(m)}$ are the same sign, so $|\varepsilon'_{m,n}| \le \pi/2 - \varepsilon$.

Because the set $\{w_m\}$ is ε-dense, at least one w_m belongs to I and so has angular distance at least $\pi/2 - \varepsilon$ from both of $\pm e^{-iMk_0 x}$. That contradicts (9.3.1) and proves that $\mathbf{E}$ is not proportional angular $(\pi/2 - \varepsilon)$-Kronecker.

However, under a suitable restriction on the order two elements, one can deduce that Sidon sets are proportional ε-Kronecker.

Let $\mathbf{\Gamma}^{(2)}$ be the set of characters in $\mathbf{\Gamma}$ whose orders are powers of 2.

Theorem 9.3.2. *Suppose $\mathbf{\Gamma}^{(2)}$ is finite and $\mathbf{E} \subseteq \mathbf{\Gamma} \smallsetminus \{1\}$ has no elements of order two. Then $\mathbf{E}$ is Sidon if and only if there are constants $\tau > 0$ and $\varepsilon < \sqrt{2}$ such that if $\mathbf{F} \subseteq \mathbf{E}$ is finite, then $\mathbf{F}$ contains a subset $\mathbf{F}'$ which is ε-Kronecker and satisfies $|\mathbf{F}'| \ge \tau |\mathbf{F}|$.*

Proof. Since an ε-Kronecker set with $\varepsilon < \sqrt{2}$ is Sidon with Sidon constant bounded by a function of ε, proportional ε-Kronecker implies proportional Sidon, which implies Sidon by Pisier's Theorem 7.2.1(5).

For the other direction, we assume that $\mathbf{E}$ is Sidon. Since $\mathbf{\Gamma}^{(2)}$ is finite, finitely many translates of its annihilator, G_2, cover G. An application of the ε-net property (Corollary 7.3.3) yields constants $\delta = \delta(\mathbf{E}) \le 1/4$ and

$\alpha = \alpha(\mathbf{E}, G_2)$ with the property that for every finite $\mathbf{F} \subseteq \mathbf{E}$ with $|\mathbf{F}| \geq \alpha$, there is a set $Y = \{g_j\} \subseteq G_2$ with $|Y| \geq 2^{\delta|\mathbf{F}|}$ and satisfying

$$\sup_{\gamma \in \mathbf{F}} |\gamma(g_j) - \gamma(g_i)| \geq \delta \text{ for } i \neq j.$$

Given such a finite set $\mathbf{F}$, apply Lemma 9.2.5 with the subset $Y \subseteq G_2$ to obtain $\mathbf{F}' \subseteq \mathbf{F}$, with $|\mathbf{F}'| \geq \tau|\mathbf{F}|$ and two intervals I_1, I_2 of equal length at most λ and separated by a gap of length at least β (with τ, λ and β positive constants depending only on $\mathbf{E}$) having the property that for all $\mathbf{A} \subseteq \mathbf{F}'$ there is some $g \in Y$ such that $\gamma(g) \in I_1$ for $g \in \mathbf{A}$ and $\gamma(g) \in I_2$ for $\gamma \in \mathbf{F}' \setminus \mathbf{A}$. We reduce the set of g slightly by putting

$$Y' = \{g \in Y : \gamma(g) \in I_1 \cup I_2 \text{ for all } \gamma \in \mathbf{F}'\}.$$

Partition each of I_1 and I_2 into s disjoint subintervals, $I_1', \ldots, I_s'$ and $I_{s+1}', \ldots, I_{2s}'$, respectively, having equal lengths at most $\min(\pi/32, \beta/4)$. Let $X^+ = \{1, \ldots, s\}$, $X^- = \{s+1, \ldots, 2s\}$ and $X = X^+ \cup X^-$.

View Y' as a subset of $X^{\mathbf{F}'}$ by identifying $g \in Y'$ with $(z_\gamma^{(g)})_{\gamma \in \mathbf{F}'}$ according to the rule $\gamma(g) \in I_{z_\gamma^{(g)}}$ for $\gamma \in \mathbf{F}'$. Define $\pi : X^{\mathbf{F}'} \to \{-1, 1\}^{\mathbf{F}'}$ by

$$\pi(z_\gamma^{(g)}) = (r_\gamma)_{\gamma \in \mathbf{F}'} \text{ where } r_\gamma = \begin{cases} +1 \text{ if } z_\gamma \in X^+, \\ -1 \text{ if } z_\gamma \in X^-. \end{cases}$$

(Equivalently, $r_\gamma = 1$ if $\gamma(g) \in I_1$ and $r_\gamma = -1$ if $\gamma(g) \in I_2$.) By taking suitable choices of g we can obtain all elements of $\{\pm1\}^{\mathbf{F}'}$, and hence $\pi(S) = \{\pm1\}^{\mathbf{F}'}$.

Appealing to Proposition 9.2.4, it follows that there exist $1 \leq t \leq s < u \leq 2s$ and $\tau_1 > 0$, depending only on δ (which in turn depends only on $\mathbf{E}$), and $\mathbf{F}_1' \subseteq \mathbf{F}'$ with $|\mathbf{F}_1'| \geq \tau_1|\mathbf{F}'|$ such that $\{t, u\}^{\mathbf{F}_1'} \subseteq P_{\mathbf{F}_1'}(S)$ (where $P_{\mathbf{F}_1'}(f) = f|_{\mathbf{F}_1'}$). This means for every $\mathbf{A} \subseteq \mathbf{F}_1'$, there exists $g \in Y'$ with $\gamma(g) \in I_t'$ if $\gamma \in \mathbf{A}$ and $\gamma(g) \in I_u'$ for $\gamma \in \mathbf{F}_1' \setminus \mathbf{A}$.

To summarize, the intervals I_t' and I_u' have length $\rho \leq \min(\pi/32, \beta/4)$ and are separated by a gap of length at least $\beta \geq 4\rho$, and both β and ρ depend only on $\mathbf{E}$. Also, $|\mathbf{F}_1'| \geq \tau\tau_1|\mathbf{F}|$ and this (new) proportionality constant $\tau_0 = \tau\tau_1$ depends only on $\mathbf{E}$.

If the gap length, β, exceeds $5\pi/8$, Corollary 2.5.5 implies $\mathbf{F}_1'$ is weak angular $3\pi/8$-Kronecker. Otherwise, there is some $k \geq 1$ such that

$$\beta \in \frac{\pi}{4} \left[\frac{1}{k+1} + \frac{1}{k}, \frac{1}{k} + \frac{1}{k-1} \right)$$

(or $[3\pi/8, 5\pi/8]$ if $k = 1$). The two gaps between the intervals $(k+1)I_t'$ and $(k+1)I_u'$ have lengths at least $(k+1)\beta$ and $2\pi - 2(k+1)\rho - (k+1)\beta$. It is an easy calculation to verify the smaller gap has length at least $\pi/2 + \varepsilon_k$ for some $\varepsilon_k > 0$.

If $g \in G_2$ has the property that $\gamma(g) \in I'_t$ if $\gamma \in \mathbf{A}$ and $\gamma(g) \in I'_u$ for $\gamma \in \mathbf{F}'_1 \smallsetminus \mathbf{A}$, then g^{k+1}, which is still in the subgroup G_2, satisfies $\gamma(g^{k+1}) \in (k+1)I'_t$ if $\gamma \in \mathbf{A}$ and $\gamma(g^{k+1}) \in (k+1)I'_u$ for $\gamma \in \mathbf{F}'_1 \smallsetminus \mathbf{A}$. Again, Corollary 2.5.5 implies $\mathbf{F}'_1$ is weak angular $\pi/2 - \varepsilon_k$-Kronecker, and hence ε-Kronecker for some $\varepsilon < \sqrt{2}$.

Finally, suppose $|\mathbf{F}| < \alpha$. Since $\mathbf{E}$ contains no elements of order two, a singleton subset of $\mathbf{E}$ is weak 1-Kronecker. Thus, simply take $\mathbf{F}'$ to be any (one) element from $\mathbf{F}$ to obtain a weak 1-Kronecker subset of size at least $|\mathbf{F}|/\alpha$. This concludes the proof that $\mathbf{E}$ is proportional ε-Kronecker for some $\varepsilon < \sqrt{2}$. $\qquad\square$

Corollary 9.3.3. *Suppose that $\mathbf{E} \subseteq \Gamma \smallsetminus \{1\}$ has no elements of order two and that the subgroup generated by $\mathbf{E}$ contains only finitely many elements of order a power of 2. Then $\mathbf{E}$ is proportional ε-Kronecker for some $\varepsilon < \sqrt{2}$ if and only if $\mathbf{E}$ is Sidon.*

Proof. Apply Theorem 9.3.2 with Γ the subgroup generated by $\mathbf{E}$ and G its dual. $\qquad\square$

9.4 The Ramsey–Wells–Bourgain Characterization of I_0 Sets

In Theorems 9.2.1 and 9.3.2 we saw that Pisier's ε-net condition implies that Sidon sets could be characterized as those which were proportional I_0 or ϵ-Kronecker. In this section, the ε-net condition will be used to prove the Ramsey–Wells–Bourgain characterization of I_0 sets as those sets $\mathbf{E}$ for which $B(\mathbf{E}) = B_d(\mathbf{E})$ (Theorem 9.4.15). Of course, every I_0 set has this property since then $\ell^\infty(\mathbf{E}) = B_d(\mathbf{E})$. The other direction is the interesting and difficult part of the theorem.

There are two main steps. First, we show that the hypothesis implies that $\mathbf{E}$, if not Sidon, would have a non-Sidon subset, $\mathbf{E}'$, with the property that for every $\varepsilon > 0$ there is an integer $M \geq 1$ and $c = (c_1, \ldots, c_M) \in \Delta^M$ such that for each $\varphi \in \mathrm{Ball}(B(\mathbf{E}'))$ there exist $x_1, \ldots, x_M \in G$ with $\|\varphi - \sum_{m=1}^M c_m \widehat{\delta_{x_m}}\|_{B(\mathbf{E}')} \leq \varepsilon$. That is done in Lemma 9.4.13 and involves sup-norm partitions.

Second, in Lemma 9.4.14, it will be shown that if $\mathbf{E}$ satisfies the conditions of Lemma 9.4.13, then $\mathbf{E}$ satisfies a Pisier ε-net condition for a suitable choice of ε. This second step uses some simple ideas from metric entropy. Theorem 9.4.15 will then follow easily from the main steps.

We note that when $B_d(\mathbf{E}) = B(\mathbf{E})$, the norms are equivalent, but the $B_d(\mathbf{E})$ norm may be larger. We will do our calculations using that larger norm.

Before those main steps, we prove a number of technical lemmas. The first is somewhat surprising.

Lemma 9.4.1 (Varopoulos's Lemma). *For every finite $Y \subseteq G$ and $\gamma \in \Gamma$,*

$$\|1 - \gamma\|_{A(Y)} \leq \pi(e-1)\|1 - \gamma\|_{\ell^\infty(Y)}. \tag{9.4.1}$$

Remark 9.4.2. It is obvious that $\|\widehat{\delta_\gamma} - \widehat{\delta_\rho}\|_{\ell^\infty(Y)} \leq \|\widehat{\delta_\gamma} - \widehat{\delta_\rho}\|_{A(Y)}$, so Lemma 9.4.1 says the two norms are equivalent for those differences.

Proof (of Lemma 9.4.1). Since Y is finite, $B_d(Y) = B(Y) = A(Y)$ and $A(Y)^*$ consists of the bounded functions on Γ with Fourier transforms supported on Y, that is the Y-polynomials. Thus,

$$\|1 - \gamma\|_{A(Y)} = \sup\{|\widehat{\nu}(1) - \widehat{\nu}(\gamma)| : \nu \in M(Y), \|\widehat{\nu}\|_\infty \leq 1\}.$$

Because $\|1 - \gamma\|_{A(Y)} \leq 2$, (9.4.1) trivially holds if $\|1 - \gamma\|_{\ell^\infty(Y)} \geq 2/\pi$. So we may assume $\varepsilon := \|1 - \gamma\|_{\ell^\infty(Y)} \leq 2/\pi$. We will be replacing γ with $e^{i\theta}$. Exercise 9.6.6 (3) says there exists $f \in A(\mathbb{T})$ such that $f(e^{i\theta}) = 1 - e^{i\theta}$ for $|\theta| \leq \delta \leq 1$ and $\|f\|_{A([-\delta,\delta])} < 2(e-1)\delta$. Thus, $f(e^{i\theta}) = \sum_m c_m e^{im\theta}$, with $\sum_m |c_m| \leq 2(e-1)\delta$.
Since $|1 - e^{i\theta}| \leq \varepsilon$ implies $|\theta| \leq \varepsilon\pi/2$, we see that

$$\|1 - e^{i\theta}\|_{A([-\varepsilon\pi/2, \varepsilon\pi/2])} \leq \pi(e-1)\varepsilon.$$

Now put $g(x) = 1 - \overline{\gamma}(x) - f(\overline{\gamma}(x))$, $x \in G$. Then $g \in A(G)$ and if $x \in Y$ then $g(x) = 0$.
Suppose $\nu \in M(Y)$ with $\|\widehat{\nu}\|_\infty = 1$. These observations imply

$$0 = \int g d\nu = \widehat{\nu}(1) - \widehat{\nu}(\gamma) - \int f \circ \overline{\gamma} d\nu,$$

and so

$$|\widehat{\nu}(1) - \widehat{\nu}(\gamma)| = \left|\int f \circ \overline{\gamma} d\nu\right| \leq \|\widehat{\nu}\|_\infty \sum_m |c_m|$$
$$\leq \pi(e-1)\varepsilon \leq \pi(e-1)\|1 - \gamma\|_\infty. \qquad \square$$

Lemma 9.4.3. *Let $\varepsilon > 0$ and assume $\mathbf{F}$ is a finite subset of Γ. Then there exists a compact e-neighbourhood $U \subseteq G$ such that, for all $\gamma \in \mathbf{F}$,*

$$\|\widehat{\delta_1} - \widehat{\delta_\gamma}\|_{A(U)} \leq \varepsilon. \tag{9.4.2}$$

Proof. Choose an e-neighbourhood U_1 such that

$$|\widehat{\delta_1} - \widehat{\delta_\gamma}(x)| < \varepsilon/(2\pi(e-1)) \text{ for } x \in U_1 \text{ and } \gamma \in \mathbf{F}.$$

We shall find an e-neighbourhood $U \subseteq U_1$ such that (9.4.2) holds. By Lemma 9.4.1, $\|1 - \widehat{\delta_\gamma}\|_{A(Y)} < \varepsilon/2$ for all finite sets $Y \subseteq U_1$. For each such Y, let

$\mu_Y \in M(\Gamma)$ be such that $\widehat{\mu_Y} = 1 - \widehat{\delta_\gamma}$ on Y and $\|\mu_Y\|_{M(\Gamma)} < \varepsilon/2$. Note that $\{\mu_Y : Y \subset U_1$ is finite$\}$ is a bounded net in $M(\overline{\Gamma})$. Let μ be any weak* cluster point of that net. Then

$$\|\mu\|_{M(\overline{\Gamma})} \leq \varepsilon/2 \quad \text{and} \quad \widehat{\mu} = 1 - \widehat{\delta_\gamma} \text{ on } U_1.$$

By the corollary to the local units theorem, Corollary C.1.7, we may choose $h \in A(G)$ such that $h = 1$ in an e-neighbourhood U, Supp $h \subseteq U_1$ and $\|h\|_{A(G)} < 2$. Thus, $h\widehat{\mu} = 1 - \widehat{\delta_\gamma}$ on U and $\|h\widehat{\mu}\|_{B(G_d)} < \varepsilon$. We note that $h\widehat{\mu}$ is continuous on G. Since the subalgebra $M(\Gamma)$ is characterized by the Bochner–Eberlein Theorem C.1.8 as the measures in $M(\overline{\Gamma})$ whose Fourier-Stieltjes transforms are continuous on G, $h\widehat{\mu} \in B(G) = A(G)$. Hence, $\|1 - \widehat{\delta_\gamma}\|_{A(U)} = \|h\widehat{\mu}\|_{A(U)} < \varepsilon$. $\qquad\square$

Corollary 9.4.4. *Let* $\mathbf{F} \subseteq \Gamma$ *be finite and* $\varepsilon > 0$. *There exists a compact e-neighbourhood* $U \subseteq G$ *such that if* $\lambda\gamma^{-1} \in \mathbf{F}$, *then*

$$|\widehat{\mu}(\gamma) - \widehat{\mu}(\lambda)| < \varepsilon\|\widehat{\mu}\|_\infty \text{ for all } \mu \in M(U).$$

Proof. Let U be given by Lemma 9.4.3. The conclusion follows from the estimate

$$|\widehat{\mu}(\gamma) - \widehat{\mu}(\lambda)| \leq \|\widehat{\mu}\|_\infty \|\widehat{\delta_\gamma} - \widehat{\delta_\lambda}\|_{A(U)} = \|\widehat{\mu}\|_\infty \|\widehat{\delta_1} - \widehat{\delta_{\lambda\gamma^{-1}}}\|_{A(U)},$$

where the first inequality comes from duality. $\qquad\square$

9.4.1 Sup-Norm Partitions

Definition 9.4.5. A decomposition $\mathbf{E} = \bigcup_{j=1}^{\infty} \mathbf{E}_j$ is a *sup-norm partition* of $\mathbf{E} \subseteq \Gamma$ if the sets $\mathbf{E}_j$ are finite and there is a constant C such that for all $\varphi \in B(\mathbf{E})$

$$\|\varphi\|_{B(\mathbf{E})} \leq C \sup_j \|\varphi|_{\mathbf{E}_j}\|_{B(\mathbf{E}_j)}. \tag{9.4.3}$$

Equivalently, whenever $J \geq 1$ and $\mu_j \in M(\mathbf{E}_j)$ for $1 \leq j \leq J < \infty$, we have

$$\sum_{j=1}^{J} \|\widehat{\mu_j}\|_\infty \leq C \|\sum_{j=1}^{J} \widehat{\mu_j}\|_\infty. \tag{9.4.4}$$

To prove the $B_d(\mathbf{E}) = B(\mathbf{E})$ theorem, we will need to find sup-norm partitioned subsets of a given set. We now develop tools to do that.

A collection of trigonometric polynomials, $p_1, \ldots, p_J$, defined on G, will be called *ε-additive* if for each choice of $t_1, \ldots, t_J \in G$, there exists $t \in G$ such that

$$\left| \sum_{j=1}^{J} p_j(t) - \sum_{j=1}^{J} p_j(t_j) \right| \leq \frac{\varepsilon}{\pi} \sum_{j=1}^{J} \|p_j\|_\infty.$$

This property is useful because it ensures that the norm of the sum of the polynomials is comparable to the sum of their norms. It can be thought of as an "individualistic" version of sup-norm partitioning.

Lemma 9.4.6. *Suppose* $p_1, \ldots, p_J$ *are* ε-*additive polynomials with Fourier transforms disjointly supported. If* $\widehat{p}_j(1) = 0$, *then*

$$\sum_{j=1}^{J} \|p_j\| \leq \frac{\pi}{1-\varepsilon} \left\| \sum_{j=1}^{J} p_j \right\|.$$

Proof. An elementary fact is that any finite set of complex numbers, X, contains a subset, Y, with $\sum_{z \in X} |z| \leq \pi |\sum_{z \in Y} z|$; see Exercise 9.6.5.

Temporarily fix $\delta > 0$ and choose $u_j \in G$ such that for each j, $|p_j(u_j)| \geq (1 - \delta) \|p_j\|$. With $X = \{p_j(u_j) : j = 1, \ldots, J\}$, choose $Y \subseteq \{1, \ldots, J\}$ such that

$$\left| \sum_{j \in Y} p_j(u_j) \right| \geq \frac{1}{\pi} \sum_{j=1}^{J} |p_j(u_j)| \geq \frac{1-\delta}{\pi} \sum_{j=1}^{J} \|p_j\|_\infty. \tag{9.4.5}$$

Put $t_j = u_j$ if $j \in Y$.

The task is to suitably redefine u_j for $j \notin Y$ so that $\left| \sum_{j=1}^{J} p_j(u_j) \right|$ is still large. For this we write $\sum_{j \in Y} p_j(u_j) = r e^{i\theta}$, where $\theta \in [0, 2\pi)$ and $r \geq 0$. The inequality (9.4.5) shows

$$r \geq \frac{1-\delta}{\pi} \sum_{j=1}^{J} \|p_j\|_\infty. \tag{9.4.6}$$

Because $0 = e^{-i\theta} \widehat{p}_j(1)$ is the average value of $\mathfrak{Re}(e^{-i\theta} p_j)$, it cannot be the case that for all $x \in G$, $\mathfrak{Re}(e^{-i\theta} p_j(x)) < -\|p_j\|_\infty \delta/\pi$. Hence, for each j, there is some $v_j \in G$ with

$$\mathfrak{Re}(e^{-i\theta} p_j(v_j)) \geq -\|p_j\|_\infty \delta/\pi. \tag{9.4.7}$$

Define $t_j = v_j$ if $j \notin Y$. Obtain $t \in G$ from the definition of ε-additive for this choice of $t_1, \ldots, t_J$. Then

$$\left\| \sum_{j=1}^{J} p_j \right\|_\infty \geq \left| \sum_{j=1}^{J} p_j(t) \right| \geq \left| \sum_{j=1}^{J} p_j(t_j) \right| - \frac{\varepsilon}{\pi} \sum_{j=1}^{J} \|p_j\|_\infty.$$

Inequalities (9.4.6) and (9.4.7) give

$$\left| \sum_{j=1}^{J} p_j(t_j) \right| = \left| e^{i\theta} \left(r + e^{-i\theta} \sum_{j \notin Y} p_j(t_j) \right) \right|$$

$$\geq \mathfrak{Re} \left(r + e^{-i\theta} \sum_{j \notin Y} p_j(t_j) \right) \geq \frac{1 - 2\delta}{\pi} \sum_{j=1}^{J} \|p_j\|_\infty.$$

Thus, $\pi\left\|\sum_{j=1}^J p_j\right\|_\infty \geq (1 - 2\delta - \varepsilon)\sum_{j=1}^J \|p_j\|_\infty$. Since $\delta > 0$ was arbitrary, the proof is complete. □

Proposition 9.4.7. *Let* $\mathbf{E} \subseteq \boldsymbol{\Gamma}$ *be a non-Sidon set. Then* $\mathbf{E}$ *contains a countable non-Sidon set* $\mathbf{F}$, *which can be sup-norm partitioned.*

The proof of Proposition 9.4.7 is not direct; we first prove a version for non-discrete groups. For that, we will need the norm relationships of functions and measures on Helson sets (defined in Remark 3.5.5). Those are given in Remarks 9.4.8.

Sup-Norm Partitions and Countable Non-Helson Sets

Remarks 9.4.8. (i) An easy application of the closed graph theorem shows that X is a Helson subset of the group H if and only if there is a constant $C > 0$ such that for every $\varphi \in C(X)$ there exists $\nu \in M(\widehat{H})$ such that $\widehat{\nu} = \varphi$ and $\|\nu\| \leq C\|\varphi\|_\infty$. And that holds if and only if for every $\mu \in M(X)$ we have $\|\widehat{\mu}\|_\infty \leq C\|\mu\|$ (same C). The infimum of such C is called the *Helson constant* of E. A duality argument shows that the dual of $A(\mathbf{E})$ is $M(\mathbf{E})$ when $\mathbf{E}$ is Helson.

(ii) When $\mathbf{E}$ is a finite or countable, compact subset of $\boldsymbol{\Lambda}$, then $\mathbf{E}$ is Helson if and only if $\mathbf{E}$ is Sidon in $\boldsymbol{\Lambda}$ when $\boldsymbol{\Lambda}$ is given the discrete topology. In that case, the Sidon and Helson constants agree. It is evident that a countable subset $\mathbf{E}$ is *non-Sidon* (or *non-Helson* if it is a compact subset of a non-discrete group) if for each $C > 0$ there exists a finitely supported measure $\mu \in M_d(\mathbf{E})$ such that $\|\mu\|_{M(E)} \geq C\|\widehat{\mu}\|_\infty$. Equivalently, there are finite subsets $\mathbf{E}_n \subset \mathbf{E}$ such that the Sidon constants satisfy $S(\mathbf{E}_n) \geq n$ for $n \geq 1$.

(iii) If $\mathbf{E}$ is a finite subset of the discrete group $\boldsymbol{\Gamma}$, then each $f \in B_d(\mathbf{E})$ can be represented using Fourier-Stieltjes transforms of discrete measures concentrated on any dense subset, H, of G. In this case,

$$\|f\|_{B_d(\mathbf{E})} = \inf\{\|\mu\| : \mu \in M_d(H), \widehat{\mu}|_\mathbf{E} = f\}.$$

That is because each character (i.e., $\widehat{\delta_x}$ for $x \in G$) on $\mathbf{E}$ may be approximated uniformly by a character $(\widehat{\delta_{x_\alpha}})$ from H. Consequently, the Sidon constant can be calculated using the discrete measures from H.

Here is the non-discrete group version of Proposition 9.4.7.

Proposition 9.4.9. *Let* G *be compact and metrizable and* $T \subseteq G$ *be non-Helson, compact and countable. Then* T *contains a compact, non-Helson subset,* Y, *such that* Y *has a unique cluster point* x_0 *and* $Y \smallsetminus \{x_0\}$ *has a sup-norm partition.*

We begin the proof of Proposition 9.4.9 with several preliminary results.

Lemma 9.4.10. *Let* $\mu \in M(G)$ *have finite support,* T. *Then* $\|\mu\|_{M(G)} \leq |T| \|\widehat{\mu}\|_{\ell^\infty(\mathbf{\Gamma})}$.

Proof. It will suffice to show that $|\mu(\{x\})| \leq \|\widehat{\mu}\|_\infty$ for all $x \in G$. Because μ is finitely supported, $\widehat{\mu}$ extends to a trigonometric polynomial on $\overline{\mathbf{\Gamma}}$, say

$$p(\gamma) = \sum_{x_j \in T} a_j \widehat{\delta_{x_j}}(\gamma), \ \gamma \in \overline{\mathbf{\Gamma}}.$$

Since the coefficients of a trigonometric polynomial on a compact abelian group are dominated by the polynomial's supremum, the lemma follows. □

Lemma 9.4.11. *Let* $T \subseteq G$ *be finite and* $\varepsilon > 0$. *Then there exists a finite set* $\mathbf{F} \subseteq \mathbf{\Gamma}$ *such that for every* $\gamma \in \mathbf{\Gamma}$, *each translate of* $\mathbf{F}$ *contains* $\lambda \in \mathbf{\Gamma}$ *such that*

$$|\widehat{\mu}(\lambda) - \widehat{\mu}(\gamma)| \leq \varepsilon \|\widehat{\mu}\|_\infty \text{ for all } \mu \in M(T).$$

Proof (of Lemma 9.4.11). By Lemma 9.4.10, $\|\mu\|_{M(G)} \leq |T| \|\widehat{\mu}\|_\infty$ for $\mu \in M(T)$. Hence, for any $\lambda \in \mathbf{\Gamma}$,

$$|\widehat{\mu}(\gamma) - \widehat{\mu}(\lambda)| = \left| \sum_{t \in T} \mu(\{t\})(\gamma(t) - \lambda(t)) \right| \leq |T| \|\widehat{\mu}\|_\infty \sup_{t \in T} |1 - (\lambda\gamma^{-1})(t)|.$$

Let $\mathbf{L} = \{\tau \in \overline{\mathbf{\Gamma}} : \sup_{t \in T} |1 - \tau(t)| < \varepsilon/|T|\}$. Then $\mathbf{L}$ is an open set in $\overline{\mathbf{\Gamma}}$, so $\overline{\mathbf{\Gamma}} = \bigcup_{\rho \in \mathbf{\Gamma}} \rho\mathbf{L}$. The compactness of $\overline{\mathbf{\Gamma}}$ implies there is a finite set $\rho_1, \ldots, \rho_J \in \mathbf{\Gamma}$, with $\bigcup_j \rho_j \mathbf{L} = \overline{\mathbf{\Gamma}}$, and so $\bigcup_j \rho_j(\mathbf{L} \cap \mathbf{\Gamma}) = \mathbf{\Gamma}$. Of course, $\bigcup_j \lambda\rho_j(\mathbf{L} \cap \mathbf{\Gamma}) = \mathbf{\Gamma}$ for each $\lambda \in \mathbf{\Gamma}$.

Put $\mathbf{F} = \{\rho_j\}_{j=1}^J$. For any $\gamma, \omega \in \mathbf{\Gamma}$, there is some j such that $\gamma \in \omega\rho_j\mathbf{L}$. Put $\lambda = \omega\rho_j$. Then $\lambda \in \omega\mathbf{F}$ and $\lambda\gamma^{-1} \in \mathbf{L}$. Hence, $|1 - \lambda\gamma^{-1}(t)| < \varepsilon/|T|$, and the lemma follows. □

Lemma 9.4.12. *Let* G *be a compact, metrizable abelian group and* $T \subseteq G$ *be a countable set which satisfies*

$$\sup\{S(T_1) : T_1 \subset T, T_1 \text{ finite}\} = \infty. \tag{9.4.8}$$

Then T *contains a convergent subsequence,* T', *which satisfies* (9.4.8) *with* T' *in place of* T.

Here, $S(T_1)$ is the Sidon constant of T_1.

Proof (of Lemma 9.4.12). We say an $x \in G$ is a *non-Helson point for* T if every neighbourhood U of x has the property that (9.4.8) holds with T replaced with $U \cap T$.

Claim. G has a non-Helson point for T.

Assuming the claim, let x be a non-Helson point for T and let $\{U_n\}$ be a neighbourhood base at x. Then for each n there exists a finite subset $T_n \subset T \cap U_n$ such that the Helson (= Sidon) constant $S(T_n) \geq n$. Let $T' = \bigcup_1^\infty T_n$. Clearly, T' is non-Sidon and T' accumulates only at x.

We now prove the claim. If the claim were false, there would exist (by the compactness of G) a finite number of open sets U_n and constant C such that

$$\sup\{S(T_1) : T_1 \subset T \cap U_n, T_1 \text{ finite}\} \leq C \qquad (9.4.9)$$

and $\bigcup_1^N U_n = G$.

Suppose $T' \subset T$ is finite. Then $S(T' \cap U_n) \leq C$ for $1 \leq n \leq N$. That means that each $T \cap U_n$ is a Sidon set in the discrete version, G_d, of G. By repeated application of Drury's union Theorem 6.3.1 for Sidon sets, $\bigcup_{n=1}^N T \cap U_n = T$ is a Sidon set in G_d. Then $S(T') \leq S(T, G_d)$ for all finite $T' \subset T$, where $S(T, G_d)$ is the Sidon constant in G_d. Since the Sidon (or Helson) constant of a finite set is independent of the topology of the group (Remarks 9.4.8), (9.4.8) cannot hold, a contradiction. □

Proofs of Propositions 9.4.9 and 9.4.7

Proof (of Proposition 9.4.9). By Lemma 9.4.12 we may assume that T has only one accumulation point, (by translating) that the accumulation point is the identity, e, and that $e \notin T$. Thus, if U is any e-neighbourhood, then $U \cap T$ contains all but finitely many points of T and hence cannot be a Helson set.

Choose a finite set $T_1 \subset T$ with $S(T_1) > 8$. By Lemma 9.4.11 there exists a finite, symmetric $\mathbf{F}_1 \subset \mathbf{\Gamma}$ which satisfies Lemma 9.4.11 with $T = T_1$ and $\varepsilon = 1/8$.

Let $\mathbf{F}_0 = \emptyset$. Suppose $J \geq 1$ and that for each $1 \leq j \leq J$ pairwise disjoint, finite sets $T_j \subset T$ and finite sets $\mathbf{F}_j \subset \mathbf{\Gamma}$ have been found such that (a) $S(T_j) \geq 8^j$, (b) $\mathbf{F}_j \supseteq \mathbf{F}_{j-1}$ and (c) $\mathbf{F}_j$ satisfy Lemma 9.4.11 with $T = \bigcup_{\ell=1}^j T_\ell$ and $\varepsilon = 8^{-j}$.

Apply Corollary 9.4.4 with $\mathbf{F} = \mathbf{F}_1 \cdot \mathbf{F}_2 \cdots \mathbf{F}_J$ and $\varepsilon = 8^{-J-1}$ to find a compact e-neighbourhood U. We may assume that $U \cap \bigcup_{j=1}^J T_j = \emptyset$. Choose a finite set $T_{J+1} \subset T \cap U$ with $S(T_{J+1}) \geq 8^{J+1}$. Finally, apply Lemma 9.4.11 to find $\mathbf{F}_{J+1}$. That completes our induction.

We claim that $\bigcup_{j=1}^\infty T_j$ is sup-norm partitioned. Of course $\{e\} \cup \bigcup_{j=1}^\infty T_j$ is a compact non-Helson set. To establish the sup-norm partitioned property, it will suffice to show that if measures $\nu_j \in M(T_j)$ are given, then the trigonometric polynomials $p_j = \hat{\nu}_j$ are $\pi/4$-additive. We may apply Lemma 9.4.6 since the identity is not in any of the T_j.

Fix $1 < J < \infty$. Let $\nu_j \in M(T_j)$ and $\gamma_j \in \mathbf{\Gamma}$ for $1 \leq j \leq J$. Set $\rho_J = \gamma_J$. By the Lemma 9.4.11, there exists ρ_{J-1} such that

$$|\widehat{\nu_{J-1}}(\rho_{J-1}) - \widehat{\nu_{J-1}}(\gamma_{J-1})| \leq 8^{-J+1}\|\widehat{\nu_{J-1}}\|_\infty \text{ and } \rho_{J-1}\rho_J^{-1} \in \mathbf{F}_{J-1}.$$

Inducting down, we find $\rho_{J-2}, \ldots, \rho_1$ such that for $1 < j \leq J$

$$|\widehat{\nu_{j-1}}(\rho_{j-1}) - \widehat{\nu_{j-1}}(\gamma_{j-1})| \leq 8^{-j+1} \|\widehat{\nu_{j-1}}\|_\infty \text{ and } \rho_{j-1}\rho_j^{-1} \in \mathbf{F}_{j-1}.$$

Then the conditions on the $\rho_j\rho_{j-1}^{-1}$ show that $\rho_1\rho_j^{-1} \in \mathbf{F}_1 \cdot \mathbf{F}_2 \cdots \mathbf{F}_{j-1}$. Since $\nu_j \in M(T_j)$, Corollary 9.4.4 gives

$$|\widehat{\nu_j}(\rho_1) - \widehat{\nu_j}(\rho_j)| \leq 8^{-j} \|\widehat{\nu_j}\|_\infty, \text{ and so}$$

$$|\widehat{\nu_j}(\rho_1) - \widehat{\nu_j}(\gamma_j)| \leq 2 \cdot 8^{-j} \|\widehat{\nu_j}\|_\infty.$$

Therefore,

$$\left| \sum_1^J \widehat{\nu_j}(\rho_1) - \sum_1^J \widehat{\nu_j}(\gamma_j) \right| \leq 2 \sum_{j=1}^J 8^{-j} \|\widehat{\nu_j}\|_\infty \leq \frac{1}{4} \sum_{j=1}^J \|\widehat{\nu_j}\|_\infty.$$

Thus, the $\widehat{\nu_j}$ are $\pi/4$-additive, completing the proof of Proposition 9.4.9. $\square$

Proof (of Proposition 9.4.7). We note first that we may assume that $\mathbf{E}$ and $\mathbf{\Gamma}$ are countable. Therefore, G will have a dense, countable subgroup H (Exercise C.4.18 (4)). We give H the discrete topology and let $\mathbf{\Lambda} = \widehat{H} = \overline{\mathbf{\Gamma}}/H^\perp$, the compact, metrizable dual group. Because H is dense in G, the natural embedding $i : \mathbf{\Gamma} \hookrightarrow \mathbf{\Lambda}$ is one-to-one into a dense subgroup of $\mathbf{\Lambda}$. If $\mathbf{F} \subseteq \mathbf{\Gamma}$ is finite, then, by Exercise 6.6.14, $S(\mathbf{F}) = S(i(\mathbf{F}))$. Here, $S(i(\mathbf{F}))$ is either the Helson constant of $i(\mathbf{F})$ in $\mathbf{\Lambda}$ or the Sidon constant of $i(\mathbf{F})$ in $\mathbf{\Lambda}$ with the discrete topology.

By Lemma 9.4.12 $i(\mathbf{E})$ has a countable subset that is a convergent sequence, whose closure is non-Helson. We may assume $i(\mathbf{E})$ is that sequence. By Proposition 9.4.9, $i(\mathbf{E})$ without its limit point has a sup-norm partitioned, non-Sidon (considered as a subset of the discrete $\mathbf{\Lambda}_d$) subset which we can write as $i(\mathbf{E}')$ for some $\mathbf{E}' \subset \mathbf{E}$. Since $i(\mathbf{E}')$ is non-Sidon, $\mathbf{E}'$ is non-Sidon. An easy computation shows that $\mathbf{E}'$ is sup-norm partitioned. $\square$

9.4.2 Completing the Proof of the $B_d(E) = B(E)$ Characterization of I_0 Sets

We have two more lemmas, these specific to the $B_d(\mathbf{E}) = B(\mathbf{E})$ context.

Lemma 9.4.13. *If $B_d(\mathbf{E}) = B(\mathbf{E})$ and $\mathbf{E} = \bigcup_{j=1}^\infty \mathbf{E}_j$ are a sup-norm partition, then for every $\varepsilon > 0$ there exist $M = M(\varepsilon)$, $c_1, \ldots, c_M \in \Delta$ and a finite $J \geq 1$ such that if $\varphi \in B_d(\mathbf{E})$ has norm 1, then there exist $x_1, \ldots, x_M \in G$ such that*

$$\left\| \varphi - \sum_{m=1}^M c_m \widehat{\delta_{x_m}} \right\|_{B_d(\bigcup_{j>J}^\infty \mathbf{E}_j)} \leq \varepsilon. \tag{9.4.10}$$

Proof. Let $C > 0$ be the product of $\sup\{\|\varphi\|_{B_d(\mathbf{E})} : \|\varphi\|_{B(\mathbf{E})} \le 1\}$ and the constant, C', for the sup-norm partition ($\|\varphi\|_{B(\mathbf{E})} \le C' \sup_j \|\varphi\|_{B(\mathbf{E}_j)}$ for all $\varphi \in B(\mathbf{E})$).

Since the $\mathbf{E}_j$ are finite, each $\mathrm{Ball}(B_d(\mathbf{E}_j))$ is isometrically isomorphic to $\mathrm{Ball}(B(\mathbf{E}_j))$ (Exercise 9.6.2) and is compact. Thus, we may include or omit the subscript d on the $B(\mathbf{E}_j)$, as we wish.

Let $X = \prod_{j=1}^{\infty} \mathrm{Ball}(B_d(\mathbf{E}_j))$. Then X is compact in the product topology. If $\varphi \in \mathrm{Ball}(B_d(\mathbf{E}))$, then we may identify φ with $(\varphi|_{\mathbf{E}_j}) \in \prod_j \mathrm{Ball}(B_d(\mathbf{E}_j))$. Also, any $(\varphi_j) \in \prod_j \mathrm{Ball}(B_d(\mathbf{E}_j))$ may be identified with $\varphi \in C\,\mathrm{Ball}(B_d(\mathbf{E}))$ where $\varphi|_{\mathbf{E}_j} = \varphi_j$. That gives $\mathrm{Ball}(B_d(\mathbf{E})) \subseteq X \subseteq C\,\mathrm{Ball}(B_d(\mathbf{E}))$. Of course, for φ and $\varphi_\alpha \in \mathrm{Ball}(B_d(\mathbf{E}))$,

$$\|\varphi_\alpha - \varphi\|_{B_d(\mathbf{E})} \to 0 \iff \sup_j \|(\varphi - \varphi_\alpha)|_{\mathbf{E}_j}\|_{B_d(\mathbf{E}_j)} \to 0, \text{ and} \quad (9.4.11)$$

$$\varphi_\alpha \to \varphi \text{ weak*} \iff \|(\varphi - \varphi_\alpha)|_{\mathbf{E}_j}\|_{B_d(\mathbf{E}_j)} \to 0 \text{ for all } j. \quad (9.4.12)$$

For each $M \ge 1$, let D_M be a countable, dense subset of Δ^M. For $M \ge 1$ and $c = (c_1, \ldots, c_M) \in D_M$, set

$$Y(M,c) = \left\{ \varphi \in X : \exists x_m \in G \text{ with } \sup_j \left\| \varphi - \sum_{m=1}^{M} c_m \widehat{\delta_{x_m}} \right\|_{B(\mathbf{E}_j)} \le \frac{\varepsilon}{C} \right\}.$$

Then $\bigcup_{M \ge 1,\, c \in D_M} Y(M,c) = X$. Because G and X are compact, (9.4.11) implies each $Y(M,c)$ is closed in X. The Baire category theorem and the definition of the product topology say at least one $Y(M,c)$ has non-empty interior in X in the product topology on X, that is, there exist an integer J and $\psi \in \prod_{j=1}^{J} \mathrm{Ball}(B(\mathbf{E}_j))$ such that $\{\psi\} \times \prod_{j>J} \mathrm{Ball}(B(\mathbf{E}_j)) \subset Y(M,c)$; see p. 209. Therefore, the lemma holds with M, J and $c = (c_1, \ldots, c_M)$. $\square$

Lemma 9.4.14. *Let $\mathbf{E} = \bigcup_j \mathbf{E}_j$ be sup-norm partitioned and $B_d(\mathbf{E}) = B(\mathbf{E})$. Then a cofinite subset of $\mathbf{E}$ satisfies a Pisier ε-net condition for some $0 < \varepsilon < 1$.*

Proof. Apply Lemma 9.4.13 to find J, $M < \infty$ and $c \in \Delta^M$ such that for every $\varphi \in B_d(\bigcup_{j>J} \mathbf{E}_j)$ there are $x_m \in G$ with

$$\left\| \sum_{1}^{M} c_m \widehat{\delta_{x_m}} - \varphi \right\|_{B_d(\mathbf{E}')} \le 1/8,$$

for $\mathbf{E}' = \bigcup_{j>J} \mathbf{E}_j$.

We will show that $\varepsilon = 1/(1000M)$ will do. By Exercise 9.6.8 each $\mathbf{F} \subset \Gamma$ with $|\mathbf{F}| \le M+1$ satisfies a Pisier $1/(M+1)$-net condition. We therefore may assume that $|\mathbf{F}| > M+1$.

Let $\mathbf{F} \subset \mathbf{E}'$ be finite and non-empty. Use Exercise 9.6.9 with $\varepsilon = 1/4$ to find $L \ge 4^{|\mathbf{F}|} = 2^{2|\mathbf{F}|}$ elements, f_ℓ, of $\mathrm{Ball}(B_d(\mathbf{F}))$ such that

$$\|f_j - f_\ell\|_{B(\mathbf{F})} \ge \frac{1}{4} \text{ for } 1 \le j \ne \ell \ne L. \quad (9.4.13)$$

Let $\mu_j = \sum_{m=1}^{M} c_m \delta_{x_{j,m}}$ be measures with $\|\widehat{\mu_j} - f_j\|_{B_d(\mathbf{F})} \leq 1/64$ for all j. Let $\{x_k : 1 \leq k \leq K\} \subseteq G$ be a subset of maximal cardinality such that

$$\|\widehat{\delta_{x_\ell}} - \widehat{\delta_{x_k}}\|_{\ell^\infty(\mathbf{F})} \geq \frac{1}{1000M} \text{ for } 1 \leq k \neq \ell \leq K. \qquad (9.4.14)$$

Let $U_k = \{x \in G : \|\widehat{\delta_x} - \widehat{\delta_{x_k}}\|_{\ell^\infty(\mathbf{F})} < 1/(1000M)\}$ for $1 \leq k \leq K$. Then the union of the sets U_k covers G. We claim that

$$K \geq 2^{2|\mathbf{F}|/(M+1)} \geq 2^{|\mathbf{F}|/(1000M)}.$$

Suppose not. By the Pigeon hole principle, there exists $k(1)$ such that $x_{j,1} \in U_{k(1)}$ for at least

$$L/K \geq 2^{2|\mathbf{F}|(1-1/(M+1))}$$

of the $x_{j,1}$. Let $\mathcal{J}(1)$ denote that set of j's, so $|\mathcal{J}(1)| \geq 2^{2|\mathbf{F}|(1-1/(M+1))}$. Then there exists an integer $k(2)$ such that for at least

$$2^{2|\mathbf{F}|(1-\frac{2}{M+1})}$$

elements $j \in \mathcal{J}(1)$ we have $x_{j,2} \in U_{k(2)}$. Proceeding inductively, we see that there exists a set $\mathcal{J}(M)$ and $k(1), \ldots, k(M)$ such that $x_{j,m} \in U_{k(m)}$ for all $j \in \mathcal{J}(M)$ and $1 \leq m \leq M$ and

$$|\mathcal{J}(M)| \geq 2^{2|\mathbf{F}|(1-\frac{M}{M+1})} = 2^{2|\mathbf{F}|/(M+1)} \geq 4$$

since $|\mathbf{F}| \geq M + 1$. We note that for $j, \ell \in \mathcal{J}(M)$, Lemma 9.4.1 implies that

$$\|\widehat{\mu_j} - \widehat{\mu_\ell}\|_{B(\mathbf{F})} \leq \sum_{m=1}^{M} |c_m| \|\widehat{\delta_{x_{j,m}}} - \widehat{\delta_{x_{\ell,m}}}\|_{B(\mathbf{F})}$$

$$\leq \pi(e-1) \sum_{m=1}^{M} |c_m| \|\widehat{\delta_{x_{j,m}}} - \widehat{\delta_{x_{\ell,m}}}\|_{\ell^\infty(\mathbf{F})}$$

$$< \frac{\pi(e-1)2M}{1000M} \leq \frac{1}{64},$$

since $x_{j,m}, x_{\ell,m} \in U_{k(M)}$. Since $|\mathcal{J}(M)| \geq 4$ we may find $j \neq \ell \in \mathcal{J}(M)$. Then the above shows that

$$\|f_j - f_\ell\|_{B(\mathbf{F})} \leq \|f_j - \widehat{\mu_j}\|_{B(\mathbf{F})} + \|\widehat{\mu_j} - \widehat{\mu_\ell}\|_{B(\mathbf{F})} + \|\widehat{\mu_\ell} - f_\ell\|_{B(\mathbf{F})}$$
$$\leq 1/64 + 1/64 + 1/64 < 1/4,$$

a contradiction of (9.4.13). This establishes the claim. Simple calculations now show the required conclusion holds. □

Here is the Ramsey–Wells–Bourgain theorem.

Theorem 9.4.15 (Ramsey–Wells–Bourgain). *A subset* $\mathbf{E}$ *of* Γ *is* I_0 *if and only if* $B_d(\mathbf{E}) = B(\mathbf{E})$.

Proof (of Theorem 9.4.15). Suppose $\mathbf{E} \subseteq \Gamma$ with $B_d(\mathbf{E}) = B(\mathbf{E})$ (the other direction being trivial). It will suffice to show that $\mathbf{E}$ is Sidon since then $\ell^\infty(\mathbf{E}) = B_d(\mathbf{E})$. We proceed by contradiction. We may assume that $1 \notin \mathbf{E}$ and that $\mathbf{E}$ is infinite.

If $\mathbf{E}$ were not Sidon, apply Proposition 9.4.7 to obtain a non-Sidon $\mathbf{E}' = \bigcup_{j=1}^{\infty} \mathbf{E}_j \subseteq \mathbf{E}$ which is sup-norm partitioned and satisfies $B_d(\mathbf{E}') = B(\mathbf{E}')$. Lemma 9.4.14 and Theorem 9.2.1 (2) show that $\bigcup_{j \geq J} \mathbf{E}_j$ is Sidon for suitable J. Because $\bigcup_{j < J} \mathbf{E}_j$ is a finite set, it too is Sidon, and hence the union of those two last sets, $\mathbf{E}'$, must have been Sidon. That contradiction proves the theorem. $\square$

9.5 Remarks and Credits

Sidon Sets Are Proportional I_0 or ϵ-Kronecker. Theorem 9.2.1 is due to Ramsey [158]; the proof given here follows his closely. The material in Sect. 9.3 is from [55]. In [55] it is also shown that Sidon sets are characterized by the property of being proportional RI_0. A different proof that Pisier's ε-net condition implies Sidonicity can be found in [119, Theorem 13.V.5].

Propositions 9.2.3 and 9.2.4 were announced by Pajor without proof in [140, p. 742]. Proofs can be found in [141, pp. 143–4] and [142, pp. 69–74]. A proof of Proposition 9.2.4 is also given in [119, pp. 418–420].

Characterization of I_0 Sets as $B_d(\mathbf{E}) = B(\mathbf{E})$. The Ramsey–Wells–Bourgain Theorem 9.4.15 appeared in [18], answering a question implicit in [50]. Ramsey and Wells earlier proved Theorem 9.4.15 for subsets of groups of bounded order and for subsets of $\mathbf{E}_1 \cdot \mathbf{E}_2$ where $\mathbf{E}_1, \mathbf{E}_2$ are disjoint and $\mathbf{E}_1 \cup \mathbf{E}_2$ is ε-Kronecker [159].

Lemma 9.4.1 appears, with proof for $G = \mathbb{T}$, in Varopoulos's [186]; variants can also be found in [13, 125], both of which reference [186]. The proof here uses the main idea of [186], while avoiding the complication of applying Bernstein's lemma to pseudomeasures on $\mathbb{T}$. Varopoulos gives no explicit constant. Bourgain stated Lemma 9.4.1 in [18] and gave a proof in [20]. A longer proof [119, p. 555] of Lemma 9.4.1, attributed to Rodríguez–Piazza, gives a better constant.

The propositions on sup-norm partitions are adapted from [13]. Lemma 9.4.3 is essentially the assertion that singletons are sets of spectral synthesis; see, for example, [167, Theorem 7.6.2(a)] or [88, Lemma 39.27]) for other proofs of that fact. Lemma 9.4.12 is a version (due to Blei [13]) of Kahane's [100, Theorem 1]; Lemma 9.4.13 is adapted from one due to Ramsey [18].

Exercises. Exercise 9.6.2 is from [65] and [158, Theorem 1]. Exercise 9.6.5 is a special case of the Kaufman–Rickert inequality [110] or [56, 12.2.3]. Other proofs are given in [2] and Asmar and Montgomery–Smith [4]. Exercise 9.6.10 is [13, Lemma 1.7]. Exercise 9.6.9 is standard. For more on the subject of metric entropy, to which it belongs, see [1], for example.

9.6 Exercises

Exercise 9.6.1. Show that $A(Y)$ is isomorphic (but not necessarily isometrically) to $B(Y)$ for all compact subsets of the locally compact abelian group G.

Exercise 9.6.2. Show that if $\mathbf{F}$ is a finite set, then $B(\mathbf{F})$ and $B_d(\mathbf{F})$ are isometrically isomorphic.

Exercise 9.6.3. For $\mathbf{F} \subseteq \Gamma$, let $\varsigma(\mathbf{F})$ denote the minimum integer N such that given any $\mathbf{A} \subseteq \mathbf{F}$, there is a measure $\mu \in M_d(G)$ of length at most N, such that $\mathfrak{Im}\widehat{\mu}(\gamma) \geq 1/2$ if $\gamma \in \mathbf{A}$ and $\mathfrak{Im}\widehat{\mu}(\gamma) \leq -1/2$ if $\gamma \in \mathbf{F} \setminus \mathbf{A}$.

1. Show that $\mathbf{E}$ is I_0 if and only if $\zeta(\mathbf{E}) < \infty$.
2. Find a bound on the I_0 constant of $\mathbf{E}$ if $\varsigma(\mathbf{E}) \leq N$.
3. Show that the set $\mathbf{E} \subseteq \Gamma$ is Sidon if and only if there exists $\tau > 0$ and integer N such that each finite subset $\mathbf{F} \subseteq \mathbf{E}$ contains a further finite subset $\mathbf{F}' \subseteq \mathbf{F}$ with $|\mathbf{F}'| \geq \tau |\mathbf{F}|$ and $\varsigma(\mathbf{F}') \leq N$.

Exercise 9.6.4. Call a subset of $\mathbb{N}$ "proportional Hadamard" if all finite subsets contain proportional-sized subsets that are Hadamard with ratios bounded away from one. Show that Sidon sets in $\mathbb{N}$, and even ε-Kronecker sets, need not be proportional Hadamard.

Exercise 9.6.5. Show that any finite set of complex numbers, X, contains a subset Y with $\sum_{z \in X} |z| \leq \pi |\sum_{z \in Y} z|$.

Exercise 9.6.6.   1. Show that $f = (1/2h)1_{[-h,h]} * 1_{[-h,h]}$ has $A(\mathbb{T})$-norm 1 and that f has graph the triangle of width $4h$ and height 1.
2. Use two triangles to find $f \in A(\mathbb{T})$ of norm at most 3, support in $[-2h, 2h]$ and equal to 1 on $[-h, h]$.
3.   (i) Use two triangles to show that $f(x) = x$ on $[-h, h]$ has $A([-h, h])$-norm at most $2h$.
    (ii) Use those same two triangles to show that $f(x) = x^n$ on $[-h, h]$ has $A([-h, h])$-norm at most $2h^n$.
    (iii) Use power series to show that $\|1 - e^{ix}\|_{A([-h,h])} \leq 2(e^h - 1) \leq 2(e - 1)h$.
4.   (i) Show that the "valley-high" function f, supported on $[-2h, 2h]$, has norm at most $\sqrt{2}h$. Here, $f(-2h) = f(0) = f(2h) = 0$, $f(-h) = -h$, $f(h) = h$ and f is linear in the gaps. Hint: Write f as a convolution of two L^2 functions.
    (ii) Improve the estimate of $3c$ to $(\sqrt{2} + 2(e - 2))h$.

Exercise 9.6.7. Let $T \subseteq G$ be finite and $\varepsilon > 0$. Show that there exists a finite subset $\mathbf{F} \subset \boldsymbol{\Gamma}$ such that for every $\rho \in \boldsymbol{\Gamma}$ there exists $\gamma \in \mathbf{F}$ such that $\|\gamma - \rho\|_{\ell^\infty(T)} < \varepsilon$.

Exercise 9.6.8. Fix $1 \leq M < \infty$. Show that every finite $\mathbf{F} \subset \boldsymbol{\Gamma} \smallsetminus \{\mathbf{1}\}$ with $|\mathbf{F}| \leq M + 1$ satisfies a Pisier $1/(M+1)$-net condition.

Exercise 9.6.9. If X is an N-dimensional Banach space, then for every $\varepsilon > 0$ there exist at least $J = \lceil 1/\varepsilon^N \rceil$ elements $x_j \in \mathrm{Ball}(X)$ such that $\|x_j - x_k\|_X \geq \varepsilon$ for all $1 \leq j \neq k \leq J$.

Exercise 9.6.10. Prove Lemma 9.4.12 using Varopoulos's Theorem 10.3.4.

Chapter 10
Sets of Zero Discrete Harmonic Density

Sets with zdhd and zhd are defined. Finite unions of I_0 sets have zdhd. A "Hadamard gap" theorem holds for sets with zhd.

10.1 Introduction

Two important themes have motivated much of the research on Sidon and related special sets: determining which classes of special sets have the property that every Sidon set is a finite union of sets from the class and understanding the "size" of Sidon sets. Much progress has been made on these themes, as discussed in Chaps. 6–9. Two specific problems which remain outstanding are:

1. Is every Sidon set a finite union of I_0 sets? [**P 1**]
2. Can a Sidon set be dense in $\overline{\Gamma}$? [**P 2**]

These questions are not independent. A finite union of I_0 sets cannot be dense in $\overline{\Gamma}$ (Theorem 3.5.1), and hence, a "yes" answer to the first question implies "no" to the second.

In this chapter we approach these two questions by introducing the notion of *zero (discrete) harmonic density*, abbreviated *z(d)hd*. A set $E \subseteq \Gamma$ is said to have property z(d)hd if for each non-empty, open $U \subseteq G$, the Fourier transform of every (discrete) measure agrees on **E** with the Fourier transform of a (discrete) measure concentrated on U. The characterizations given of z(d)hd in Sect. 10.2 will make it easy to show that property zdhd implies zhd.

The definition is motivated by the facts that when G is connected every Sidon (or I_0) set is Sidon(U) (resp., $I_0(U)$) (see Corollary 6.3.7 and Theorem 5.3.6) for all non-empty, open U. It follows that Sidon sets have the property zhd and I_0 sets have the property zdhd. In particular, every finite set has zdhd. But there are also non-Sidon sets with property zdhd; several examples

C.C. Graham and K.E. Hare, *Interpolation and Sidon Sets for Compact Groups*,   175
CMS Books in Mathematics, DOI 10.1007/978-1-4614-5392-5_10,
© Springer Science+Business Media New York 2013

are given in Sect. 10.4. It is unknown if every Sidon set has zdhd or even if every ε-Kronecker (for $\varepsilon \geq \sqrt{2}$) or dissociate set has zdhd [**P 11**].

It is easy to see that a set with the zdhd property cannot be dense in $\overline{\Gamma}$ (Proposition 10.2.6). A deeper result, Theorem 10.3.5, is that finite unions of I_0 sets have property zdhd. Thus, if it could be resolved whether every Sidon set has zdhd (either way), then one of the two questions stated in the opening paragraph could be answered: If every Sidon set has zdhd, then a Sidon set cannot be dense in $\overline{\Gamma}$. If, instead, there is a Sidon set which does not have zdhd, then that Sidon set is not a finite union of I_0 sets.

Another motivation for the study of the z(d)hd property is a "globalization principle", which is illustrated in Sect. 10.4.1 by a novel proof of the classical Hadamard gap Theorem 1.2.2.

10.2 Characterizations and Closure Properties

Throughout this chapter, the compact group G is assumed to be connected. That is a natural assumption to make because a set of two elements of finite order will not even have zhd; just take for U any open subset of G on which the two elements coincide.

Definition 10.2.1. Let $U \subseteq G$. We say that $\mathbf{E} \subseteq \Gamma$ has U-hd (respectively, U-dhd) if for every $\mu \in M(G)$ (respectively, $\mu \in M_d(G)$) there exists $\nu \in M(U)$ (resp., $\nu \in M_d(U)$) satisfying $\widehat{\mu} = \widehat{\nu}$ on $\mathbf{E}$.

Clearly, $\mathbf{E}$ has zhd (respectively, zdhd) if and only if $\mathbf{E}$ has U-hd (resp., U-dhd) for every non-empty, open set $U \subseteq G$. A translation argument shows that $\mathbf{E}$ has zhd (zdhd) if and only if $\mathbf{E}$ has U-hd (U-dhd) for every e-neighbourhood $U \subseteq G$.

Sets with the U-dhd property can be characterized in an analogous fashion to Kalton's characterization of $I_0(U)$ sets, Theorem 3.2.5. Similar statements characterize U-hd but with $M_d(G)$ and $M_d(U)$ replaced by $M(G)$ and $M(U)$ and $B_d(\mathbf{E})$ replaced by $B(\mathbf{E})$.

Theorem 10.2.2. Let $U \subseteq G$ be open and $\mathbf{E} \subseteq \Gamma$. The following are equivalent:

1. $\mathbf{E}$ has U-dhd.
2. There is a constant N such that for all $\mu \in M_d(G)$ there exists $\nu \in M_d(U)$ with $\|\nu\|_{M(G)} \leq N \|\mu\|_{M(G)}$ and $\widehat{\nu}(\gamma) = \widehat{\mu}(\gamma)$ for all $\gamma \in \mathbf{E}$.
3. There is a constant N such that for all $x \in G$ there exists $\nu \in M_d(U)$ with $\|\nu\|_{M(G)} \leq N$ and $\widehat{\nu}(\gamma) = \widehat{\delta_x}(\gamma)$ for all $\gamma \in \mathbf{E}$.
4. There exists $0 < \varepsilon < 1$ and constant $N = N(\varepsilon)$ such that for every $\mu \in M_d(G)$ there exists $\nu \in M_d(U)$ with $\|\nu\|_{M(G)} \leq N \|\mu\|_{M(G)}$ and

$$\|\widehat{\mu} - \widehat{\nu}\|_{B_d(\mathbf{E})} < \varepsilon \|\mu\|_{M(G)}.$$

As in the characterizations of $I_0(U)$ sets, the proof will show that the phrase, "There exists $0 < \varepsilon < 1$", can be replaced by "For every $0 < \varepsilon < 1$".

Proof. (1) $\Rightarrow$ (2) is the closed graph theorem. The implications (2) $\Rightarrow$(3) $\Rightarrow$(4) are clear. (4) $\Rightarrow$(1) is a variation on the standard iteration argument. The details are left as Exercise 10.6.1. $\qquad\square$

The proof of the analogous theorem for zdhd is also left as Exercise 10.6.2.

Theorem 10.2.3. *The subset* $\mathbf{E} \subseteq \Gamma$ *has zdhd if and only if any of the following conditions hold:*

1. *Any one of properties Theorem 10.2.2 (1)–(4) is satisfied for all e-neighbourhoods $U \subseteq G$.*
2. *For every $x \in G$ and for every e-neighbourhood $U \subseteq G$, there exists $\nu \in M_d(U)$ with $\widehat{\nu}(\gamma) = \widehat{\delta}_x(\gamma)$ for all $\gamma \in \mathbf{E}$.*
3. *For every e-neighbourhood $U \subseteq G$ and $0 < \varepsilon < 1$ there exists $N = N(U, \varepsilon)$ such that for every $x \in G$ there are scalars $c_n \in \Delta$ and elements $u_n \in U$ such that*

$$\left\| \widehat{\delta}_x - \sum_{n=1}^{N} c_n \widehat{\delta_{u_n}} \right\|_{B_d(\mathbf{E})} < \varepsilon. \tag{10.2.1}$$

Remark 10.2.4. In Theorem 10.2.3(3), δ_x cannot be replaced by arbitrary $\mu \in \mathrm{Ball}(M_d(G))$. Here is why. Suppose $\omega \in \mathrm{Ball}(M(G))$. Let $\mu_\alpha \in \mathrm{Ball}(M_d(G))$ converge weak* to ω, and let $c_{n,\alpha} \in \Delta$ and $u_{n,\alpha} \in U$ have $\|\widehat{\mu_\alpha} - \sum_{n=1}^{N} c_{n,\alpha} \widehat{\delta_{u_{n,\alpha}}}\|_{B(\mathbf{E})} < \varepsilon$ for all α. Letting c_n be a cluster point of $c_{n,\alpha}$ and u_n a cluster point of $u_{n,\alpha}$, we see that $\|\widehat{\omega} - \sum_1^N c_n \widehat{\delta_{u_n}}\|_{B(\mathbf{E})} \le \varepsilon$. That shows that $\mathbf{E}$ satisfies the conclusion of Lemma 9.4.13. By the completion (Sect. 9.4.2) of the proof of the Ramsey–Wells–Bourgain Theorem 9.4.15, $\mathbf{E}$ is I_0. But there are non-I_0 sets (even non-Sidon sets) that have zdhd; see Sect. 10.4 for several examples.

Corollary 10.2.5. *A set with zdhd has zhd.*

Proof. It will be enough to verify that if $\mathbf{E}$ has U-dhd, then $\mathbf{E}$ has $\overline{U}$-hd. Let $\mu \in M(G)$ and choose $\nu_\alpha \in M_d(G)$ with $\|\nu_\alpha\|_{M(G)} \le \|\mu\|_{M(G)}$ and $\nu_\alpha \to \mu$ weak* in $M(G)$. Since $\mathbf{E}$ has U-dhd, by Theorem 10.2.2(2), there are discrete measures $\sigma_\alpha \in M_d(U)$ and a constant N such that $\widehat{\sigma_\alpha} = \widehat{\nu_\alpha}$ on $\mathbf{E}$ and $\|\sigma_\alpha\|_{M(G)} \le N \|\nu_\alpha\|_{M(G)} \le N \|\mu\|_{M(G)}$. Being norm bounded, the net $\{\sigma_\alpha\}$ has a weak* cluster point $\sigma \in M(G)$. Because the measures σ_α are supported on $\overline{U}$, the same is true of σ and since $\widehat{\nu_\alpha}(\gamma) \to \widehat{\mu}(\gamma)$ for all $\gamma \in \Gamma$, $\widehat{\sigma} = \widehat{\mu}$ on $\mathbf{E}$. Thus, $\mathbf{E}$ has $\overline{U}$-hd. $\qquad\square$

Here are some easy facts about the "size" of a set with zdhd. In particular, Proposition 10.2.6 (1) implies a set with zdhd is not dense in $\overline{\Gamma}$.

Proposition 10.2.6. *Suppose* $\mathbf{E} \subseteq \boldsymbol{\Gamma}$ *has zdhd.*

1. *If* $\boldsymbol{\Lambda}$ *is a non-trivial, closed subgroup of* $\overline{\boldsymbol{\Gamma}}$, *then* $\overline{\mathbf{E} \cap \boldsymbol{\Lambda}} \neq \boldsymbol{\Lambda}$.
2. *The interior of* $\overline{\mathbf{E}}$ *in* $\overline{\boldsymbol{\Gamma}}$ *is empty.*

Proof. (1) Suppose $\mathbf{E} \cap \boldsymbol{\Lambda}$ is dense in the non-trivial, closed subgroup $\boldsymbol{\Lambda}$. Let $H \subseteq G$ be the annihilator of $\boldsymbol{\Lambda}$. Since $\boldsymbol{\Lambda}$ is non-trivial, its dual group G/H contains a proper open subset UH. Choose $x \in G$ such that the coset $xH \notin UH$. The set UH can also be viewed as an open subset of G. Hence, there is a discrete measure $\nu = \sum c_j \delta_{x_j} \in M_d(UH)$ with $\widehat{\nu} = \widehat{\delta_x}$ on $\mathbf{E}$. Put $\mu = \sum c_j \delta_{x_j H}$. Then μ is a discrete measure on G/H concentrated on UH.

If $\gamma \in \boldsymbol{\Lambda}$, then, since $\gamma(H) = 1$,

$$\widehat{\mu}(\gamma) = \sum c_j \widehat{\delta_{x_j H}}(\gamma) = \widehat{\nu}(\gamma) = \widehat{\delta_x}(\gamma) = \widehat{\delta_{xH}}(\gamma).$$

Thus, the Fourier–Stieltjes transforms of μ and δ_{xH} agree on $\boldsymbol{\Lambda}$. But $xH \notin UH$, so this is not possible.

(2) Suppose $\overline{\mathbf{E}}$ contains a non-empty, open set in $\overline{\boldsymbol{\Gamma}}$. There is no loss of generality in assuming this open set is a neighbourhood of the identity since translates of sets with property zdhd also have zdhd. Thus, $\overline{\mathbf{E}}$ contains a set of the form $\{\gamma : |\gamma(x_j) - 1| < \varepsilon \text{ for } j = 1, \ldots, J\}$. In particular, $\overline{\mathbf{E}}$ will contain $H^{\perp}$, where $H = \langle \{x_1, \ldots, x_J\} \rangle$. By the first part of the proposition, this subgroup must be trivial and so H must be dense in G.

Consider the map $T : \overline{\boldsymbol{\Gamma}} \to \mathbb{T}^J$ given by $T(\gamma) = (\gamma(x_1), \ldots, \gamma(x_J))$. The map T is clearly continuous and, since H is dense, T is $1-1$. Thus, $T : \overline{\boldsymbol{\Gamma}} \to T(\overline{\boldsymbol{\Gamma}})$ is a homeomorphism. This shows $\overline{\boldsymbol{\Gamma}}$ is homeomorphic to a compact subgroup of $\mathbb{T}^J$, and so G is countable. But there are no countably infinite, compact abelian groups (Exercise C.4.17 (1)). □

Remark 10.2.7. It is known that, unlike I_0 sets, a set $\mathbf{E}$ with zdhd can cluster at a continuous character; see Corollary 10.4.5, but it is unknown if $\overline{\mathbf{E}}$ is a U_0 set [**P 12**], for example.

Proposition 10.2.8. *A subset* $\mathbf{E} \subseteq \mathbb{Z}$ *which has zhd cannot contain arbitrarily long arithmetic progressions of fixed step length.*

Proof. Since translation and dilation do not affect the property zhd, or even the zhd constants, there is no loss of generality in assuming $\mathbf{E}$ contains arbitrarily long arithmetic progressions of step length 1.

Let $U \subseteq \mathbb{T}$ be a non-empty, open subset that is not dense and choose $x \notin \overline{U}$. Let N be the U-hd constant of $\mathbf{E}$, that is, for every $\mu \in M(G)$ there is some $\nu \in M(U)$ such that $\|\nu\|_{M(G)} \leq N \|\mu\|_{M(G)}$ and $\widehat{\nu} = \widehat{\mu}$ on $\mathbf{E}$. By translating $\mathbf{E}$ repeatedly, we can obtain measures $\nu_j \in M_d(U)$ such that $\widehat{\nu_j} = \widehat{\delta_x}$ on $[-j, j]$ and $\|\nu_j\|_{M(G)} \leq N$. Let ν be a weak* limit. This measure is supported on $\overline{U}$ and its transform agrees with δ_x on all of $\mathbb{Z}$. But this is impossible because $x \notin \overline{U}$. □

10.3 Union Results

It is unknown if the union of two sets with zdhd has zdhd [**P 11**]. In this section it will be shown that a finite union of zdhd sets has property zdhd under additional assumptions. In particular, we will prove that a finite union of I_0 sets, although not necessarily I_0, has zdhd.

10.3.1 Unions of Zdhd Sets with Disjoint Closures

We begin by showing that a finite union of zdhd sets with disjoint closures has zdhd.

Proposition 10.3.1. *Suppose* $\mathbf{E}, \mathbf{F} \subseteq \Gamma$ *have* V*-dhd for some symmetric* e*-neighbourhood* V *and that* $\overline{\mathbf{E}} \cap \overline{\mathbf{F}}$ *is empty. Then* $\mathbf{E} \cup \mathbf{F}$ *has* V^6*-dhd.*

Proof. We claim it will be enough to find $\nu \in M_d(V^5)$ with $\widehat{\nu} = 1$ on $\mathbf{E}$ and $\widehat{\nu} = 0$ on $\mathbf{F}$. To see this, note that for each $\mu \in M_d(G)$ there exist $\omega_{\mathbf{E}}, \omega_{\mathbf{F}} \in M_d(V)$ such that $\widehat{\omega_{\mathbf{E}}} = \widehat{\mu}$ on $\mathbf{E}$ and $\widehat{\omega_{\mathbf{F}}} = \widehat{\mu}$ on $\mathbf{F}$. Set $\omega = \nu * \omega_{\mathbf{E}} + (1 - \nu) * \omega_{\mathbf{F}}$. We have $\omega \in M_d(V^6)$ and $\widehat{\omega} = \widehat{\mu}$ on $\mathbf{E} \cup \mathbf{F}$.

We turn to finding ν. That is done exactly as in the proof of Proposition 5.2.2, up to the point at which the $I_0(V)$ property for $\mathbf{F}$ is called upon. In the notation of the proof of Proposition 5.2.2, for each $\gamma \in \overline{\mathbf{E}}$ there is a measure $\tau_1 \in M_d(V)$, with $\widehat{\tau_1} \geq 1/2$ on $\overline{\mathbf{F}}$ and $\widehat{\tau_1}(\gamma) = 0$. In the present context, we call upon Gel'fand's Theorem C.1.12. Applying that theorem to $A(\overline{\mathbf{F}})$ and $\widehat{\tau_1}$, we see that there exists $\widehat{\tau_0} \in A(\overline{\mathbf{F}})$ such that $\widehat{\tau_0} = 1/\widehat{\tau_1}$ on $\mathbf{F}$. We may assume $\tau_0 \in M_d(G)$. Because $\mathbf{F}$ has V-dhd, there exists $\tau \in M_d(V)$ with $\widehat{\tau} = \widehat{\tau_0}$ on $\mathbf{F}$. Then $\omega_\gamma = (1 - \tau_1 * \tau) * (1 - \widetilde{\tau_1} * \widetilde{\tau})$ has $\widehat{\omega_\gamma}(\gamma) = 1$, $\widehat{\omega_\gamma} = 0$ on $\mathbf{F}$ and $\widehat{\omega_\gamma} \geq 0$ everywhere. Also, $\omega_\gamma \in M_d(V^4)$.

By the compactness of $\overline{\mathbf{E}}$, there are $\gamma_1, \ldots, \gamma_M$ such that $\tau_1' := \sum_1^M \omega_{\gamma_m}$ has $\widehat{\tau_1'} \geq 1/2$ on $\overline{\mathbf{E}}$ (and 0 on $\mathbf{F}$). Again, by Gel'fand's theorem, there exists $\tau_0' \in M_d(G)$ such that $\widehat{\tau_0'} = 1/\widehat{\tau_1'}$ on $\mathbf{E}$. Because $\mathbf{E}$ has V-dhd, there exists $\tau' \in M_d(V)$ such that $\widehat{\tau'} = \widehat{\tau_0'}$ on $\mathbf{E}$. Then $\nu = \tau_1' * \tau'$ has $\widehat{\nu} = 1$ on $\mathbf{E}$ and $\widehat{\nu} = 0$ on $\mathbf{F}$. Also, $\nu \in M_d(V^5)$. $\qquad\square$

Corollary 10.3.2. *If* $\mathbf{E}, \mathbf{F}$ *have zdhd and* $\overline{\mathbf{E}} \cap \overline{\mathbf{F}}$ *is empty, then* $\mathbf{E} \cup \mathbf{F}$ *has zdhd.*

Corollary 10.3.3. *If* $\mathbf{E}$ *has zdhd and* $\mathbf{F}$ *is finite, then* $\mathbf{E} \cup \mathbf{F}$ *has zdhd.*

Proof. There is no loss in assuming $\mathbf{F} = \{\gamma\}$. If μ and ν are discrete measures whose Fourier transforms agree on $\mathbf{E}$, then by continuity $\widehat{\mu} = \widehat{\nu}$ on $\overline{\mathbf{E}}$. Thus, if $\gamma \in \overline{\mathbf{E}}$, then $\mathbf{E} \cup \{\gamma\}$ has zdhd. Otherwise, $\overline{\mathbf{E}} \cap \overline{\mathbf{F}}$ is empty and we may apply the previous corollary. $\qquad\square$

10.3.2 A Finite Union of I_0 Sets Has Zdhd

The goal of this section is to prove the theorem stated in its title. The proof will require the notion of a Helson set (Remark 3.5.5). Recall that the Helson constant (p. 165) of a closed set $\mathbf{S} \subset \overline{\Gamma}$ is the infimum of the numbers C such that $\|f\|_{B_d(\mathbf{S})} \le C\|f\|_\infty$ for all $f \in B_d(\mathbf{S})$.

Helson sets are relevant here because, as was observed in Remark 3.5.5, the closure of an I_0 set is a Helson set. Like Sidon sets, a finite union of Helson sets is Helson. This deep result, stated below, will be used in what follows.

Theorem 10.3.4 (Varopoulos's union theorem).
The union of two Helson sets is Helson.

Here is our union theorem for I_0 sets.

Theorem 10.3.5. *If* $\mathbf{E}, \mathbf{F} \subseteq \overline{\Gamma}$ *are* I_0 *sets, then* $\mathbf{E} \cup \mathbf{F}$ *has zdhd.*

One can immediately deduce that there are non-I0 sets with zdhd (such as Example 1.5.2). We will see later (Proposition 10.4.4) that there are non-Sidon sets that have zdhd.

First we prove a technical lemma.

Lemma 10.3.6. *Suppose* $\mathbf{E}, \mathbf{F} \subseteq \overline{\Gamma}$ *are* I_0 *sets. Let* $V \subseteq G$ *be an e-neighbourhood. Then there is an open set* $\Omega \subseteq \overline{\Gamma}$, *containing* $\overline{\mathbf{E}} \cap \overline{\mathbf{F}}$, *such that* $(\mathbf{E} \cup \mathbf{F}) \cap \Omega$ *has* V-*dhd.*

Proof (of Lemma 10.3.6). The characterization of V-dhd given in Theorem 10.2.2 (4) implies that it will suffice to show that there is an open set $\Omega \supseteq \overline{\mathbf{E}} \cap \overline{\mathbf{F}}$ and constant N such that for every $\mu \in M_d(G)$ with $\|\mu\|_{M(G)} \le 1$ there exists $\nu \in M_d(V)$ such that $\|\nu\|_{M(G)} \le N$ and $\|\hat{\mu} - \hat{\nu}\|_{B_d((\mathbf{E} \cup \mathbf{F}) \cap \Omega)} \le 1/2$.

Choose an e-neighbourhood $W \subseteq G$ such that $W^2 \subseteq V$. By compactness, there are finitely many points $x_k, k = 1, \ldots, K$, such that $\bigcup_{k=1}^{K} x_k W = G$. Being I_0, the set $\mathbf{E}$ has zdhd, and hence there are measures $\nu_k \in M_d(W)$ and a constant $C_{\mathbf{E}}$ such that $\|\nu_k\|_{M(G)} \le C_{\mathbf{E}}$ and $\widehat{\nu_k} = \widehat{\delta_{x_k}}$ on $\mathbf{E}$. Continuity implies that this equality continues to hold on $\overline{\mathbf{E}}$.

The sets $\mathbf{E}, \mathbf{F}$ are both Helson, and therefore so is their union. Let C_0 be the Helson constant of $\overline{\mathbf{E}} \cup \overline{\mathbf{F}}$. Put

$$\Omega = \bigcap_{k=1}^{K} \{\gamma \in \overline{\Gamma} : \left| \widehat{\nu_k}(\gamma) - \widehat{\delta_{x_k}}(\gamma) \right| < 1/(4C_0)\} \subseteq \overline{\Gamma}.$$

The set Ω is open as each of the sets $\{\gamma : \left| \widehat{\nu_k}(\gamma) - \widehat{\delta_{x_k}}(\gamma) \right| < 1/(4C_0)\}$ is open in $\overline{\Gamma}$. Since $\widehat{\nu_k} = \widehat{\delta_{x_k}}$ on $\overline{\mathbf{E}}$, Ω contains all of $\overline{\mathbf{E}}$.

Let μ be any discrete measure on G with $\|\mu\|_{M(G)} \leq 1$. Then μ can be written as

$$\mu = \sum_{j,k} c_{j,k} \delta_{x_k w_{j,k}} = \sum_{k=1}^{K} \sum_{j=1}^{\infty} c_{j,k} \delta_{x_k} * \delta_{w_{j,k}},$$

where $w_{j,k} \in W$ and $\sum_{j,k} |c_{j,k}| = \|\mu\|_{M(G)} \leq 1$.

Consider the measure $\nu = \sum_{j,k} c_{j,k} \nu_k * \delta_{w_{j,k}} \in M_d(W^2) \subseteq M_d(V)$. This measure satisfies $\|\nu\|_{M(G)} \leq \sum_{j,k} |c_{j,k}| \|\nu_k\|_{M(G)} \leq C_{\mathbf{E}}$. Furthermore, the definition of Ω ensures that for each $\gamma \in \overline{\Omega}$,

$$|\widehat{\nu}(\gamma) - \widehat{\mu}(\gamma)| = \Big| \sum_{j,k} c_{j,k} (\widehat{\nu_k}(\gamma) - \widehat{\delta_{x_k}}(\gamma)) \widehat{\delta_{w_{j,k}}}(\gamma) \Big|$$

$$\leq \sum_{j,k} |c_{j,k}| \, |\widehat{\nu_k}(\gamma) - \widehat{\delta_{x_k}}(\gamma)| \leq \frac{1}{4C_0} \|\mu\| \leq \frac{1}{4C_0}.$$

Thus, the function $\widehat{\nu} - \widehat{\mu}$, viewed as an element of $C((\overline{\mathbf{E}} \cup \overline{\mathbf{F}}) \cap \overline{\Omega})$, has norm at most $1/(4C_0)$. Since $\overline{\mathbf{E}} \cup \overline{\mathbf{F}}$ is Helson, it follows that there is a measure $\sigma \in M_d(G)$ such that $\widehat{\sigma} = \widehat{\nu} - \widehat{\mu}$ on $(\overline{\mathbf{E}} \cup \overline{\mathbf{F}}) \cap \overline{\Omega}$ and $\|\sigma\|_{M(G)} \leq 2C_0/(4C_0) = 1/2$. Therefore, $\|\widehat{\nu} - \widehat{\mu}\|_{B_d((\mathbf{E} \cup \mathbf{F}) \cap \Omega)} \leq \|\sigma\|_{M(G)} \leq 1/2$, which proves that ν has the desired properties. □

Proof (of Theorem 10.3.5). Let $U \subseteq G$ be an e-neighbourhood and let $V \subseteq G$ be an e-neighbourhood with $V^5 \subseteq U$. As in Lemma 10.3.6, choose $\Omega \subseteq \overline{\Gamma}$, an open set containing $\overline{\mathbf{E}} \cap \overline{\mathbf{F}}$, such that $(\mathbf{E} \cup \mathbf{F}) \cap \Omega$ has V-dhd. The regularity of the topology of $\overline{\Gamma}$ implies there is an open set $\Omega_1 \supseteq \overline{\mathbf{E}} \cap \overline{\mathbf{F}}$, such that $\overline{\Omega_1} \subseteq \Omega$.

Because $\mathbf{E}$ has zdhd, there is a measure $\mu_{\mathbf{E}} \in M_d(V)$ such that $\widehat{\mu_{\mathbf{E}}} = 0$ on $\overline{\mathbf{E}} \smallsetminus \Omega_1$ and $\widehat{\mu_{\mathbf{E}}} = 1$ on the disjoint closed set $\overline{\mathbf{E}} \cap \overline{\mathbf{F}}$. Similarly, there exists a measure $\mu_{\mathbf{F}} \in M_d(V)$ such that $\widehat{\mu_{\mathbf{F}}} = 0$ on $\overline{\mathbf{F}} \smallsetminus \Omega_1$ and $\widehat{\mu_{\mathbf{F}}} = 1$ on the disjoint set $\overline{\mathbf{E}} \cap \overline{\mathbf{F}}$. Put $\sigma = \mu_{\mathbf{E}} * \mu_{\mathbf{F}} \in M_d(V^2)$. Then $\widehat{\sigma} = 0$ on $(\overline{\mathbf{E}} \cup \overline{\mathbf{F}}) \smallsetminus \Omega_1$ and $\widehat{\sigma} = 1$ on $\overline{\mathbf{E}} \cap \overline{\mathbf{F}}$. Let

$$\Omega_2 = \{\gamma \in \overline{\mathbf{E}} \cup \overline{\mathbf{F}} : |\widehat{\sigma}(\gamma)| > 1/2\}.$$

The choice of σ ensures that $\overline{\mathbf{E}} \cap \overline{\mathbf{F}} \subseteq \Omega_2 \subseteq \overline{\Omega_2} \subseteq \Omega_1$.

Since $\widehat{\sigma}|_{\Omega_2}$ is bounded away from 0, an application of Gel'fand's theorem (as in Proposition 10.3.1) implies there exists $\sigma_1 \in M_d(G)$ such that $\widehat{\sigma} \cdot \widehat{\sigma_1} = 1$ on Ω_2. Since $(\mathbf{E} \cup \mathbf{F}) \cap \Omega$ has V-dhd, the same is true for its subset $(\mathbf{E} \cup \mathbf{F}) \cap \Omega_2$. Hence, we can choose $\sigma_2 \in M_d(V)$ such that $\widehat{\sigma_2} = \widehat{\sigma_1}$ on $(\mathbf{E} \cup \mathbf{F}) \cap \Omega_2$.

Because the closed sets $\overline{\Omega_2}$ and Ω_1^c are disjoint, there is a discrete measure ν such that $\widehat{\nu} = 1$ on Ω_2, $\widehat{\nu} = 0$ on Ω_1 and $0 \leq \widehat{\nu} \leq 1$. Again, because $(\mathbf{E} \cup \mathbf{F}) \cap \Omega$ has V-dhd, there exists $\nu_1 \in M_d(V)$ such that $\widehat{\nu_1} = \widehat{\nu}$ on $(\mathbf{E} \cup \mathbf{F}) \cap \Omega$. Finally, put

$$\mu = \sigma * \sigma_2 * \nu_1 \in M_d(V^4).$$

By construction, $\widehat{\mu} = 1$ on $(\mathbf{E} \cup \mathbf{F}) \cap \Omega_2$ and $\widehat{\mu} = 0$ on $(\mathbf{E} \cup \mathbf{F}) \smallsetminus \Omega_1$.

Let $\rho \in M_d(G)$ be given. To prove that $\mathbf{E} \cup \mathbf{F}$ has U-dhd it will be enough to prove there exists $\rho_0 \in M_d(V^5)$ such that $\widehat{\rho_0} = \widehat{\rho}$ on $\mathbf{E} \cup \mathbf{F}$.

The closed sets $\overline{\mathbf{E}} \setminus \Omega_2$ and $\overline{\mathbf{F}} \setminus \Omega_2$ are disjoint since $\overline{\mathbf{E}} \cap \overline{\mathbf{F}} \subseteq \Omega_2$. By Corollary 10.3.2, $(\mathbf{E} \setminus \Omega_2) \cup (\mathbf{F} \setminus \Omega_2) = (\mathbf{E} \cup \mathbf{F}) \setminus \Omega_2$ has zdhd and this ensures there is a discrete measure $\rho_1 \in M_d(V)$ such that $\widehat{\rho_1} = \widehat{\rho}$ on $(\mathbf{E} \cup \mathbf{F}) \setminus \Omega_2$. Since $(\mathbf{E} \cup \mathbf{F}) \cap \Omega$ has V-dhd, there is also a measure $\rho_2 \in M_d(V)$ such that $\widehat{\rho_2} = \widehat{\rho}$ on $(\mathbf{E} \cup \mathbf{F}) \cap \Omega$. Set

$$\rho_0 = \rho_1 * (\delta_e - \mu) + \rho_2 * \mu \in M_d(V^5).$$

We claim this measure interpolates $\widehat{\rho}$ on $\mathbf{E} \cup \mathbf{F}$. To see this, observe the following:

- On $(\mathbf{E} \cup \mathbf{F}) \cap (\Omega_1 \setminus \Omega_2)$, $\widehat{\rho_1} = \widehat{\rho} = \widehat{\rho_2}$; hence, $\widehat{\rho_0} = \widehat{\rho_1}(1 - \widehat{\mu}) + \widehat{\rho_2}\widehat{\mu} = \widehat{\rho}$.
- On $(\mathbf{E} \cup \mathbf{F}) \setminus \Omega_1$, $\widehat{\rho_1} = \widehat{\rho}$ and $\widehat{\mu} = 0$; thus, $\widehat{\rho_0} = \widehat{\rho_1} = \widehat{\rho}$.
- Finally, on $(\mathbf{E} \cup \mathbf{F}) \cap \Omega_2$, $\widehat{\rho_2} = \widehat{\rho}$ and $\widehat{\mu} = 1$, and so $\widehat{\rho_0} = \widehat{\rho_2} = \widehat{\rho}$.

These observations demonstrate that $\widehat{\rho_0} = \widehat{\rho}$ on $\mathbf{E} \cup \mathbf{F}$, as claimed. □

10.3.3 Other Union Results for Zdhd Sets

The union of two zdhd sets can also be shown to have zdhd if their closures have finite intersection, though we know of no non-trivial instances of this. This is a consequence of a more general result which relies on the property of spectral synthesis (p. 213).

Theorem 10.3.7. *Suppose* $\mathbf{E}, \mathbf{F} \subseteq \Gamma$ *have zdhd and* $\overline{\mathbf{E}} \cap \overline{\mathbf{F}}$ *obeys spectral synthesis. Then* $\mathbf{E} \cup \mathbf{F}$ *has zdhd.*

Corollary 10.3.8. *If* $\mathbf{E}, \mathbf{F}$ *have zdhd and* $\overline{\mathbf{E}} \cap \overline{\mathbf{F}}$ *is finite, then* $\mathbf{E} \cup \mathbf{F}$ *has zdhd.*

Proof. This follows since finite sets have spectral synthesis (see p. 213). □

The theorem is deduced from a technical lemma similar to Lemma 10.3.6. We remark that the definition of zdhd extends to subsets of $\overline{\Gamma}$ in the obvious fashion and the characterization results, Theorems 10.2.2 and 10.2.3, continue to hold in this setting.

Lemma 10.3.9. *Suppose* $\mathbf{E}$ *has zdhd,* $\overline{\mathbf{E}} \cap \overline{\mathbf{F}}$ *obeys spectral synthesis and* $V \subseteq G$ *is an* e-*neighbourhood. Then there is an open set* $\Omega \supseteq \overline{\mathbf{E}} \cap \overline{\mathbf{F}}$ *such that* Ω *has* V-*dhd.*

Proof (of Lemma 10.3.9). We begin in the same manner as the proof of Lemma 10.3.6. Let $W^2 \subseteq V$ be open and assume $\bigcup_{k=1}^{K} x_k W = G$. Use the zdhd property of $\mathbf{E}$ to obtain a constant C and measures $\nu_k \in M_d(W)$

such that $\|\nu_k\|_{M(G)} \le C$ and $\widehat{\nu_k} = \widehat{\delta_{x_k}}$ on $\mathbf{E}$ for $k = 1,\ldots,K$. Since $\overline{\mathbf{E}} \cap \overline{\mathbf{F}}$ obeys spectral synthesis, there is a neighbourhood $\boldsymbol{\Omega}$ of $\overline{\mathbf{E}} \cap \overline{\mathbf{F}}$ such that $\|\widehat{\nu_k} - \widehat{\delta_{x_k}}\|_{B_d(\boldsymbol{\Omega})} < 1/(2K)$ for each $k = 1,\ldots,K$ (see Lemma C.1.14). This inequality is enough to prove that if $\mu = \sum_{j,k} c_{j,k}\delta_{x_k w_{j,k}}$ is any norm one, discrete measure, with $w_{j,k} \in W$, then $\nu = \sum_{j,k} c_{j,k}\nu_k * \delta_{w_{j,k}} \in M_d(V)$ has measure norm at most C and satisfies $\|\widehat{\mu} - \widehat{\nu}\|_{B_d(\boldsymbol{\Omega})} \le 1/2$. □

Proof (of Theorem 10.3.7). The proof is like that of Theorem 10.3.5 but without the intersections with $\mathbf{E} \cup \mathbf{F}$ and using Lemma 10.3.9. □

10.4 Examples and Applications

Proposition 10.4.1. *Let $\mathbf{E} \subseteq \boldsymbol{\Gamma}$ and suppose the cluster points of $\mathbf{E}$ are contained in the intersection of cosets of the form $\gamma_n H_n^\perp \subseteq \overline{\boldsymbol{\Gamma}}$, where H_n are finite subgroups of G and $\bigcup_{n=1}^\infty H_n$ is dense in G. Then $\mathbf{E}$ has zdhd.*

Proof. Fix an e-neighbourhood $U \subseteq G$ and let V be a symmetric e-neighbourhood such that $V^{12} \subseteq U$. Assume $G = \bigcup_{k=1}^K x_k V$. Since $\bigcup_n H_n$ is dense in G, for each k, there are elements $h_k \in x_k V \cap (\bigcup_n H_n)$, say $h_k = x_k v_k \in H_{n_k}$ with $v_k \in V$. Because V is symmetric,

$$\bigcup_{k=1}^K h_k V^2 = \bigcup_{k=1}^K x_k v_k V^2 \supseteq \bigcup_{k=1}^K x_k V = G.$$

We first check that $\boldsymbol{\Lambda} = \bigcap_{k=1}^K \gamma_{n_k} H_{n_k}^\perp$ has V^2-dhd by verifying Theorem 10.2.2 (3). Let $x \in G$. Then $x = h_k v$ for some $k = 1,\ldots,K$ and $v \in V^2$. Fix $\gamma_0 \in \gamma_{n_k} H_{n_k}^\perp$ and let σ be the discrete, norm one measure $\sigma = \gamma_0(h_k)\delta_{v^{-1}} \in M_d(V^2)$. If $\gamma \in \boldsymbol{\Lambda}$, then $\gamma \in \gamma_{n_k} H_{n_k}^\perp$, so $\gamma(h_k) = \gamma_{n_k}(h_k) = \gamma_0(h_k)$. Thus, $\widehat{\sigma}(\gamma) = \gamma_0(h_k)\gamma(v) = \gamma(h_k v) = \gamma(x) = \widehat{\delta_{x^{-1}}}(\gamma)$, showing that Theorem 10.2.2 (3) is satisfied with $N = 1$.

Since H_n is a finite group, $\boldsymbol{\Lambda}$ is open. By assumption, all the cluster points of $\mathbf{E}$ belong to $\boldsymbol{\Lambda}$. Thus, $\mathbf{E} \smallsetminus \boldsymbol{\Lambda}$ must be finite. Since $\boldsymbol{\Lambda}$ is also closed, $\mathbf{E}$ has V^{12}-dhd by Proposition 10.3.1. □

Definition 10.4.2. Sets $\mathbf{E} \subseteq \boldsymbol{\Gamma}$ which satisfy the hypotheses of the proposition will be said to have *strong zdhd*.

Example 10.4.3. Hadamard sets of the form $\{r^n\}_{n=1}^\infty$, with integer $r \ge 2$, are strong zdhd. Just take H_n to be the subgroup of the r^nth roots of unity. Similarly, the set $\{k \cdot 100^{j!} : k = 1,\ldots,j, 1 \le j < \infty\}$ is another example of a strong zdhd set, and this set is not Sidon since it contains arbitrarily long arithmetic progressions.

The next proposition gives another example of a non-Sidon, zdhd set.

Proposition 10.4.4. *If* $\mathbf{E} = \{r^n\}_{n=1}^{\infty} \subset \mathbb{Z}$ *with* $r \geq 3$ *an integer, then* $\mathbf{E} + \mathbf{E}$ *and* $\mathbf{E} - \mathbf{E}$ *have zdhd.*

Proof. Let H_n be the subgroup consisting of the r^nth roots of unity and let $\Lambda = \bigcap_n H_n^{\perp} \subseteq \overline{\mathbb{Z}}$. Fix a non-empty, open set $U \subseteq \mathbb{T}$ and choose $h_1, \ldots, h_K \in \bigcup_{n=1}^{\infty} H_n$ such that $\bigcup_{k=1}^{K}(h_k + U) = \mathbb{T}$.

Being a compact subgroup, Λ is a set of spectral synthesis (p. 213) and, since $\widehat{\delta_{h_k}} = 1$ on Λ, it follows that there is a neighbourhood $\Omega \subseteq \overline{\mathbb{Z}}$ containing Λ such that $\left\| \widehat{\delta_{h_k}} - \widehat{\delta_1} \right\|_{B_d(\Omega)} < 1$ for each $k = 1, \ldots, K$. By an argument similar to the proof that Theorem 10.2.3(3) implies zdhd (see Exercise 10.6.3), we conclude that Ω has U-dhd.

Notice that $(\overline{\mathbf{E} + \mathbf{E}}) \smallsetminus \Omega$ consists of a finite number of elements of $\mathbf{E} + \mathbf{E}$, plus a finite number of sets of the form $r^n + \overline{\mathbf{E}}$. Each of these finitely many sets has zdhd and their closures in $\overline{\mathbb{Z}}$ are pairwise disjoint (see Exercise 10.6.6). Therefore, their union has zdhd by Proposition 10.3.1.

The argument is similar for $\mathbf{E} - \mathbf{E}$. □

Corollary 10.4.5. *A zdhd set can cluster at a continuous character.*

Proof. 1 is a cluster point of $\mathbf{E} \cdot \mathbf{E}^{-1}$ whenever $\mathbf{E}$ is an infinite set. □

Remark 10.4.6. Being Hadamard, the $\mathbf{E}$ of Proposition 10.4.4 is $I_0(U)$ with bounded length (Remark 3.2.15 (i)). In [30] it is shown that if $\mathbf{E} = \{n_j\}$ is any Hadamard set with ratio at least 3, then for each k the set $\{n_{j_1} \pm \cdots \pm n_{j_k} : j_1 < \cdots < j_k\}$ has zhd. It is unknown, in general, if $\mathbf{E} \cdot \mathbf{E}^{\pm 1}$ has zdhd whenever $\mathbf{E}$ has bounded length (or bounded constants) and if all such "sums" of Hadamard sets have zdhd [**P 11**].

10.4.1 The Hadamard Gap Theorem for Sets with Zhd

We began this book by introducing Hadamard sets and giving examples of some of the unusual properties possessed by power series and trigonometric series with frequencies supported on a Hadamard set. Throughout the book, we have seen various generalizations these properties. To conclude, we give a short proof of a generalization of the classical Hadamard gap Theorem 1.2.2 for sets with property zhd.

Proposition 10.4.7. *Suppose* $\{n_j\}_{j=1}^{\infty}$ *is an increasing sequence of positive integers and has property zhd. The function* $f(z) = \sum_{j=1}^{\infty} c_j z^{n_j}$ *cannot be analytically continued, at any point, across the circle of convergence.*

Remark 10.4.8. This proves the classical Hadamard gap theorem since Hadamard sets, being Sidon, have property zhd.

Proof. Suppose that f could be analytically continued at z_0 on the circle of convergence. There is no loss of generality in assuming the circle of convergence has radius 1 and that $z_0 = 1$. Then f can be continued to be analytic in the open set $U_\varepsilon = \{z \in \mathbb{C} : |z - 1| < \varepsilon\}$ for some $\varepsilon > 0$.

Let $t \in [0, 2\pi]$. Since $\{n_j\}$ has zhd, it is possible to obtain a measure ν on $\mathbb{T}$, concentrated on $\mathbb{T} \cap U_{\varepsilon/3}$ (an open subset of $\mathbb{T}$), such that $\widehat{\nu}(n_j) = \delta_t(n_j)$ for all j.

The function $g(z) = \int f(e^{-i\theta}z)d\nu(\theta)$ is analytic in the interior of the unit disk, as well as in the set $U_{\varepsilon/3}$. Since

$$\int e^{-i\theta n_j}d\nu(\theta) = \widehat{\nu}(n_j) = e^{itn_j},$$

the Taylor coefficients of g are the same as those of the function $z \mapsto f(e^{it}z)$. Thus, f has an analytic continuation to $\{z \in \mathbb{C} : |z - e^{it}| < \varepsilon/3\}$. Since t was arbitrary, f can be continued to $\{z : |z| < 1 + \varepsilon/3\}$, which contradicts the assumption that 1 is the radius of convergence.   □

10.5 Remarks and Credits

The term "zero harmonic density" seems to have first appeared in Déchamps-Gondim's 1976 note [28]. In [29], she gave the proof that if $\{\gamma_j\}$ is dissociate, then $\{\pm\gamma_j \pm \gamma_k : 1 \le j < k < \infty\}$ has zhd. Déchamps and Selles in 1996 [30] gave a lengthy construction to establish that the sets $\{n_{j_1} \pm \cdots \pm n_{j_k} : j_1 < \cdots < j_k\}$, where $\{n_j\}$ is Hadamard with ratio at least 3, have zhd. Lust [125] showed that if $\mathbf{E} \subseteq \mathbb{Z}$ is such that for each $k \ge 1$, $C_{k\mathbf{E}}$ (where $k\mathbf{E} = \mathbf{E} + \cdots + \mathbf{E}$ k times) does not contain a subspace isomorphic to c_0, then $\mathbf{E}$ has zhd.

The concept of zero discrete harmonic density was introduced in [58] and most of the results of this chapter can be found there. The property zdhd with bounded constants, defined analogously to that of $I_0(U)$ with bounded constants, was also studied in that paper. It is shown there that a set which has zdhd with bounded constants has at most one cluster point in Γ and that $M_0(\overline{\mathbf{E}}) = \{0\}$ if $\mathbf{E}$ is a finite union of sets having zdhd with bounded constants. Like Sidon sets, a set which contains the sum of two disjoint infinite sets cannot be a finite union of sets with zdhd with bounded constants. Thus, $\{3^n\} + \{3^n\}$ is a zdhd set that is not a finite union of sets with zdhd with bounded constants. The proof of Proposition 10.4.1 actually shows that strong zdhd sets have zdhd with bounded constants.

Theorem 10.3.4 was proved by Varopoulos for compact Helson sets, one of which was metrizable [188, 189]. It was extended to the non-metrizable (but still compact) case by Lust [124] and then to the general case by Saeki [170]. Other proofs can be found in [56, Chapter 2], [86, 120]. See also [177], which improves on the original constants.

The connection between zero harmonic density and zero density for subsets of $\mathbb{Z}$ (meaning $\limsup |\mathbf{E} \cap [-N, N]|/(2N+1) = 0$) is unclear [**P 13**]. It is only known that not all sets of zero density have zero harmonic density. The set $\{100^{j!} + k : k \leq j\}$ is such an example since it contains arbitrarily long arithmetic progressions of fixed step length (see Proposition 10.2.8).

10.6 Exercises

Exercise 10.6.1. Prove Theorem 10.2.2 (4) $\Rightarrow$ (1).

Exercise 10.6.2. Prove Theorem 10.2.3.

Exercise 10.6.3. Let $V \subseteq G$ be open and $\varepsilon < 1$. Show that if there exist finitely many points $\{x_j\}_{j=1}^N \subseteq G$ with $\bigcup_{j=1}^N x_j V = G$ and $\nu_j \in M_d(V)$ with $\left\| \widehat{\nu_j} - \widehat{\delta_{x_j}} \right\|_{B_d(E)} < 1$, then $\mathbf{E} \subseteq \mathbf{\Gamma}$ has V^2-dhd.

Exercise 10.6.4. Suppose that $U \subseteq [-\pi/k, \pi/k) \subseteq \mathbb{T}$. Show that $\mathbf{E} \subseteq \mathbb{Z}$ is kU-dhd with constant N (as in Theorem 10.2.2 (2)) if and only if $k\mathbf{E}$ is U-dhd with constant N.

Exercise 10.6.5. Show that if $\mathbf{E}$ has U-dhd for some proper open subset $U \subseteq G$, then $\mathbf{E}$ is not dense in $\mathbf{\Gamma}$.

Exercise 10.6.6. Suppose $\mathbf{E} = \{r^n\}_{n=1}^\infty$ with $r \geq 3$ an integer. Show that the sets $r^n + \mathbf{E}$ and $r^m + \mathbf{E}$ for $m \neq n$ have disjoint closures.

Exercise 10.6.7. One can define the notion of *zdhd with bounded constants* analogously to that of I_0 sets with bounded constants.

1. Prove that a set which has zdhd with bounded constants has at most one cluster point in $\mathbf{\Gamma}$.
2. Prove that a set with strong zdhd has zdhd with bounded constants.

Exercise 10.6.8. Prove that the union of a set with strong zdhd and a set with zdhd has zdhd.

Appendix A

Interpolation and Sidon Sets for Groups That Are Not Compact and Abelian

Interpolation and Sidon subsets of non-discrete abelian groups. Distinction between metrizable and non-metrizable Γ. Perturbations of I_0 and Sidon sets. A survey of I_0 and Sidon sets for compact non-abelian groups. Characterization in terms of FTR sets. Central lacunarity.

In the main part of this book, Sidon and interpolation sets in duals of compact abelian groups were investigated. Here we briefly review what is known about these special sets in duals of groups that are either abelian and locally compact, but not compact (Sect. A.1), or compact, but not abelian (Sect. A.2). While there are similarities, there are also many differences and these will be highlighted. Only some proofs will be given, but extensive references to the literature will be provided.

A.1 Interpolation and Sidon Sets in Non-compact Abelian Groups

In this section G will denote a locally compact abelian group that is not (necessarily) compact. We also assume G is not discrete. Thus, the dual group, Γ, is locally compact abelian, non-compact and not (necessarily) discrete. One such example is the group $\mathbb{R}$ with the usual topology, whose dual is itself.

We begin by generalizing the definition of an I_0 set.

Definition A.1.1. A set $\mathbf{E} \subseteq \Gamma$ is an *interpolation* or *I-set* if every uniformly continuous function f on $\mathbf{E}$ extends to an almost periodic function on Γ. In this case there exists a continuous function on the Bohr compactification, $\overline{\Gamma}$, of Γ that agrees on $\mathbf{E}$ with f.

C.C. Graham and K.E. Hare, *Interpolation and Sidon Sets for Compact Groups*,
CMS Books in Mathematics, DOI 10.1007/978-1-4614-5392-5_11,
© Springer Science+Business Media New York 2013

Definition A.1.2. $\mathbf{E} \subseteq \mathbf{\Gamma}$ is an I_0 *set* if every bounded function on $\mathbf{E}$ extends to an almost periodic function.

A.1.1 Interpolation Sets When $\mathbf{\Gamma}$ Is Metrizable

The properties of I-sets and I_0 sets in a non-discrete group $\mathbf{\Gamma}$ depend on whether G is σ-compact (equivalently, $\mathbf{\Gamma}$ is metrizable; see Exercise C.4.18 (1)). The metrizability of $\mathbf{\Gamma}$ will be assumed throughout this subsection.

Since G is σ-compact, all the results about I_0 subsets of discrete groups $\mathbf{\Gamma}$ that used the Baire category theorem in their proofs have essentially *identical versions* for I_0 subsets of metrizable groups. These results include:

- Kalton's characterization of I_0 sets (Theorem 3.2.5)
- Kahane's AP theorem (Theorem 3.3.1)
- Méla's decomposition of an I_0 set into a finite union of $I_0(U)$ sets with bounded length (Theorem 5.3.1)
- The assertion that an I_0 set does not cluster at a continuous character (Theorem 3.5.1)
- The assertion that if $\mathbf{E}$ is I_0 with bounded length, then the only continuous character at which $\mathbf{E} \cdot \mathbf{E}^{-1}$ clusters is $\mathbf{1}$ (Theorem 5.3.9).

The only difference in those results is the addition to the conclusions of a compact set $K \subset G$ on which all the interpolating measures are supported, and the addition to the proofs of a sequence of compact subsets K_n with $\bigcup_n K_n = G$.

We have seen that the closure of an I_0 set in a discrete group supports no non-zero measures in $M_0(\overline{\mathbf{\Gamma}})$ (Proposition 3.5.3). Even more is true when $\mathbf{\Gamma}$ is not discrete.

Proposition A.1.3. *Let* $\mathbf{E} \subset \mathbf{\Gamma}$ *be an* I_0 *set and* $\mathbf{V} \subseteq \mathbf{\Gamma}$ *compact. Then* $\overline{\mathbf{E}} \cdot \mathbf{V}$ *supports no non-zero measure in* $M_0(\overline{\mathbf{\Gamma}})$.

It is unknown if $m_{\overline{\mathbf{\Gamma}}}(\overline{\mathbf{E}} \cdot \mathbf{\Gamma}) = 0$ whenever $\mathbf{E} \subseteq \mathbf{\Gamma}$ is an I_0 set and $\mathbf{\Gamma}$ is metrizable, discrete or not [**P 15**].

Proof. The proof will make good use of the fact that if $\mu \in M_0$ and $\nu \ll \mu$, then $\nu \in M_0$ (Lemma C.1.9).

Suppose $\mu \in M_0(\overline{\mathbf{\Gamma}})$ has support in $\overline{\mathbf{E}} \cdot \mathbf{V}$. Let $\mathbf{E} = \bigcup_1^N \mathbf{E}_n$ be a decomposition of $\mathbf{E}$ into $I_0(U)$ sets with bounded length (Exercise A.4.3). Since the restriction of μ to each $\overline{\mathbf{E}}_n \cdot \mathbf{V}$ is also in $M_0(\overline{\mathbf{\Gamma}})$, it will suffice to show that each such restriction is the zero measure. Thus, we may assume that $\mathbf{E}$ is $I_0(U)$ with bounded length.

We may assume that μ is a probability measure and has no mass on any set of the form $\gamma \mathbf{V}$, where $\gamma \in \mathbf{\Gamma}$. Indeed, let ν be the restriction of μ to $\gamma \mathbf{V}$.

If $\mu \in M_0(\overline{\Gamma})$, then also $\nu \in M_0(\overline{\Gamma})$. Now, $\nu \in M(\Gamma)$, so $\widehat{\nu}$ is continuous on G with its original topology. If there existed $x \in G$ with $\widehat{\nu}(x) \neq 0$, then $\widehat{\nu}$ would have an uncountable number of non-zero values. Since $\nu \in M_0(\overline{\Gamma})$, that is impossible, unless $\nu = 0$.

Let C be the bound for the length and let $U \subseteq G$ be an open e-neighbourhood such that $|\widehat{\mu}| < 1/(10C)$ on $U \setminus \{e\}$. That is possible because $\widehat{\mu} \in c_0(G_d)$. Let U also be so small that $|1 - \gamma(u)| < 1/(10C)$ for all $\gamma \in V$ and $u \in U$. The openness of $U' = U \setminus \{e\}$ implies the existence of $\nu \in M_d(U')$ and a finite $F \subset E$ such that $\widehat{\nu}(\gamma) = 1$ for all $\gamma \in E \setminus F$ and $\|\nu\| < 2C$. Hence, $|1 - \widehat{\nu}| < 1/5$ on $(\overline{E} \setminus F) \cdot V$ and, as μ is a probability measure, $|1 - \int \widehat{\nu} d\mu| < 1/5$. But

$$\left| \int \widehat{\nu} d\mu \right| = \left| \int \widehat{\mu} d\nu \right| \leq 2C \sup_{u \in U} |\widehat{\mu}(u)| < \frac{1}{5},$$

a contradiction. $\qquad \square$

Another similarity with the discrete Γ case is that the elements of an I_0 set are discrete subsets of Γ, as the next proposition shows.

Lemma A.1.4. *If* $E \subseteq \Gamma$ *is* I_0, *then there exists a* 1-*neighbourhood* $U \subset \Gamma$ *such that* $(\gamma U) \cap (\chi U) = \emptyset$ *for all* $\gamma \neq \chi \in E$.

Proof. Let d be a metric for Γ. Suppose to the contrary that for each $n \geq 1$ there existed $\gamma_n \neq \chi_n \in E$ such that $\lim_{n \to \infty} d(\gamma_n, \chi_n) = 0$.

Case I. There is a subsequence $n(j)$ and $\gamma \in \Gamma$ such that $\gamma_{n(j)} = \gamma$ for all j (or, alternatively, $\chi_{n(j)} = \chi$ for all j). Without loss of generality we may assume the $\chi_{n(j)}$ are distinct. Then the sequence $\chi_{n(j)}$ also converges to γ. Let $\varphi : E \to \{-1, 1\}$ have $\varphi(\chi_{n(j)}) = (-1)^j$ for all j. Since φ extends to an almost periodic function, φ is continuous on Γ, contradicting $\varphi(\chi_{n(j)}) = (-1)^j$.

Case II. Otherwise, there is a subsequence $n(j)$ such that the $\gamma_{n(j)}$ are distinct, the $\chi_{n(j)}$ are distinct and no $\gamma_{n(j)}$ is a $\chi_{n(k)}$. Let $\varphi : E \to \{-1, 1\}$ have $\varphi(\gamma_{n(j)}) = 1$ and $\varphi(\chi_{n(j)}) = -1$ for all j. Let ψ be any extension of φ to an almost periodic function. Since ψ is uniformly continuous on Γ, there exists $\delta > 0$ such that $d(\gamma, \chi) < \delta$ implies $|\psi(\gamma) - \psi(\chi)| < 1/2$. But $d(\gamma_{n(j)}, \chi_{n(j)}) \to 0$, so for sufficiently large j, $|\psi(\gamma_{n(j)}) - \psi(\chi_{n(j)})| < 1/2$, a contradiction. Therefore, there exists $\delta > 0$ such that $d(\gamma, \chi) > \delta$ for all $\gamma \neq \chi \in E$. $\qquad \square$

Remarks A.1.5. (i) A discrete (in Γ) I-set E is I_0 if and only if every bounded function on E is uniformly continuous, and that occurs only if there is a 1-neighbourhood $W \subset \Gamma$ such that $\gamma \neq \rho \in E$ implies $(\gamma W) \cap (\rho W) = \emptyset$.

(ii) A compact, infinite $E \subset \Gamma$ is always an I-set and never an I_0 set (Exercise A.4.1 (2)).

This property allows us to show that given an I-set E, one can extract a "thick" I_0 subset.

Proposition A.1.6. *Let* $\mathbf{E}$ *be an* I-*set and suppose* $\mathbf{V} \subset \mathbf{\Gamma}$ *is a symmetric* 1-*neighbourhood. Then there exists an* I_0 *set* $\mathbf{F} \subseteq \mathbf{E}$ *such that* $\mathbf{E} \subseteq \mathbf{F} \cdot \mathbf{V}^2$.

Proof. Let $\mathcal{E}$ be the set of subsets $\mathbf{F}$ of $\mathbf{E}$ such that for all $\gamma \neq \rho \in \mathbf{F}$, $(\gamma \mathbf{V}) \cap (\rho \mathbf{V}) = \emptyset$ for $\gamma \neq \rho \in \mathbf{F}$. By Remark A.1.5(i), each such $\mathbf{F}$ is I_0. We note that $\mathcal{E}$ is not empty since it contains singletons. Zorn's lemma gives us a maximal subset $\mathbf{F} \in \mathcal{E}$. We claim that $\mathbf{E} \subseteq \mathbf{F} \cdot \mathbf{V}^2$. Indeed, if $\rho \in \mathbf{E} \setminus \mathbf{F} \cdot \mathbf{V}^2$, then for all $\gamma \in \mathbf{F}$, $\gamma \rho^{-1} \notin \mathbf{V}^2$, that is, $(\gamma \mathbf{V}) \cap (\rho \mathbf{V})$ is empty. Hence, we may adjoin ρ to $\mathbf{F}$ and $\mathbf{F}$ would not have been maximal. □

Corollary A.1.7. *Suppose* $\mathbf{E} \subseteq \mathbf{\Gamma}$ *is an* I-*set and* $\mathbf{W} \subseteq \mathbf{\Gamma}$ *is compact. Then* $\overline{\mathbf{E}} \cdot \mathbf{W}$ *supports no non-zero measure in* $M_0(\overline{\mathbf{\Gamma}})$. *In particular,* $\overline{\mathbf{E}}$ *has zero Haar measure in* $\overline{\mathbf{\Gamma}}$.

Proof. Fix a compact 1-neighbourhood $\mathbf{V} \subseteq \mathbf{\Gamma}$. Use Proposition A.1.6 to find an I_0 set $\mathbf{F} \subseteq \mathbf{E}$ such that $\mathbf{E} \subseteq \mathbf{F} \cdot \mathbf{V}^2$. Proposition A.1.3 shows that $\overline{\mathbf{F}} \cdot \mathbf{V}^2 \mathbf{W}$ supports no non-zero measure in $M_0(\overline{\mathbf{\Gamma}})$. Since $\overline{\mathbf{E}} \cdot \mathbf{W} \subseteq \overline{\mathbf{F}} \cdot \mathbf{V}^2 \mathbf{W}$, the corollary follows. □

There is also a partial converse to the proposition.

Theorem A.1.8 (Hartman–Ryll-Nardzewski extension). *If* $\mathbf{E}$ *is an* I_0 *set, then there exists a compact* 1-*neighbourhood* $\mathbf{V} \subseteq \mathbf{\Gamma}$ *such that* $\mathbf{E} \cdot \mathbf{V}$ *is an* I-*set.*

Exercise A.4.4 shows that $\mathbf{V}$ may be limited in size.

Proof. The σ-compact G version of Kalton's Theorem 3.2.5, gives us a compact set $K \subset \mathbf{\Gamma}$ and integer N such that for every $\varphi : \mathbf{E} \to \Delta$, there exists $\mu \in M_d(K)$ of length at most N such that $|\varphi - \widehat{\mu}| < 1/5$ on $\mathbf{E}$. Let $\mathbf{V} \subset \mathbf{\Gamma}$ be a compact 1-neighbourhood such that $|1 - \gamma(x)| < 1/(5N)$ for all $x \in K$.

We claim that $\chi \mathbf{V} \cap \rho \mathbf{V} = \emptyset$ for all distinct $\chi, \rho \in \overline{\mathbf{E}}$. Indeed, if $\chi \neq \rho$ are given in $\overline{\mathbf{E}}$, then there exists $\mathbf{E}_1 \subset \mathbf{E}$ such that $\chi \in \overline{\mathbf{E}}_1$ and $\rho \notin \overline{\mathbf{E}}_1$. Let $\gamma, \lambda \in \mathbf{V}$ and use Kalton's theorem to find $\mu \in M_d(K)$ with length at most N, $|1 - \widehat{\mu}| < 1/5$ on $\mathbf{E}_1$ and $|\widehat{\mu}| < 1/5$ on $\mathbf{E} \setminus \mathbf{E}_1$. Then for all $\gamma, \lambda \in \mathbf{V}$, $|1 - \widehat{\mu}(\chi \gamma)| < 2/5$ and $|\widehat{\mu}(\rho \lambda)| < 2/5$, so $\chi \gamma \notin \rho \mathbf{V}$, proving the claim.

Now suppose that $\chi_\alpha \gamma_\alpha$ and $\chi'_\alpha \gamma'_\alpha$ are nets in $\mathbf{E} \cdot \mathbf{V}$ with common limit in $\overline{\mathbf{E}} \cdot \mathbf{V}$ and that $f : \mathbf{E} \cdot \mathbf{V} \to \Delta$ is uniformly continuous in the topology of $\mathbf{\Gamma}$. We must show that the limits $\lim f(\chi_\alpha \gamma_\alpha)$ and $\lim f(\chi'_\alpha \gamma'_\alpha)$ exist and are equal. By passing to subnets, we may assume $\lim f(\chi_\alpha \gamma_\alpha)$ and $\lim f(\chi'_\alpha \gamma'_\alpha)$ exist and that $\gamma_\alpha, \gamma'_\alpha$ converge, respectively to $\gamma, \gamma' \in \mathbf{V}$.

The claim above tells us that if $\chi = \lim \chi_\alpha$ and $\chi' = \lim \chi'_\alpha$ were different, then the nets $\chi_\alpha \gamma_\alpha$, $\chi'_\alpha \gamma'_\alpha$ could not have a common limit point.

Let $\varepsilon > 0$. The uniform continuity of f gives us cofinal nets $\chi_\beta \gamma$ and $\chi'_\beta \gamma'$ such that for all β, $|f(\chi_\beta \gamma) - f(\chi_\beta \gamma_\beta)| < \varepsilon$ and $|f(\chi'_\beta \gamma') - f(\chi'_\beta \gamma'_\beta)| < \varepsilon$.

Because $\gamma \mathbf{E}$ is I_0 and χ_β, χ'_β have a common limit point, $\lim f(\chi_\beta \gamma) = \lim f(\chi'_\beta \gamma)$. The continuity of multiplication in $\overline{\mathbf{E}} \cdot \mathbf{V}$ implies $\gamma = \gamma'$ so $\lim f(\chi_\beta \gamma) = \lim f(\chi_\beta \gamma') = \lim f(\chi'_\beta \gamma) = \lim f(\chi'_\beta \gamma')$. Thus,

$$|\lim f(\chi_\beta\gamma_\beta) - \lim f(\chi'_\beta\gamma'_\beta)| < 2\varepsilon.$$

That holds for all $\varepsilon > 0$ and so $\mathbf{E} \cdot \mathbf{V}$ is an I-set.   □

Finally, we note that the elements of an I_0 set can be perturbed and the resulting set will still be I_0.

Corollary A.1.9. *If* $\mathbf{E}$ *is an* I_0 *set, then there exists a* 1-*neighbourhood* $\mathbf{W} \subseteq \Gamma$ *such that if* $f : \mathbf{E} \to \Gamma$ *has the property that* $f(\gamma) \in \gamma\mathbf{W}$ *for all* $\gamma \in \mathbf{E}$, *then* $\mathbf{F} = f(\mathbf{E})$ *is* I_0.

Proof. Let symmetric 1-neighbourhoods $\mathbf{U}, \mathbf{V} \subseteq \Gamma$ be given by Lemma A.1.4 and Theorem A.1.8, respectively, both applied to $\mathbf{E}$. Let $\mathbf{W} \subseteq \Gamma$ be a symmetric 1-neighbourhood such that $\mathbf{W}^4 \subseteq \mathbf{U} \cap \mathbf{V}$. Then $\mathbf{E} \cdot \mathbf{W}$ is an I-set.

Suppose $f(\gamma) \in \gamma\mathbf{W}$ for all $\gamma \in \mathbf{E}$ and $\mathbf{F} = f(\mathbf{E})$. Then $\mathbf{F} \subseteq \mathbf{E} \cdot \mathbf{W}$ and therefore $\mathbf{F}$ is an I-set. By Remark A.1.5, $\mathbf{F}$ is I_0 if $(\gamma'\mathbf{W}) \cap (\rho'\mathbf{W}) = \emptyset$ for all $\gamma' \neq \rho' \in \mathbf{F}$. To see that this is the case, assume otherwise. Let $\gamma, \rho \in \mathbf{E}$ be such that $\gamma'\gamma^{-1}, \rho'\rho^{-1} \in \mathbf{W}$. Then $\gamma\rho^{-1} \in \mathbf{W}^4 \subseteq \mathbf{U}^2$, contradicting the choice of $\mathbf{U}$. Therefore, $\mathbf{F}$ is I_0.   □

A.1.2 Interpolation Sets When Γ Is Non-metrizable

In the absence of metrizability many of the preceding results fail. Here is an example.

Example A.1.10. Let $\Gamma = \mathbb{R} \times \overline{\mathbb{Z}}$. Fix an I_0 set, $\mathbf{E}_1 = \{n_j\}_1^\infty \subseteq \mathbb{Z}$, and let $\{x_j\}_1^\infty$ be a dense subset of $\mathbb{R}$. Set $\mathbf{E} = \{(x_j, n_j) : j = 1, 2, \dots\}$. Then $\mathbf{E}$ is I_0 since the interpolation can be done in the second factor. This example has the following properties, whose verification we leave to Exercise A.4.2:

1. If $x_{j(\ell)}$ converges to $x \in \mathbb{R}$, then each cluster point of $\{(x_{j(\ell)}, n_{j(\ell)})\}$ is a continuous character (Theorem 3.5.1 fails).
2. There is *no* compact $U \subseteq \mathbb{R} \times \mathbb{T}_d$ for which $\mathbf{E} \setminus \mathbf{F}$ is $I_0(U)$, whatever the finite set $\mathbf{F}$.
3. $\overline{\mathbf{E}} \cdot \Gamma$ has non-zero $\overline{\Gamma}$-Haar measure and therefore supports non-zero measures in $M_0(\overline{\Gamma})$ (Proposition A.1.3 fails).
4. $\mathbf{E} \cdot \Gamma$ is not an I-set.
5. There does not exist any $\varepsilon < 1$, integer N and σ-compact $U \subseteq G$ such that $\mathbf{E}$ is $I_0(U, N, \varepsilon)$ (Kalton's characterization Theorem 3.2.5 of I_0 sets fails in this sense).
6. $\mathbf{E}$ is not a finite union of sets with step tending to infinity (Corollary 5.3.5 fails).

We do not know if there exist a non-metrizable Γ and subset $\mathbf{E}$ such that every bounded function on $\mathbf{E}$ extends to an almost periodic function, but not all bounded functions extend to elements of $B(\overline{\Gamma})$ (Kahane's characterization fails) [P 14].

A.1.3 Sidon Sets in Non-compact Abelian Groups

Definition A.1.11. A subset $\mathbf{E} \subseteq \Gamma$ is a *topological Sidon set* if and only if $\ell^{\infty}(\mathbf{E}) = B(\mathbf{E})$.

Since elements of $B(\Gamma)$ are uniformly continuous, we see that if $\mathbf{E}$ is a topological Sidon set, then there exists a 1-neighbourhood $\mathbf{V} \subset \Gamma$ such that $\gamma \neq \rho \in \mathbf{E}$ implies $\gamma\rho^{-1} \notin \mathbf{V}$.

Again, most of the results concerning Sidon sets have topological Sidon set versions when Γ is metrizable. The examples of I_0 sets in non-metrizable Γ apply here, of course. In the remainder of this section we assume Γ is metrizable.

Theorem A.1.12. *The union of two disjoint topological Sidon sets, $\mathbf{E}$ and $\mathbf{F}$, is a topological Sidon set provided that there exists a 1-neighbourhood $\mathbf{V} \subset \Gamma$ such that $\gamma \in \mathbf{E}$, $\rho \in \mathbf{F}$ implies $\gamma\rho^{-1} \notin \mathbf{V}$.*

There is a connection between Sidon sets in Γ_d (Γ with the discrete topology) and topological Sidon sets.

Theorem A.1.13. *A subset $\mathbf{E} \subset \Gamma$ is a topological Sidon set if and only if $\mathbf{E}$ is a Sidon set in Γ_d and there exists a compact subset $K \subset G$ such that $\ell^{\infty}(\mathbf{E}) = M(K)^{\curlyvee}|_{\mathbf{E}}$.*

Topological Sidon sets have an *enlargement* property. A consequence (Exercise A.4.5) is that if $\mathbf{E}$ is topological Sidon and $\mathbf{F}$ is "close" in the sense of Corollary A.1.9, then the perturbed set $\mathbf{F}$ is also topological Sidon.

Theorem A.1.14. *If a subset $\mathbf{E}$ is Sidon(U), where $U \subset G$ is compact, then there exists a 1-neighbourhood $\mathbf{V} \subset \Gamma$ such that for each $\varphi \in \ell^{\infty}(\mathbf{E})$ and $f \in A(\mathbf{V})$, the function $\psi(\lambda\gamma) = \varphi(\lambda)f(\gamma\lambda^{-1})$ belongs to $B(\mathbf{E} \cdot \mathbf{V})$.*

A.2 Interpolation and Sidon Sets in Compact, Non-abelian Groups

A.2.1 Introduction to Compact, Non-abelian Groups

Sidon and I_0 sets have also been defined and investigated in the setting of a compact, non-abelian group G. Although there are similarities with the abelian theory, there are very significant differences. For example, there are infinite non-abelian groups whose dual contains no infinite Sidon set.

In fact, the general harmonic analysis theory is quite different in the non-abelian case. We begin with a brief overview of some of the most important aspects of this theory and refer the reader to [88] for a more thorough explanation.

It is customary to let $\widehat{G}$ denote the set[1] of irreducible, inequivalent, unitary representations of G. When G is abelian, all irreducible representations of G are one dimensional, and hence $\widehat{G}$ is the dual group Γ. When G is non-abelian, $\widehat{G}$ is no longer a group and is called the *dual object*. All irreducible representations of a compact group are finite dimensional, and thus we may view $\widehat{G}$ as a set of matrix-valued functions $\sigma : G \to U(d_\sigma)$, where d_σ denotes the degree of σ and $U(d_\sigma)$ is the group of unitary matrices of size $d_\sigma \times d_\sigma$.

Important examples of compact, non-abelian groups include the classical Lie groups such as the unitary groups, $U(n)$, the group of orthogonal matrices, $O(n)$, and their subgroups consisting of the matrices of determinant 1, $SU(n)$ and $SO(n)$, respectively. Much is understood about the representations of these classical matrix groups; see, for example, [185].

Compact groups all admit a left Haar measure, m_G, and consequently one can speak of the integrable functions on G, the spaces $L^p(G) = L^p(G, m_G)$ and the absolutely continuous measures on G, identified with $L^1(G)$ (all with respect to m_G).

Given an integrable function, f, defined on G, the *Fourier transform* of f is the function $\widehat{f}$ defined on $\widehat{G}$ by

$$\widehat{f}(\sigma) = \int_G f(x)\sigma(x)\mathrm{d}m_G.$$

The integration should be understood coordinatewise, and thus $\widehat{f}(\sigma)$ is a matrix of size $d_\sigma \times d_\sigma$. The *Fourier series* of f is given by

$$\sum_{\sigma \in \widehat{G}} d_\sigma \operatorname{Tr}\left(\widehat{f}(\sigma)\sigma(x)\right)$$

where "Tr" denotes the trace of the matrix. When $\widehat{f}$ is non-zero for only finitely many σ, then f is called a trigonometric polynomial.

The *Fourier–Stieltjes transform* of a measure on G is defined similarly.

Given an $n \times n$ matrix A, let $\|A\|_{op}$ be the maximum eigenvalue of $|A|$, where $|A|$ is the operator which is the positive square root of AA^*. Then $\|A\|_{op}$ is the operator norm of A when viewed as a map on $\mathbb{C}^n$.

---

[1] This should be the set of *equivalence classes*, but, as usual, we shall assume that a convenient representative is chosen from each class.

Let $\ell^\infty(\widehat{G})$ denote the Banach space of all $(A_\sigma)_{\sigma \in \widehat{G}}$, where A_σ is a $d_\sigma \times d_\sigma$ matrix, with norm $\|(A_\sigma)_\sigma\|_\infty = \sup_{\sigma \in \widehat{G}} \|A_\sigma\|_{op} < \infty$. We define $\ell^\infty(\mathbf{E})$ similarly for $\mathbf{E} \subseteq \widehat{G}$ by restricting the representations to $\mathbf{E}$.

There is a non-abelian version of the Riemann–Lebesgue lemma and an analogue of Parseval's theorem, known as the Peter–Weyl theorem.

Lemma A.2.1 (Riemann–Lebesgue). *If $f \in L^1(G)$, then $\left\|\widehat{f}(\sigma)\right\|_{op} \to 0$ as $\sigma \to \infty$.*

Theorem A.2.2 (Peter–Weyl). *Let $f : G \to \mathbb{C}$. Then*

$$\|f\|_2^2 = \sum_{\sigma \in \widehat{G}} d_\sigma \operatorname{Tr}\left(\widehat{f}(\sigma)\widehat{f}(\sigma)^*\right).$$

A measure μ is called *central* if μ commutes with all other measures under convolution. Equivalently, $\mu(gXg^{-1}) = \mu(X)$ for all Borel sets $X \subseteq G$ and $g \in G$. Similarly, *central functions* satisfy $f(gxg^{-1}) = f(x)$ for all $g, x \in G$. Central measures are characterized by the property that their Fourier transforms are central matrices, that is, $\widehat{\mu}(\sigma) = c_\sigma I_{d_\sigma}$ for all σ, where I_{d_σ} denotes the identity matrix of size $d_\sigma \times d_\sigma$ and c_σ is a scalar. In this case, $\|\widehat{\mu}(\sigma)\|_{op} = |c_\sigma|$ and $\|\widehat{\mu}\|_\infty = \sup_\sigma |c_\sigma|$.

A.2.2 Sidon and I_0 Sets

Definition A.2.3. A subset $\mathbf{E} \subseteq \widehat{G}$ is called a *Sidon set* if whenever $(A_\sigma)_{\sigma \in \mathbf{E}} \in \ell^\infty(\mathbf{E})$, there is a measure μ on G satisfying $\widehat{\mu}(\sigma) = A_\sigma$ for all $\sigma \in \mathbf{E}$. If μ can be chosen to be discrete, then $\mathbf{E}$ is said to be an I_0 *set*.

Sidon(U), $I_0(U)$, $RI_0(U)$ and $FZI_0(U)$ are all defined analogously, with the understanding that $\varphi \in \ell^\infty(\mathbf{E})$ is *Hermitian* if $\varphi(\overline{\sigma}) = \overline{\varphi(\sigma)}$ whenever $\sigma, \overline{\sigma} \in \mathbf{E}$ are not equivalent and if σ and $\overline{\sigma}$ are equivalent, say $\sigma = P\overline{\sigma}P^{-1}$ with P unitary, then $\varphi(\sigma) = P\overline{\varphi(\sigma)}P^{-1}$. Of course, every set that is $I_0(U)$ is Sidon(U).

There are many equivalent characterizations of Sidon and I_0 in the same spirit as Theorems 6.2.2 and 3.2.5.

Theorem A.2.4. *Let $\mathbf{E} \subseteq \widehat{G}$ and U be a compact subset of G. The following are equivalent:*

1. $\mathbf{E}$ *is a Sidon(U) set.*
2. *There is a constant $S(\mathbf{E}, U)$ such that given any $(A_\sigma)_{\sigma \in \mathbf{E}} \in \ell^\infty(\mathbf{E})$, there is a measure $\mu \in M(U)$ such that $\widehat{\mu}(\sigma) = A_\sigma$ for all $\sigma \in \mathbf{E}$ and*

$$\|\mu\|_{M(G)} \leq S(\mathbf{E}, U) \left\|(A_\sigma)_\sigma\right\|_\infty.$$

3. For every $(A_\sigma)_{\sigma \in \mathbf{E}} \in \ell^\infty(\mathbf{E})$, with A_σ unitary, there is some $\mu \in M(U)$ such that $\widehat{\mu}(\sigma) = A_\sigma$ for all $\sigma \in \mathbf{E}$.

4. There exists $0 < \varepsilon < 1$ (equivalently, for every $0 < \varepsilon < 1$) such that for each $(A_\sigma)_{\sigma \in \mathbf{E}} \in Ball(\ell^\infty(\mathbf{E}))$ there is some $\mu \in M(U)$ such that

$$\sup\{\|\widehat{\mu}(\sigma) - A_\sigma\|_{op} : \sigma \in \mathbf{E}\} < \varepsilon.$$

5. There is a constant $S(\mathbf{E}, U)$ such that if $f(x) = \sum_{\sigma \in E} d_\sigma \operatorname{Tr}(A_\sigma \sigma(x))$ is a trigonometric polynomial with Fourier transform supported on $\mathbf{E}$, then

$$\sum_{\sigma \in E} d_\sigma \operatorname{Tr}|A_\sigma| \leq S(\mathbf{E}, U)\|f|_U\|_\infty.$$

A proof for Sidon sets can be found in [88, 37.2], where many other basic facts about Sidon sets are also demonstrated. A proof for Sidon(U) sets can be given by making the obvious modifications. In addition to the usual iteration argument, a key idea in the equivalence of (1) and (3) is the observation that every matrix of norm at most one is half the sum of four unitary matrices.

Theorem A.2.5. *Let* $\mathbf{E} \subseteq \widehat{G}$ *and* U *be a compact subset of* G. *The following are equivalent:*

1. $\mathbf{E}$ *is* $I_0(U)$.

2. *There is a constant* $S_d(\mathbf{E}, U)$ *such that given any* $(A_\sigma)_{\sigma \in \mathbf{E}} \in \ell^\infty(\mathbf{E})$, *there is a measure* $\mu \in M_d(U)$ *such that* $\widehat{\mu}(\sigma) = A_\sigma$ *for all* $\sigma \in \mathbf{E}$ *and*

$$\|\mu\|_{M(G)} \leq S_d(\mathbf{E}, U)\|(A_\sigma)_\sigma\|_\infty.$$

3. *For every* $(A_\sigma)_{\sigma \in \mathbf{E}} \in \ell^\infty(\mathbf{E})$, *with* A_σ *unitary, there is some* $\mu \in M_d(U)$ *such that* $\widehat{\mu}(\sigma) = A_\sigma$ *for all* $\sigma \in \mathbf{E}$.

4. *There exists* $0 < \varepsilon < 1$ *(equivalently, for every* $0 < \varepsilon < 1$*) such that for each* $(A_\sigma)_{\sigma \in \mathbf{E}} \in Ball(\ell^\infty(\mathbf{E}))$ *there is some* $\mu \in M_d(U)$ *such that*

$$\|\widehat{\mu}(\sigma) - A_\sigma\|_{op} < \varepsilon \text{ for all } \sigma \in \mathbf{E}.$$

5. *There exists* $0 < \varepsilon < 1$ *(equivalently, for every* $0 < \varepsilon < 1$*) and integer* $N = N(\varepsilon)$ *such that for each* $(A_\sigma)_{\sigma \in \mathbf{E}} \in \ell^\infty(\mathbf{E})$ *with* A_σ *unitary, there is some* $\mu = \sum_{j=1}^N c_j \delta_{x_j} \in M_d(U)$ *with* $|c_j| \leq 1$ *and*

$$\|\widehat{\mu}(\sigma) - A_\sigma\|_{op} < \varepsilon \text{ for all } \sigma \in \mathbf{E}.$$

I_0 and $I_0(U)$ sets for non-abelian groups were introduced in [54, 75], respectively, where proofs can be found of the above characterizations. In [54] a similar characterization is given for $FZI_0(U)$ sets. It is also shown there that $FZI_0(U)$ sets are $I_0(U)$ and that finite sets are $I_0(U)$.

As in the abelian case, every finite union of Sidon sets is Sidon; proofs have been given by Rider [162], Wilson [196], Marcus and Pisier [127] and Baur [8]. It is unknown if Sidon (or I_0) sets in duals of compact, connected groups are necessarily Sidon(U) (resp., $I_0(U)$) for every non-empty, open U [P 16].

$\Lambda(p)$ sets are defined in the non-abelian setting exactly as in abelian groups, and, as in the abelian case, Sidon sets are $\Lambda(p)$ for all $p < \infty$ with $\Lambda(p)$ constant $O(\sqrt{p})$ for every $p > 2$ [38].

Let $G = \prod_{n=1}^{\infty} U(d_n)$ be the product of unitary groups with the product topology. If $\sup d_n = \infty$, then $\widehat{G}$ contains an infinite Sidon set with unbounded degree, as follows. Let π_n denote the projection of G onto $U(d_n)$ and put $\mathbf{E} = \{\pi_n : n = 1, 2, \dots\}$. Each π_n is an irreducible, unitary representation of G of degree d_n. It is easy to see that $\mathbf{E}$ is an I_0 set. Let $(A_n)_n \in \ell^{\infty}(\mathbf{E})$ with A_n unitary and take $x = (A_n) \in G$. Then $\widehat{\delta_x}(\pi_n) = \pi_n(x) = A_n$, and thus condition (3) is satisfied in both theorems.

More generally, groups which have infinitely many irreducible representations of bounded degree contain infinite Sidon sets [96].

However, in contrast to the situation for abelian groups, there are infinite, compact, non-abelian groups whose dual object contains no infinite Sidon set. Even those groups whose dual admits infinite Sidon sets need not have the property that every infinite subset of the dual contains an infinite Sidon subset. Cartwright and McMullen [23] used the relationship between Sidon and $\Lambda(p)$ sets to effectively describe the infinite Sidon sets. The typical example of a Sidon set with unbounded degree, called the FTR set (defined below), is a generalization of the set of projections onto the product group $\prod U(d_n)$.

Definition A.2.6. Suppose G is one of the matrix groups $SU(n)$, $O(n)$, $SO(n)$ or $Sp(n)$. Let $\sigma : G \to U(n)$ be the self-representation, $\sigma(x) = x$. If $G = Spin(n)$ let $q : G \to SO(n)$ be the canonical covering map and let σ be the composition of the self-representation with q. The *Figà–Talamanca and Rider set* (denoted $FTR(G)$) is $\{\sigma, \overline{\sigma}, 1\}$.

Suppose for each j that G_j is one of the groups $SU(n_j)$, $O(n_j)$, $SO(n_j)$, $Sp(n_j)$ or $Spin(n_j)$ and $G = \prod_j G_j$. Let $\pi_j : G \to G_j$ denote the projection maps. Then the FTR set of G is defined as

$$FTR(G) = \bigcup_j \{\sigma \circ \pi_j : \sigma \in FTR(G_j)\}.$$

In the abelian case, every singleton is an I_0 set with the I_0 constant equal to 1 since the interpolating measure can be taken to be a suitable multiple of the point mass measure at the identity. That is not true for non-abelian groups, and hence it is of interest to consider sets where singletons have uniformly bounded Sidon or I_0 constants.

Definition A.2.7. A set $\mathbf{E} \subseteq \widehat{G}$ is called a *local Sidon* (resp., *local I_0*) set if there is a constant $S_1(\mathbf{E})$ such that whenever $\sigma \in \mathbf{E}$ and A_σ is a $d_\sigma \times d_\sigma$ matrix of norm one, then there is a (resp., discrete) measure μ on G with $\|\mu\|_{M(G)} \le S_1(\mathbf{E})$ and $\widehat{\mu}(\sigma) = A_\sigma$.

Characterizations similar to Theorems A.2.4 and A.2.5 can be proved for local Sidon and local I_0 sets.

Obviously, a Sidon (I_0) set is local Sidon (I_0). The converse is false, even in abelian groups, since the dual of an infinite compact abelian group is local I_0, but not Sidon. In contrast, in the non-abelian setting, the dual object need not even be local Sidon. Indeed, in [23, Prop 5.5], Cartwright and McMullen gave an elegant characterization of local Sidon sets in terms of the FTR set.

Theorem A.2.8 (Cartwright–McMullen). *Let $G = \prod G_j$ where each G_j is one of the groups $SU(n_j)$, $SO(n_j)$, $Sp(n_j)$ or $Spin(n_j)$. Let $\mathbf{E} \subseteq \widehat{G}$ be a local Sidon set. Then there are a partition $J = J_1 \cup J_2 \cup J_3$ and subsets $\mathbf{E}_j \subseteq \widehat{H}_j$ where $H_j = \prod_{i \in J_j} G_i$ for $j = 1, 2, 3$, such that:*

1. *$\mathbf{E} \subseteq \mathbf{E}_1 \times \mathbf{E}_2 \times \mathbf{E}_3$*
2. *$\mathbf{E}_1 = \{1\}$*
3. *$\sup\{\operatorname{rank} G_i : i \in J_2\} < \infty$ and $\sup\{d_\sigma : \sigma \in \mathbf{E}_2\} < \infty$*
4. *$\mathbf{E}_3 = FTR(H_3)$*

Conversely, all such sets are local Sidon.

This theorem is useful because a structure theorem states that all compact connected groups are homomorphic to a product of the form $A \times \prod_j G_j$ where G_j are classical, simple, simply connected Lie groups [153, 6.5.6] and A is abelian. Consequently, a compact connected group admits infinite Sidon sets if and only if it admits infinite local Sidon sets.

It is known that $FTR(G)$ is an I_0 set [75] and $FTR(G) \setminus \{1\}$ is FZI_0 [54]. Moreover, given any non-empty, open set $U \subseteq G$, there is a finite subset $\mathbf{F} \subseteq FTR(G)$ such that $FTR(G) \setminus \mathbf{F}$ is $FZI_0(U)$. From these observations the following can be deduced.

Theorem A.2.9. *If G is a compact, connected group, then every infinite local Sidon set in $\widehat{G}$ contains an infinite subset $\mathbf{E}$ that is FZI_0 and has the property that given any non-empty, open set $U \subseteq G$, there is a finite subset $\mathbf{F} \subseteq \mathbf{E}$ such that $\mathbf{E} \setminus \mathbf{F}$ is $FZI_0(U)$.*

Remark A.2.10. It is unknown if the set $\mathbf{E}$ can be chosen with the same cardinality as the local Sidon set [**P 17**].

Corollary A.2.11. *Every infinite local Sidon set contains an infinite I_0 set.*

Just as sums of Hadamard sets in $\mathbb{Z}$ are examples of $\Lambda(p)$ sets that are not Sidon (Example 6.3.14), "products" of FTR sets provide examples of $\Lambda(p)$ sets that are not Sidon. Given $\mathbf{E} \subseteq \widehat{G}$, let

$$\mathbf{E}^m = \{\sigma \in \widehat{G} : \sigma \leq \gamma_1 \otimes \cdots \otimes \gamma_m, \gamma_1, \ldots, \gamma_m \in \mathbf{E}\}$$

where $\sigma \leq \gamma$ means σ is a subrepresentation of γ. It is shown in [78] that if $\mathbf{P}$ is an asymmetric subset of the FTR set of the product group G (as in Definition A.2.6), then $\mathbf{P}^m$ is a $\Lambda(p)$ set for all $p < \infty$.

A.2.3 Central Lacunarity

The weaker notions of central Sidon and central $\Lambda(p)$, where the function and measure spaces are replaced by their centres, were introduced by Parker [143].

Definition A.2.12. The subset $\mathbf{E} \subseteq \widehat{G}$ is said to be *central Sidon* if whenever $(a_\sigma I_{d_\sigma})_{\sigma \in \mathbf{E}} \in \ell^\infty(\mathbf{E})$, there is a central measure $\mu \in M(G)$ such that $\widehat{\mu}(\sigma) = a_\sigma I_{d_\sigma}$ for all $\sigma \in \mathbf{E}$. Equivalently, there is a constant S_c such that if f is a central trigonometric polynomial with transform supported on $\mathbf{E}$, then

$$\sum_{\sigma \in \mathbf{E}} d_\sigma \, \text{Tr} \, |\widehat{f}(\sigma)| \leq S_c \|f\|_\infty \, .$$

Central $\Lambda(p)$ sets are defined similarly.

It is clear from the second statement in the definition that Sidon implies central Sidon and similarly that $\Lambda(p)$ implies central $\Lambda(p)$. But the converse does not hold; for example, see [143, 161]. In fact, Rider proved in [161] that a set $\mathbf{E}$ is $\Lambda(p)$ if and only if $\mathbf{E}$ is both local $\Lambda(p)$ (meaning singleton subsets of $\mathbf{E}$ are $\Lambda(p)$ with uniform constant) and central $\Lambda(p)$. He gave an example of a set that is central Sidon and central $\Lambda(p)$ for all $p < \infty$, but not local $\Lambda(p)$ for any p, and also showed that the union property fails for central Sidon sets. Arithmetic properties of central $\Lambda(p)$ sets, similar to those of $\Lambda(p)$ sets in abelian groups (see Sect. 6.3.2), were established in [10].

Although central Sidon is a weaker property than Sidon, it is still not true that all infinite groups admit infinite central Sidon sets. This follows from a geometric result of Ragozin [155] which states that if G is a compact, simple, simply connected Lie group and ν is a continuous, central measure on G, then $\nu^{\dim G}$ is absolutely continuous with respect to m_G. Consequently, the Riemann–Lebesgue lemma implies $\widehat{\nu} \to 0$. If μ is a central measure, then $\mu = \mu_c + \mu_d$, where μ_c is central and continuous and μ_d is central and discrete. Since the central, discrete measures are supported on a finite set, namely the centre of G, the vector space consisting of their transforms is a finite dimensional space. Hence, when $\mathbf{E}$ is infinite, it is not possible to interpolate all central elements in $\ell^\infty(\mathbf{E})$ with the transforms of central measures, and therefore G does not admit an infinite central Sidon set.

More generally, with these ideas and the structure theorem, Rider [161] proved the following.

Theorem A.2.13. *Let G be compact and connected. Then G has an infinite central Sidon set if and only if G is not a semi-simple Lie group.*

One could analogously define *central I_0 sets*, but since central, discrete measures must be supported on the centre of G, there would be groups for which not even all finite sets would be central I_0. Thus, instead of requiring interpolation by central, discrete measures in the definition of central I_0 we ask, instead, for interpolation from the subspace generated by the orbital measures. The *orbital measures* are the central, probability measures, μ_x, supported on the conjugacy class containing $x \in G$ and defined by

$$\int_G f d\mu_x = \int_G f(gxg^{-1}) dm_G(g)$$

for all continuous functions f on G. Clearly, any central I_0 set is central Sidon. Any I_0 set can be shown to be central I_0 and all finite sets are central I_0 [67]. Parker's [143] independent-like so-called I-sets are examples of infinite central I_0 sets in the product group setting.

A.3 Remarks and Credits

Interpolation and Sidon Sets in Non-compact Abelian Groups. Sect. A.1 is adapted from [81–83]. Example A.1.10 was suggested by [156, Theorem 2]. For more on topological Sidon sets, including the proofs of Theorems A.1.12–A.1.14, see [27, 128, 132].

Thin Sets in Other Settings. The notion of interpolation has been extended in a variety of other directions, as well. For instance, the problem of approximating random choices of signs by characters was addressed in [67]. This is related to the study of "weighted" Sidon or I_0 sets; see, for example, [68]. Thin sets in the setting of discrete non-abelian groups have been studied using ideas from operator spaces in [71], while in [68, 191], Sidon and I_0-like sets in duals of compact abelian hypergroups are investigated.

A.4 Exercises

In the following exercises Γ is a locally compact abelian group.

Exercise A.4.1.   1. Let $\mathbf{E}$ be an infinite, compact subset of Γ with non-empty interior. Show that

(i) For each $j \geq 1$, $\mathbf{E}$ contains sets of the form $\mathbf{F}_1 \cdot \mathbf{F}_2$ with $|\mathbf{F}_1| = |\mathbf{F}_2| = j$.

(ii) $\mathbf{E}$ supports a non-zero measure in $M_0(\mathbf{\Gamma})$.

(iii) $\mathbf{E}$ is neither Helson nor I_0.

2. Show that an infinite compact subset $\mathbf{E}$ of a metrizable $\mathbf{\Gamma}$ is always an I-set and never an I_0 set.

Exercise A.4.2. Prove the assertions of Example A.1.10.

Exercise A.4.3. Prove that if $\mathbf{E}$ is an I_0 subset of a metrizable group $\mathbf{\Gamma}$, then there exists a finite number of I_0 sets $\mathbf{E}_n$ such that each $\mathbf{E}_n$ is I_0 with bounded length and $\mathbf{E} = \bigcup_1^N \mathbf{E}_n$.

Exercise A.4.4. Show that one cannot always make $\mathbf{V}$ arbitrarily large in Theorem A.1.8. Hint: Consider a set $\mathbf{E}$ contained in two disjoint subsets of $\mathbb{Z}$ and $\mathbf{V} = [-1, 1]$.

Exercise A.4.5. Let $\mathbf{\Gamma}$ be a metrizable group and let $\mathbf{E} \subset \mathbf{\Gamma}$ be topological Sidon.

1. Let $\mathbf{F} \subset \mathbf{\Gamma}$ be finite. Show that $\mathbf{E} \cup \mathbf{F}$ is topological Sidon without using the union theorem for topological Sidon sets.

2. Show that there exist $C > 0$, a compact $K \subseteq G$ and finite $\mathbf{F} \subseteq \mathbf{E}$ such that for every $\varphi : \mathbf{E} \to \Delta$ there exists $\mu \in M(K)$ such that $\|\mu\| \leq C$ and $\widehat{\mu} = \varphi$ on $\mathbf{E} \smallsetminus \mathbf{F}$.

3. Let $\mathbf{V} \subset \mathbf{\Gamma}$ be a compact 1-neighbourhood. Show that there exist $C > 0$, a compact 1-neighbourhood $\mathbf{D} \subseteq \mathbf{V}$ and finite $\mathbf{F} \subseteq \mathbf{E}$ such that for every $\varphi : \mathbf{E} \cdot \mathbf{D} \to \Delta$, that is constant on each $\gamma \mathbf{D}$ with $\gamma \in \mathbf{E} \smallsetminus \mathbf{F}$, there exists $\mu \in M(K)$ such that $\|\mu\| \leq C$ and $|\widehat{\mu} - \varphi| \leq 1/2$ on $(\mathbf{E} \smallsetminus \mathbf{F}) \cdot \mathbf{D}$.

4. Formulate and prove a Sidon set version of Corollary A.1.9.

Appendix B
Combinatorial Results

Combinatorial results needed for the proportional quasi-independent characterization of Sidon sets.

B.1 Words and Quasi-independent Sets

We recall Definition 6.2.9: For a subset $\mathbf{F}$ of Γ and positive integer k, let

$$W_k(\mathbf{F}) = \prod_{\gamma \in \mathbf{F}} \left\{ \gamma^{\varepsilon_\gamma} : \gamma \in \mathbf{F}, \varepsilon_\gamma \in \mathbb{Z} \text{ and } \sum_\gamma |\varepsilon_\gamma| \leq k \right\} \text{ and}$$

$$Q_k(\mathbf{F}) = \left\{ \prod_{\gamma \in \mathbf{F}} \gamma^{\varepsilon_\gamma} : \gamma \in \mathbf{F}, \varepsilon_\gamma \in \{0, -1, 1\} \text{ and } \sum_\gamma |\varepsilon_\gamma| = k \right\}.$$

The elements of $W_k(\mathbf{F})$ are called *words from* $\mathbf{F}$ *of length at most* k and the elements of $Q_k(\mathbf{F})$ are called *quasi-words from* $\mathbf{F}$ *of presentation length* k. Of course, $Q_k(\mathbf{F}) \subseteq W_k(\mathbf{F})$. We note that $|Q_k(\mathbf{F})| \leq 3^{|\mathbf{F}|}$.

The number of words of length k in $\mathbf{F} \subseteq \Gamma$ can be estimated as follows.

Lemma B.1.1. *Let* $\mathbf{F} \subseteq \Gamma$ *be finite and* $k \geq 1$. *Then*

$$|W_k(\mathbf{F})| \leq \begin{cases} \left(4e\frac{|\mathbf{F}|}{k}\right)^k & \text{if } k \leq |\mathbf{F}| \\ \left(4e\frac{k}{|\mathbf{F}|}\right)^{|\mathbf{F}|} & \text{if } |\mathbf{F}| \leq k. \end{cases} \qquad \text{(B.1.1)}$$

C.C. Graham and K.E. Hare, *Interpolation and Sidon Sets for Compact Groups*,
CMS Books in Mathematics, DOI 10.1007/978-1-4614-5392-5_12,
© Springer Science+Business Media New York 2013

Proof. Let $N = |\mathbf{F}|$ and put $\mathbb{N}_0 = \mathbb{N} \cup \{0\}$. Then

$$|W_k(\mathbf{F}))| \leq |\{(k_1, \ldots, k_N) \in \mathbb{Z}^N : \sum_{n=1}^N |k_n| \leq k\}|$$

$$\leq \min(2^k, 2^N) \, |\{(k_1, \ldots, k_N) \in \mathbb{N}_0^N : \sum_{n=1}^N k_n \leq k\}|$$

$$= \min(2^k, 2^N) \, |\{(k_1, \ldots, k_{N+1}) \in \mathbb{N}_0^{N+1} : \sum_{n=1}^{N+1} k_n = k\}|.$$

Let $a_k = |\{(k_1, \ldots, k_{N+1}) \in \mathbb{N}_0^{N+1} : \sum k_n = k\}|$. Then $a_0 = 1$ and

$$\sum_{k=0}^\infty a_k x^k = \left(\sum_{k=0}^\infty x^k\right)^{N+1} = \frac{1}{(1-x)^{N+1}}.$$

Therefore $a_k = \binom{N+k}{k}$ for all $k \geq 0$. We use Stirling's formula [37, p. 52] in the form $\sqrt{2\pi}\, n^{n+\frac{1}{2}} e^{-n} \leq n! \leq \sqrt{2\pi}\, n^{n+\frac{1}{2}} e^{-n} e^{\frac{1}{12n}}$. Applying this to a_k, we have

$$a_k = \frac{(N+k)!}{k!N!} \leq \frac{e^{1/(12N)}}{\sqrt{2\pi}} \sqrt{2} \left(1 + \frac{k}{N}\right)^N \left(1 + \frac{N}{k}\right)^k.$$

Now suppose $N \leq k$. Then $\left(1 + \frac{k}{N}\right)^N \leq 2^N (\frac{k}{N})^N$ and $\left(1 + \frac{N}{k}\right)^k \leq e^N$, so

$$|W_k(\mathbf{F})| \leq \min(2^k, 2^N)\, 2^N \left(\frac{k}{N}\right)^N e^N \leq \left(4e\frac{k}{N}\right)^N,$$

and (B.1.1) holds in this case. The case $k \leq N$ is similar. $\square$

We now use these results to find quasi-independent subsets.

Proposition B.1.2. *There is a constant $A > 10$ such that whenever the finite sets $\mathbf{E}_1, \ldots, \mathbf{E}_J$ are quasi-independent, pairwise disjoint and satisfy $|\mathbf{E}_{j+1}|/|\mathbf{E}_j| \geq A$ for $1 \leq j < J$, we can find subsets $\mathbf{E}'_j \subseteq \mathbf{E}_j$ such that $\bigcup_{j=1}^J \mathbf{E}'_j$ is quasi-independent and $|\mathbf{E}'_j| \geq |\mathbf{E}_j|/10$ for $1 \leq j \leq J$.*

Proposition B.1.2 is used on page 121. The proof of Proposition B.1.2 is first reduced to a combinatorial criterion in Lemma B.1.3 and then further technical probabilistic and combinatorial arguments, involving Riesz products, are used to show that under the assumptions of the proposition, the hypotheses of the lemma can be satisfied.

Lemma B.1.3. *Let $\mathbf{E}_1, \ldots, \mathbf{E}_J$ be finite, disjoint subsets of Γ. Suppose that for each j there is a subset $\mathbf{F}_j \subseteq \mathbf{E}_j$, with $|\mathbf{F}_j| \geq \frac{1}{5}|\mathbf{E}_j|$, and such that*

$$\prod_{\ell=1}^{J} \lambda_\ell \neq 1 \qquad (\text{B.1.2a}_j)$$

$$\text{whenever } \lambda_j \in Q_{p_j}(\mathbf{F}_j) \text{ with } p_j \geq |\mathbf{F}_j|/2 \text{ and} \qquad (\text{B.1.2b}_j)$$

$$\lambda_k \in W_{d_{j,k}}(\mathbf{E}_k) \text{ with } d_{j,k} = \frac{|\mathbf{E}_j|}{|\mathbf{E}_k|} p_j \text{ for } k \neq j. \qquad (\text{B.1.2c}_j)$$

Then there are subsets $\mathbf{E}'_j \subseteq \mathbf{E}_j$ *such that* $\bigcup_{j=1}^{J} \mathbf{E}'_j$ *is quasi-independent and* $|\mathbf{E}'_j| \geq |\mathbf{E}_j|/10$ *for* $1 \leq j \leq J$.

Proof (of Lemma B.1.3). We first construct the $\mathbf{E}'_j \subseteq \mathbf{E}_j$ and then show that their union is quasi-independent. Temporarily fix $j \in \{1, \ldots, J\}$. There are two cases to consider.

Case I. Assume there are words λ_ℓ for $\ell = 1, \ldots, J$, such that $\lambda_j \in Q_{p_j}(\mathbf{F}_j)$, $\lambda_k \in W_{d_{j,k}}(\mathbf{E}_k)$ for each $k \neq j$, $k = 1, \ldots, J$ (where $d_{j,k}$ is as in (B.1.2c$_j$)) and $\prod_{\ell=1}^{J} \lambda_\ell = 1$.

Since the sets $\mathbf{E}_1, \ldots, \mathbf{E}_J$ are finite and since $\mathbf{F}_j \subseteq \mathbf{E}_j$, there can be only finitely many such collections of (quasi-)words. Among that set of collections, let $\sigma_1^{(j)}, \ldots, \sigma_J^{(j)}$ be a choice with the presentation length, $p(\sigma_j^{(j)})$, of the quasi-word, $\sigma_j^{(j)}$, maximal. Of course, $p_j := p(\sigma_j^{(j)}) \leq |\mathbf{F}_j|/2$.

Because $\sigma_j^{(j)}$ is a quasi-word in $Q_{p_j}(\mathbf{F}_j)$, there are characters $\gamma \in \mathbf{F}_j$ and scalars $\epsilon_\gamma \in \{0, -1, 1\}$ such that $\sigma_j^{(j)} = \prod_{\gamma \in \mathbf{F}_j} \gamma^{\epsilon_\gamma}$. Let $\mathbf{F}'_j = \{\gamma \in \mathbf{F}_j : \epsilon_\gamma \neq 0\}$ and $\mathbf{E}'_j = \mathbf{F}_j \smallsetminus \mathbf{F}'_j$. Then $|\mathbf{E}'_j| \geq |\mathbf{F}_j| - |\mathbf{F}'_j| \geq |\mathbf{F}_j|/2 \geq |\mathbf{E}_j|/10$, as desired.

Case II. Otherwise, there is no such collection of words. In this case, we set $\mathbf{E}'_j = \mathbf{F}_j$. Clearly, $|\mathbf{E}'_j| \geq |\mathbf{E}_j|/10$.

We repeat the above to define the sets $\mathbf{E}'_1, \ldots, \mathbf{E}'_J$ for each $1 \leq j \leq J$.

We now prove that $\mathbf{E}' = \bigcup_{j=1}^{J} \mathbf{E}'_j$ is quasi-independent. Assume not. Then there is a non-trivial quasi-relation in $\mathbf{E}'$, say $\prod_{j=1}^{J} \lambda_j = 1$, where for each j, λ_j is a quasi-word from $\mathbf{E}'_j$ and not all presentation lengths, $p(\lambda_j)$, are zero. Choose the index j_0 such that $|\mathbf{E}_{j_0}| p(\lambda_{j_0}) = \max_j |\mathbf{E}_j| p(\lambda_j)$.

The choice of j_0 ensures that for each $k \neq j_0$, λ_k is a quasi-word from $\mathbf{E}_k$ of presentation length $p(\lambda_k) \leq |\mathbf{E}_{j_0}| p(\lambda_{j_0})/|\mathbf{E}_k|$. In particular, Case I applies to the index $j = j_0$. For notational ease, we set $\sigma_k = \sigma_k^{(j_0)}$, $1 \leq k \leq J$.

Consider the words $\omega_\ell = \sigma_\ell \lambda_\ell$, $1 \leq \ell \leq J$. Clearly $\prod_{\ell=1}^{J} \omega_\ell = 1$. Since σ_{j_0} is a quasi-word from $\mathbf{F}'_{j_0}$ and λ_{j_0} is a quasi-word from the disjoint set $\mathbf{E}'_{j_0}$, ω_{j_0} is a quasi-word from $\mathbf{F}'_{j_0} \cup \mathbf{E}'_{j_0} = \mathbf{F}_{j_0}$ with presentation length $p(\omega_{j_0}) = p(\sigma_{j_0}) + p(\lambda_{j_0}) > p(\sigma_{j_0})$. Furthermore, for $k \neq j_0$, ω_k is a word from $\mathbf{E}_k$ of length at most

$$\frac{|\mathbf{E}_{j_0}|}{|\mathbf{E}_k|} p(\sigma_{j_0}) + \frac{|\mathbf{E}_{j_0}|}{|\mathbf{E}_k|} p(\lambda_{j_0}) = \frac{|\mathbf{E}_{j_0}|}{|\mathbf{E}_k|} p(\omega_{j_0}).$$

That shows that the collection of words $\omega_1, \ldots, \omega_J$ satisfies the criteria of Case I with the index $j = j_0$. But the presentation length of ω_{j_0} is greater than that of σ_{j_0}, contradicting the choice of σ_{j_0}. $\qquad\qquad\square$

Proof (of Proposition B.1.2). The strategy will be to construct sets $\mathbf{F}_j \subseteq \mathbf{E}_j$ satisfying the three conditions of $(\mathrm{B.1.2}_j)$. We note that equations $(\mathrm{B.1.2}_j)$ depend on the set $\mathbf{F}_j$, but not on any other sets $\mathbf{F}_k$ for $k \neq j$. Thus, we can construct the sets $\mathbf{F}_j$ independently of each other and verify $(\mathrm{B.1.2}_j)$ for each index j separately.

We put $A = (\frac{480}{\log 2})^2$. Fix an index j and first suppose $|\mathbf{E}_j| \geq A$. The subset $\mathbf{F}_j$ will be defined probabilistically. To begin, choose independent, Bernoulli random variables ξ_γ, for $\gamma \in \mathbf{E}_j$, such that

$$\mathbb{P}(\xi_\gamma = 1) = \frac{1}{4} \quad \text{and} \quad \mathbb{P}(\xi_\gamma = 0) = \frac{3}{4}.$$

For each ω define subsets of $\boldsymbol{\Gamma}$ by $\mathbf{F}_j(\omega) = \{\gamma \in \mathbf{E}_j : \xi_\gamma(\omega) = 1\}$. For $\ell \in \mathbb{N}$ and $1 \leq k \leq J$, $k \neq j$ (and remembering that j is temporarily fixed) we put $d_k(\ell) = \ell |\mathbf{E}_j|/|\mathbf{E}_k|$ and define functions

$$g_j(\omega, x) = \sum_{\ell = \lceil \frac{|\mathbf{E}_j|}{10} \rceil}^{|\mathbf{E}_j|} \left(\sum_{\substack{\mathbf{H} \subseteq \mathbf{E}_j \\ |\mathbf{H}| = \ell}} \prod_{\gamma \in \mathbf{H}} \xi_\gamma(\omega) q_\gamma(x) \prod_{\substack{1 \leq k \leq J \\ k \neq j}} \sum_{\lambda \in W_{d_k(\ell)}(\mathbf{E}_k)} \lambda(x) \right),$$

where $q_\gamma(x) = 2\mathfrak{Re}\gamma(x)$ if γ is not of order two and $q_\gamma = \gamma$ otherwise.

Then $\int_G g_j(\omega, x) \mathrm{d}m_G(x)$ is the number of J-tuples $(\lambda_1, \ldots, \lambda_J)$ with $\prod_{k=1}^{J} \lambda_k = 1$ for which $\lambda_j \in Q_\ell(\mathbf{F}_j(\omega))$ for some $\ell \geq |\mathbf{E}_j|/10$ and $\lambda_k \in W_{d_k(\ell)}(\mathbf{E}_k)$ for each $k \neq j$.

Thus, to verify that the three conditions of $(\mathrm{B.1.2}_j)$ hold for this index j, it will be enough to show that there exists ω such that $\int_G g_j(\omega, x) \mathrm{d}m_G(x) = 0$ and $|\mathbf{F}_j(\omega)| \geq \frac{1}{5}|\mathbf{E}_j|$. Let

$$\Omega_1 = \left\{ \omega \in \Omega : \int_G g_j(\omega, x) \mathrm{d}m_G(x) = 0 \right\} \quad \text{and}$$

$$\Omega_2 = \left\{ \omega : |\mathbf{F}_j(\omega)| \geq \frac{1}{5}|\mathbf{E}_j| \right\}.$$

It will suffice to show that $\mathbb{P}(\Omega_1 \cap \Omega_2) > 0$. We argue by contradiction and assume that $\mathbb{P}(\Omega_1 \cap \Omega_2) = 0$. Then Ω_2 would be a subset of the complement Ω_1^c (up to a $\mathbb{P}$-null set), and so $\mathbb{P}(\Omega_2) \leq \mathbb{P}(\Omega_1^c)$.

Let us first obtain a lower bound for $\mathbb{P}(\Omega_2)$. Because $\mathbb{P}(\xi_\gamma \neq 0) = 1/4$,

$$\frac{|\mathbf{E}_j|}{4} = \mathbb{E}\left(\sum_{\gamma \subseteq \mathbf{E}_j} \xi_\gamma \right) = \int_\Omega |\mathbf{F}_j(\omega)| \mathrm{d}\mathbb{P}(\omega)$$

$$\leq \frac{|\mathbf{E}_j|}{5} \mathbb{P}(\Omega_2^c) + |\mathbf{E}_j| \mathbb{P}(\Omega_2) \leq \frac{|\mathbf{E}_j|}{5} + \frac{4|\mathbf{E}_j|}{5} \mathbb{P}(\Omega_2).$$

Hence, $\frac{1}{16} \leq \mathbb{P}(\Omega_2) \leq \mathbb{P}(\Omega_1^c)$. Because $\int_G g_j(\omega, x)\mathrm{d}x \geq 1$ on Ω_1^c,

$$\mathbb{P}(\Omega_1^c) \leq \int_{\Omega_1^c} \int_G g_j(\omega, x)\mathrm{d}x\,\mathrm{d}\mathbb{P}(\omega) \leq \int_\Omega \int_G g_j(\omega, x)\,\mathrm{d}x\,\mathrm{d}\mathbb{P}(\omega).$$

It will now suffice to show that $\int_\Omega \int_G g_j(\omega, x)\,\mathrm{d}x\,\mathrm{d}\mathbb{P}(\omega) < 1/16$ to obtain a contradiction. Note that the independence of the functions ξ_γ, $\gamma \in \mathbf{E}_j$, implies that for each $\mathbf{H} \subseteq \mathbf{E}_j$ with $|\mathbf{H}| = \ell$, $\int \prod_{\gamma \in \mathbf{H}} \xi_\gamma(\omega)\mathrm{d}\mathbb{P}(\omega) = 4^{-\ell}$. We interchange the order of integration and use this fact to get

$$\int_\Omega \int_G g_j(\omega, x)\,\mathrm{d}x\,\mathrm{d}\mathbb{P}(\omega) = \int_G \int_\Omega g_j(\omega, x)\,\mathrm{d}\mathbb{P}(\omega)\,\mathrm{d}x$$

$$= \sum_{\ell = \lceil \frac{|\mathbf{E}_j|}{10} \rceil}^{|\mathbf{E}_j|} \frac{1}{4^\ell} \int_G \Big(\sum_{\substack{\mathbf{H} \subseteq \mathbf{E}_j \\ |\mathbf{H}|=\ell}} \prod_{\gamma \in \mathbf{H}} q_\gamma \Big) \Big(\prod_{\substack{1 \leq k \leq J \\ k \neq j}} \sum_{\lambda \in W_{d_k(\ell)}(\mathbf{E}_k)} \lambda \Big)\,\mathrm{d}x,$$

where we have suppressed the variable x in the integrands. The expansion of the Riesz product $p(x) := \prod_{\gamma \in \mathbf{E}_j} p_\gamma(x)$, where $p_\gamma(x) = 1 + \frac{1}{2}(\gamma(x) + \overline{\gamma}(x))$ if γ is not of order two or $p_\gamma(x) = 1 + \frac{1}{2}\gamma(x)$ if γ is of order two, includes all the terms of the sum

$$\frac{1}{2^\ell} \sum_{\substack{\mathbf{H} \subseteq \mathbf{E}_j \\ |\mathbf{H}|=\ell}} \prod_{\gamma \in \mathbf{H}} q_\gamma,$$

as well as other terms. Since each term in the expansion of $p(x)$ has a real, non-negative contribution to the integral over G,

$$\int_\Omega \int_G g_j(\omega, x)\,\mathrm{d}x\,\mathrm{d}\mathbb{P}(\omega) \leq \sum_{\ell = \lceil \frac{|\mathbf{E}_j|}{10} \rceil}^{|\mathbf{E}_j|} \frac{1}{2^\ell} \int_G \Big(\prod_{\gamma \in \mathbf{E}_j} p_\gamma \prod_{k \neq j} \sum_{\lambda \in W_{d_k(\ell)}(\mathbf{E}_k)} \lambda \Big)\,\mathrm{d}x.$$

Since $\mathbf{E}_j$ is quasi-independent, $p(x)$ has $L^1(G)$-norm 1, and so

$$\int_G \Big(\prod_{\gamma \in \mathbf{E}_j} p_\gamma \prod_{k \neq j} \sum_{\lambda \in W_{d_k(\ell)}(\mathbf{E}_k)} \lambda \Big) \mathrm{d}m_G(x) \leq \Big\| \prod_{k \neq j} \sum_{\lambda \in W_{d_k(\ell)}(\mathbf{E}_k)} \lambda \Big\|_\infty$$

$$\leq \prod_{k \neq j} |W_{d_k(\ell)}(\mathbf{E}_k)| \leq \prod_{k \neq j} |W_{d_k}(\mathbf{E}_k)|,$$

where $d_k = |\mathbf{E}_j|^2/|\mathbf{E}_k|$. The sets $\mathbf{E}_k$ are increasing in cardinality, and hence $d_k \leq |\mathbf{E}_k|$ if $j \leq k$ and $d_k \geq |\mathbf{E}_k|$ if $j \geq k$. Therefore, Lemma B.1.1 gives

$$|W_{d_k}(\mathbf{E}_k)| \leq \begin{cases} \Big(4e\frac{|\mathbf{E}_k|^2}{|\mathbf{E}_j|^2}\Big)^{|\mathbf{E}_j|^2/|\mathbf{E}_k|} & \text{if } j < k \\[2mm] \Big(4e\frac{|\mathbf{E}_j|^2}{|\mathbf{E}_k|^2}\Big)^{|\mathbf{E}_k|} & \text{if } k < j. \end{cases}$$

Now use the fact that $\log x \leq 2\sqrt{x}$ for $x \geq 1$ to obtain

$$
\log | W_{d_k}(\mathbf{E}_k)| \leq
\begin{cases}
\dfrac{|\mathbf{E}_j|^2}{|\mathbf{E}_k|} \times 8\sqrt{\dfrac{|\mathbf{E}_k|}{|\mathbf{E}_j|}} = 8\dfrac{|\mathbf{E}_j|^{3/2}}{|\mathbf{E}_k|^{1/2}} & \text{if } j < k \\[3mm]
|\mathbf{E}_k| \times 8\sqrt{\dfrac{|\mathbf{E}_j|}{|\mathbf{E}_k|}} = 8|\mathbf{E}_j|^{1/2}|\mathbf{E}_k|^{1/2} & \text{if } k < j.
\end{cases}
$$

Since $|\mathbf{E}_{j+1}|/|\mathbf{E}_j| \geq 10$ and $A > 10$,

$$
\log \prod_{k \neq j} | W_{d_k}(\mathbf{E}_k)| \leq 8|\mathbf{E}_j| \left[\sum_{k<j} \left(\frac{|\mathbf{E}_k|}{|\mathbf{E}_j|} \right)^{1/2} + \sum_{k>j} \left(\frac{|\mathbf{E}_j|}{|\mathbf{E}_k|} \right)^{1/2} \right]
$$

$$
\leq 8|\mathbf{E}_j| \left[\sum_{k<j} \left(\frac{1}{A^{j-k}} \right)^{1/2} + \sum_{k>j} \left(\frac{1}{A^{k-j}} \right)^{1/2} \right]
$$

$$
\leq 8|\mathbf{E}_j| \times \frac{2}{\sqrt{A}} \times \frac{1}{1 - 1/\sqrt{A}} \leq 8|\mathbf{E}_j| \frac{3}{\sqrt{A}}.
$$

Thus,

$$
\int_\Omega \int_G g_j(\omega, x)\mathrm{d}x\, \mathrm{d}\mathbb{P}(\omega) \leq \exp\left(24|\mathbf{E}_j|/\sqrt{A} \right) \sum_{\ell \geq |\mathbf{E}_j|/10} \frac{1}{2^\ell}
$$

$$
\leq \exp\left(\frac{24|\mathbf{E}_j|}{\sqrt{A}} \right) \times 2 \times 2^{-|\mathbf{E}_j|/10}
$$

$$
= 2\exp\left(|\mathbf{E}_j| \left(-\frac{\log 2}{10} + \frac{24}{\sqrt{A}} \right) \right)
$$

$$
= 2\exp\left(|\mathbf{E}_j| \left(\frac{-\log 2}{20} \right) \right),
$$

because $A = \left(\frac{480}{\log 2} \right)^2$. Since we have assumed $|\mathbf{E}_j| \geq A$, we conclude that $\int_\Omega \int_G g_j(\omega, x)\mathrm{d}x\, \mathrm{d}\mathbb{P}(\omega) < 1/20$, a contradiction.

Finally, suppose $|\mathbf{E}_j| < A$. Of course, the condition $|\mathbf{E}_{\ell+1}|/|\mathbf{E}_\ell| \geq A$ for all ℓ ensures that $j = 1$. In this case, set $\mathbf{F}_1 = \mathbf{E}_1$. Since the presentation length of a quasi-word from $\mathbf{F}_1$ is at most $|\mathbf{F}_1|$, an easy computation shows that (in the notation of Lemma B.1.3) $d_{1,k} < 1$ for $k \geq 2$. Thus, $W_{d_{1,k}}(\mathbf{E}_k) = \{\mathbf{1}\}$ for all $k \geq 2$. Consequently, if $\lambda_1 \in Q_{p_1}(\mathbf{F}_1)$ with $p_1 \geq |\mathbf{F}_1|/2 > 0$ and $\lambda_k \in W_{d_{1,k}}(\mathbf{E}_k)$ for $k \geq 2$, then $\prod_{k=1}^J \lambda_k = \lambda_1 \neq \mathbf{1}$. Therefore, the three conditions of $(B.1.2_j)$ are satisfied in this case, as well. $\square$

This material is from [119, pp. 496–8]. Proposition B.1.2 is in Bourgain [19].

Appendix C
Background Material

Background material required from abstract harmonic analysis and probability.

C.1 Overview of Harmonic Analysis on Locally Compact Abelian Groups

C.1.1 Topological Groups

A topological group is a Hausdorff space, G, that is also a group, and for which the map $(x, y) \mapsto xy^{-1}$ is continuous from the product space $G \times G$ to G. If the topology is compact, G is called a compact group. G is called locally compact if there is a neighbourhood base of compact sets.

The definition of a topological group implies that neighbourhoods of $x \in G$ are those sets of the form xU where U is a neighbourhood of the identity $e \in G$. Every e-neighbourhood U contains a symmetric e-neighbourhood V (i.e. $V = V^{-1}$) with $V^2 \subseteq U$. Thus, every topological group G is a regular topological space, meaning that if A is a closed subset of G and $x \notin A$, then there are disjoint open sets U, V such that $x \in U$ and $A \subseteq V$. It also follows that every compact group is locally compact.

Every locally compact abelian group G or compact group G admits a *Haar measure* m_G – a non-negative, regular, Borel measure, not identically 0 and satisfying $m_G(E) = m_G(xE)$ for all $x \in G$ and Borel sets $E \subseteq G$. A Haar measure is also invariant under inversion ($m_G(E) = m_G(E^{-1})$) for all Borel sets E), the measure of any non-empty, open set is strictly positive and the measure of any compact set is finite. Up to scaling, the Haar measure is unique. When the group is compact it is customary to normalize m_G so that $m_G(G) = 1$ and when it is discrete to normalize so the measure of a singleton is 1.

C.C. Graham and K.E. Hare, *Interpolation and Sidon Sets for Compact Groups*,   207
CMS Books in Mathematics, DOI 10.1007/978-1-4614-5392-5_13,
© Springer Science+Business Media New York 2013

For the remainder of this section G will denote a locally compact abelian group.

Dual Groups

A complex-valued function $\gamma : G \to \mathbb{T}$ is called a *character* if $\gamma(xy) = \gamma(x)\gamma(y)$ for all $x, y \in G$. The set of all continuous characters on G forms a special abelian group, denoted $\widehat{G}$ or Γ, known as the *dual group* of G. It is a separating family, that is, if $\gamma(xy^{-1}) = 1$ for all $\gamma \in \Gamma$, then $x = y$. We note that $\gamma^{-1} = \overline{\gamma}$.

When G is compact, the continuous characters are *orthonormal* in the sense that $\int_G \gamma\overline{\chi}\,dm_G = 0$ if $\gamma \neq \chi$ and 1 otherwise.

The group Γ is given the *compact-open* topology as follows. Let $\varepsilon > 0$ and $K \subseteq G$ be compact. A basic neighbourhood of $\gamma_0 \in \Gamma$ is the set

$$\{\gamma \in \Gamma : |\gamma(x) - \gamma_0(x)| < \varepsilon \text{ for all } x \in K\}.$$

With this topology, Γ is also a locally compact abelian group.

If $x \in G$, then x can be viewed as a continuous character, $\alpha(x)$ on Γ defined by $\alpha(x)(\gamma) = \gamma(x)$. Thus, there is a natural map $\alpha : G \to \widehat{\Gamma}$ given by $\alpha : x \mapsto \alpha(x)$. The Pontryagin duality theorem says that under this identification the dual of Γ is G.

Theorem C.1.1 (Pontryagin duality theorem). *The map α, described above, is a homeomorphism of G onto the dual group of Γ.*

When $G = \mathbb{T}$, then $\Gamma = \mathbb{Z}$, and the Haar measure on G is normalized Lebesgue measure. When $G = \mathbb{R}$, $\Gamma = \mathbb{R}$, and Lebesgue measure is Haar measure. More generally, we have the following.

Proposition C.1.2. *G is compact if and only if its dual group Γ is discrete.*

Many other properties of the group have "dual" properties in the dual group.

Theorem C.1.3. *Suppose G is a compact group.*

1. *G is metrizable if and only if Γ is countable.*
2. *G is connected if and only if Γ has no non-trivial elements of finite order if and only if Γ is divisible.*
3. *$G = \{x^2 : x \in G\}$ if and only if Γ has no elements of order two.*
4. *If every element of Γ has finite order, then G is totally disconnected.*

If H is a closed subgroup of a (locally) compact abelian group G, then the quotient group G/H is also a (locally) compact abelian group. The quotient group is discrete if and only if H is open.

The *annihilator* of H, denoted $H^{\perp}$, is the set of all $\gamma \in \Gamma$ such that $\gamma(x) = 1$ for all $x \in H$. $H^{\perp}$ is a closed subgroup of Γ and $(H^{\perp})^{\perp} = H$. If $q : G \to G/H$ is the quotient map, then for each $\gamma \in H^{\perp}$ the function Φ_{γ}: $G/H \to \mathbb{T}$ given by $\Phi_{\gamma}(q(x)) = \gamma(x)$ belongs to $\widehat{G/H}$. The correspondence $\gamma \mapsto \Phi_{\gamma}$ is a homeomorphism between $H^{\perp}$ and $\widehat{G/H}$. By duality, the dual of $\widehat{H}$ is the quotient $\Gamma/H^{\perp}$. This shows that every continuous character on H extends to a continuous character on G.

Product Groups

Let X_{α}, $\alpha \in B$, be compact, Hausdorff spaces. The direct product is the set

$$\prod_{\alpha \in B} X_{\alpha} = \left\{ \psi : B \to \bigcup_{\alpha} X_{\alpha} : \psi(\alpha) \in X_{\alpha} \text{ for all } \alpha \in B \right\}.$$

A base for the product topology consists of the sets $\prod_{\alpha \in B} U_{\alpha}$, where U_{α} is open in X_{α} for each $\alpha \in B$ and for all but finitely many coordinates, $U_{\alpha} = X_{\alpha}$. This makes the product space a compact, Hausdorff space.

Theorem C.1.4 (Tychonoff's theorem). *If X_{α} are compact spaces, then* $\prod_{\alpha \in B} X_{\alpha}$ *is compact.*

By definition, each non-empty, open set contains a set of the form $V \times \prod_{\alpha \in B \backslash F} X_{\alpha}$, where F is a finite subset of B and $V \subseteq \prod_{\alpha \in F} X_{\alpha}$ is open. If all the sets X_{α} are finite, then each open set in the direct product will contain an open set of the form $\{\psi\} \times \prod_{\alpha \in B \backslash F} X_{\alpha}$, where F is finite and $\psi \in \prod_{\alpha \in F} X_{\alpha}$. Furthermore, each open set in the product contains a set of that form, even if the sets X_{α} are not finite.

When the X_{α} are all compact abelian groups, then $\prod_{\alpha \in B} X_{\alpha}$ is again a compact abelian group and its dual is the direct sum, $\bigoplus_{\alpha \in B} \widehat{X_{\alpha}}$, the set of all $\varphi : B \to \bigcup_{\alpha} \widehat{X_{\alpha}}$ such that $\varphi(\alpha) \in \widehat{X_{\alpha}}$ for all $\alpha \in B$ and $\varphi(\alpha) = 1$ for all but finitely many α.

C.1.2 Fourier and Fourier–Stieltjes Transforms

The *Fourier transform* of $f \in L^1(G) := L^1(G, m_G)$ is the function $\widehat{f}$ defined on Γ by

$$\widehat{f}(\gamma) = \int_G f(x)\overline{\gamma(x)} dm_G.$$

$\widehat{f}$ is a continuous function on $\mathbf{\Gamma}$ and $\|\widehat{f}\|_\infty \leq \|f\|_1$. The complex numbers $\{\widehat{f}(\gamma)\}_{\gamma \in \mathbf{\Gamma}}$ are called the *Fourier coefficients* of f and they uniquely determine $f \in L^1(G)$.

Whenever $f, g \in L^1(G)$, we can define their convolution

$$f * g(x) = \int_G f(xy^{-1})g(y)\mathrm{d}m_G.$$

The function $f * g$ belongs to L^1 with $\|f * g\|_1 \leq \|f\|_1 \|g\|_1$. An application of Fubini's theorem proves $\widehat{f * g} = \widehat{f}\,\widehat{g}$.

Theorem C.1.5 (Plancherel). *The Fourier transform restricted to $L^1(G) \cap L^2(G)$ is an isometry (with respect to the L^2 norm) onto a dense subspace of $L^2(\mathbf{\Gamma})$ and thus extends uniquely to an isometry of $L^2(G)$ onto $L^2(\mathbf{\Gamma})$.*

One can deduce from this *Parseval's formula*: For all $f, g \in L^2(G)$ and suitable normalizations of the Haar measures,

$$\int_G f(x)\overline{g(x)}\mathrm{d}m_G = \int_{\mathbf{\Gamma}} \widehat{f}(\gamma)\overline{\widehat{g}(\gamma)}\mathrm{d}m_{\mathbf{\Gamma}}.$$

When G is compact, the formal series $\sum_{\gamma \in \mathbf{\Gamma}} \widehat{f}(\gamma)\gamma$ is known as the *Fourier series* of f. A finite linear combination of continuous characters, $P = \sum_{j=1}^N a_j \gamma_j$, is called a *trigonometric polynomial*. The orthogonality property of characters implies $\widehat{P}(\gamma_j) = a_j$ for all j. The Stone–Weierstrass theorem implies the set of trigonometric polynomials is dense in $C(G)$ and hence also dense in $L^p(G)$ for all $p < \infty$. Plancherel's theorem implies the partial sums of the Fourier series of f converge to f in L^2 norm.

The *Riemann–Lebesgue lemma* says $\widehat{f}(x) \to 0$ as $x \to \infty$ for all $f \in L^1(\mathbf{\Gamma})$, and hence $A(G) = \{\widehat{f} : f \in L^1(\mathbf{\Gamma})\} \subseteq C_0(G)$. Being a separating, self-adjoint subalgebra, it is dense in $C_0(G)$. The property known as "local units" is one illustration of the richness of the set $A(G)$. It applies whether G is compact, discrete or neither.

Theorem C.1.6 (Local units theorem). *Let C, V be compact subsets of G with the interior of V non-empty. Then there exists $f \in A(G)$ such that*

1. $f(x) = 1$ *on* C, $f(x) = 0$ *off* $C \cdot V \cdot V^{-1}$ *and* $0 \leq f(x) \leq 1$ *for all* $x \in G$
2. $\|f\|_{A(G)} \leq \left(m_G(C \cdot V^{-1})/m_G(V)\right)^{1/2}$

Corollary C.1.7. *If $U \subseteq G$ is an e-neighbourhood, then there exists $h \in A(G)$ such that* Supp $h \subseteq U$, $h = 1$ *on an e-neighbourhood and* $\|h\|_{A(G)} \leq 2$.

The Fourier transform has an extension, the *Fourier–Stieltjes transform*, to $\mu \in M(G)$ given by

$$\widehat{\mu}(\gamma) = \int_G \overline{\gamma(x)}\mathrm{d}\mu.$$

This function is uniformly continuous and is bounded by the measure norm of μ. The *uniqueness theorem* states that if $\widehat{\mu}(\gamma) = 0$ for all $\gamma \in \Gamma$, then $\mu = 0$. When $f \in L^1(G)$, then f can be identified with the absolutely continuous measure $d\mu_f = f dm_G$. One can easily see that $\widehat{f}(\gamma) = \widehat{\mu_f}(\gamma)$ for all $\gamma \in \Gamma$.

If $\mu, \nu \in M(G)$, then their convolution, $\mu * \nu$, is defined as the linear transformation on $C_0(G)$ that maps

$$f \mapsto \int_G f d\mu * \nu = \int_G \int_G f(xy) d\mu(x) d\nu(y).$$

Thus, $\mu * \nu$ is another finite, regular Borel measure on G, with $\|\mu * \nu\|_{M(G)} \leq \|\mu\|_{M(G)} \|\nu\|_{M(G)}$ and $\widehat{\mu * \nu} = \widehat{\mu} \widehat{\nu}$.

The space $B(\Gamma) = \{\widehat{\mu} : \mu \in M(G)\}$ is a self-adjoint algebra of $C(\Gamma)$, which is invariant under the group action and multiplication by $\gamma(x)$ for any $x \in G$, and contains an identity, $\widehat{\delta_e}$. Since $L^1(G) \subseteq M(G)$, $A(\Gamma) \subseteq B(\Gamma)$. The Bochner–Eberlein theorem gives one way to conclude that a function belongs to $B(\Gamma)$.

Theorem C.1.8 (Bochner–Eberlein). *Let* Γ *be a locally compact abelian group. Suppose that* $f \in B(\Gamma_d)$, *where* Γ_d *is* Γ *with the discrete topology. If* f *is continuous on* Γ, *then* $f \in B(\Gamma)$.

Given a non-discrete, abelian group Λ, denote by

$$M_0(\Lambda) = \{\mu \in M(\Lambda) : \widehat{\mu} \in C_0(\widehat{\Lambda})\}.$$

Similarly, for a Borel set $\mathbf{E} \subseteq \Lambda$, write $M_0(\mathbf{E})$ for the set of $\mu \in M_0(\Lambda)$ that are concentrated on $\mathbf{E}$.

Lemma C.1.9. *1.* $M_0(\Lambda)$ *is a norm-closed ideal in* $M(\Lambda)$.
  2. If $\mu \in M_0(\Lambda)$ *and* $\nu \ll \mu$, *then* $\nu \in M_0(\Lambda)$.
  3. $M_0(\Lambda)$ *is a subspace of the continuous measures on* Λ.

Proof. (1) is easy. For (2), it is also easy to see that $\gamma\mu \in M_0(\Lambda)$ whenever $\mu \in M_0(\Lambda)$ and $\gamma \in \widehat{\Lambda}$. Hence, $p\mu \in M_0(\Lambda)$ for all trigonometric polynomials p. If $\nu \ll \mu$, then $\nu = f\mu$ for some $f \in L^1(|\mu|)$. Being a finite measure, there are compactly supported measures $\mu_n \to \mu$ and $f\mu_n \to f\mu$ in measure norm. Thus, by (1), there is no loss in assuming μ is compactly supported. Then for every $\varepsilon > 0$ there exists a trigonometric polynomial $p \in \text{Trig}(\Lambda)$ such that $\|f\mu - p\mu\|_{M(\Lambda)} < \varepsilon$. Again, because $M_0(\Lambda)$ is norm closed, $\nu \in M_0(\Lambda)$.

(3) If $\mu \in M_0(\Lambda)$ had a discrete component, then there would be a point mass $\delta_x \ll \mu$. But, of course, a point mass cannot be in $M_0(\Lambda)$. $\square$

More generally, a continuous measure can be characterized by the average decay in its Fourier transform.

Lemma C.1.10 (Wiener). *Let* $\{V_\alpha\}$ *be a neighbourhood base at* $e \in G$ *and suppose for each* α *that* f_α *is a positive-definite function[1], compactly*

---

[1] That is, $\widehat{f}_\alpha \geq 0$ with $\int_\Gamma \widehat{f}_\alpha d\gamma < \infty$.

supported on V_α, such that $f_\alpha(e) = 1$. Then for every $\mu \in M(G)$,

$$\lim_\alpha \int_\Gamma \widehat{f_\alpha}(\gamma) |\widehat{\mu}(\gamma)|^2 \mathrm{d}\gamma = \sum_{x \in G} |\mu(\{x\})|^2.$$

C.1.3 The Bohr Compactification and the Bohr Topology

Suppose G is a compact abelian group but now considered as a topological group with the discrete topology, denoted G_d. The dual of G_d is a compact topological group known as the *Bohr compactification* of Γ and denoted $\overline{\Gamma}$. Since all functions defined on G are continuous functions on G_d, $\overline{\Gamma}$ consists of all characters (continuous or otherwise) on G. If G is infinite, $\overline{\Gamma}$ is not metrizable and $|\overline{\Gamma}| = 2^{|G|}$. (See Exercises C.4.17 and 4.7.8.)

The compact-open topology on $\overline{\Gamma}$ is known as the *Bohr topology*. The basic neighbourhoods of $\gamma_0 \in \overline{\Gamma}$ are the sets of the form

$$\{\gamma \in \overline{\Gamma} : |\gamma(x_j) - \gamma_0(x_j)| < \varepsilon \text{ for } j = 1, \ldots, N\},$$

where $\varepsilon > 0$ and $\{x_1, \ldots, x_N\}$ is any finite set (equivalently, compact set) in G_d. Thus, the Bohr topology restricted to Γ is the weakest topology that makes all the maps $\gamma \mapsto \gamma(x)$ continuous for $x \in G$. These maps are all the continuous characters on $\overline{\Gamma}$ since the dual group of $\overline{\Gamma}$ is equal to G_d by the Pontryagin duality theorem.

The subgroup Γ naturally embeds into $\overline{\Gamma}$ as a dense subgroup because if $x \in \widehat{\overline{\Gamma}}$ satisfies $\gamma(x) = 1$ for all $\gamma \in \Gamma$, then $x = e \in G$. When $\overline{\Gamma}$ is not discrete (equivalently, when Γ is infinite), then every element of $\overline{\Gamma}$ is a cluster point. It follows that every element of $\overline{\Gamma}$ is a limit of a net from Γ.

Being a compact topological group, $\overline{\Gamma}$ admits a normalized Haar measure. The $m_{\overline{\Gamma}}$-measure of Γ must be zero (except in the trivial case that Γ is a finite group); for otherwise, being a subgroup Γ would be open and hence closed in $\overline{\Gamma}$, and therefore $\Gamma = \overline{\Gamma}$ (as topological groups), a contradiction.

The Fourier–Stieltjes transform of the point mass measure, $\delta_x \in M_d(G)$, naturally extends to a continuous function on $\overline{\Gamma}$ by defining $\widehat{\delta_x}(\gamma) = \gamma(x)$ for $\gamma \in \overline{\Gamma}$. If $\mu = \sum_{j=1}^\infty c_j \delta_{x_j}$ is any finite discrete measure on G, then $\widehat{\mu}$ is a continuous function on $\overline{\Gamma}$ being the uniform limit of the partial sums, $\sum_{j=1}^N c_j \widehat{\delta_{x_j}}$. By the Stone–Weierstrass theorem, the algebra $\{\widehat{\mu} : \mu \in M_d(G)\}$ is dense in $C(\overline{\Gamma})$. Since $M(G_d) = M_d(G) = \ell^1(G) = L^1(\widehat{\overline{\Gamma}})$, this subalgebra is equal to both $A(\overline{\Gamma})$ and $B(\overline{\Gamma})$.

C.1.4 Consequences of the Local Units Theorem

In Exercise C.4.20 we ask that local units be used to prove the next theorem.

Theorem C.1.11 (Wiener–Levy). *Let* $\mathbf{E} \subset \mathbf{\Gamma}$ *be compact. If* $f \in A(\mathbf{E})$ *and* F *is analytic in a neighbourhood of* $\overline{f(\mathbf{E})}$, *then* $F \circ f \in A(\mathbf{E})$. *In particular, if* $f \geq 1$ *on* $\mathbf{E}$, *then* $1/f \in A(\mathbf{E})$.

A more general version of the Wiener–Levy theorem is due to Gel'fand. We use $\widehat{x}$ to denote the Gel'fand transform of an element x of a commutative Banach algebra.

Theorem C.1.12 (Gel'fand). *Let* $\mathcal{A}$ *be a commutative, semi-simple, unital Banach algebra with maximal ideal space* $\mathcal{M}$. *Let* $x \in \mathcal{A}$. *If* F *is analytic in a neighbourhood of the spectrum of* x, *then there exists* $y \in \mathcal{A}$ *such that* $\widehat{y} = F(\widehat{x})$ *on* $\mathcal{M}$.

Spectral Synthesis

Local units can be avoided by using a Cauchy integral formula proof of Theorem C.1.11, but local units are essential when discussing ideals of $A(G)$. Every $X \subseteq G$ defines a closed ideal (the hull) $I(X)$, consisting of all $f \in A(G)$ which are zero on X. By continuity, X and its closure in G define the same ideal, so we shall assume X is closed in this discussion.

Because of the local units property, for each $x \notin X$, there exists $f \in I(X)$ with $f(x) = 1$. Therefore, we may identify $A(G)/I(X)$ with the set of restrictions of elements of $A(G)$ to X. We call this quotient $A(X)$ and give elements the quotient norm. We can, of course, do this whether G is compact or non-compact, discrete or non-discrete.

Each closed ideal $I \subset A(G)$ gives rise to a closed set (the kernel of I) consisting of those $x \in G$ for which $f(x) = 0$ for all $f \in I$. For some sets there are other closed ideals of interest. We let $J_0(X)$ be the set of elements of $A(G)$ that are zero in a neighbourhood of X and $J(X)$, the norm closure of $J_0(X)$.

Definition C.1.13. The set X is said to be a *set of spectral synthesis* if $I(X) = J(X)$.

Important examples of sets of spectral synthesis include singletons and compact subgroups. In fact if X is a compact subgroup, then X is a *Ditkin set*: there exists a bounded net $g_\alpha \in J_0(X)$ such that $g_\alpha f \to f$ in $A(G)$ norm for all $f \in I(X)$. That singletons are sets of spectral synthesis can be deduced from Lemma 9.4.3. The argument goes as follows. Suppose $f \in A(G)$, $x \in G$, $f(x) = 0$ and $\varepsilon > 0$. By translating f, we may assume $x = e$. There exists a trigonometric polynomial, $g = \sum_1^N c_n \gamma_n$, with $\|f - g\|_{A(G)} < \varepsilon/4$ and

$g(e) = 0$. Apply Lemma 9.4.3 to obtain a compact neighbourhood U of x such that $\|1 - \gamma_n\|_{A(U)} < \varepsilon/4N$ for each n. Then $\|g\|_{A(U)} < \varepsilon/4$. Let $h \in A(G)$ agree with g on U and have $\|h\|_{A(G)} < \varepsilon/4$. Then $g - h = 0$ on U and $\|f - (g - h)\|_{A(G)} < \varepsilon$.

The following property of spectral synthesis is used in Sect. 10.3.3.

Lemma C.1.14. *Suppose the compact set* $\mathbf{V} \subset \overline{\boldsymbol{\Gamma}}$ *obeys spectral synthesis and* $f, g \in A(\overline{\boldsymbol{\Gamma}})$ *with* $f = g$ *on* $\mathbf{V}$. *Then for each* $\varepsilon > 0$ *there exists a closed neighbourhood* $\mathbf{U} \supset \mathbf{V}$ *such that* $\|f - g\|_{A(\mathbf{U})} < \varepsilon$.

Proof. Let $\varepsilon > 0$. Since $f - g \in I(\mathbf{V})$, the spectral synthesis property of $\mathbf{V}$ implies that there exists $h \in J_0(\mathbf{V})$ with $\|f - g - h\|_{A(\overline{\boldsymbol{\Gamma}})} < \varepsilon$. Let $\mathbf{U} = \{\chi \in \overline{\boldsymbol{\Gamma}} : h(\chi) = 0\}$. Then $\mathbf{U}$ is a compact neighbourhood of $\mathbf{V}$.

Since $h \in I(\mathbf{U})$, $\|h\|_{A(\mathbf{U})} = 0$ by the definition of the quotient norm in $A(\mathbf{U}) = A(\overline{\boldsymbol{\Gamma}})/I(\mathbf{U})$. Therefore, $\|f - g\|_{A(\mathbf{U})} = \|f - g - h\|_{A(\mathbf{U})} \leq \|f - g - h\|_{A(\overline{\boldsymbol{\Gamma}})} < \varepsilon$. □

C.1.5 Elements of Order Two

In the study of ε-Kronecker and I_0 sets, elements of order two cause complications. Recorded here are some useful facts.

Lemma C.1.15. *1. Let* G_2 *be the annihilator of* $\boldsymbol{\Gamma}^{(2)}$, *the subgroup consisting of the set of characters of order* 2^k *for some* k. *Then every element of* G_2 *is a square.*
2. *If* $\boldsymbol{\Gamma}$ *has only finitely many elements of order two, then each character on* G *of order two is continuous.*

Proof. (1) Apply Exercise C.4.11(1) because $\widehat{G_2} = \boldsymbol{\Gamma}/\boldsymbol{\Gamma}^{(2)}$ has no elements of order two.

(2) Let G_s be the set of squares in G. As $\boldsymbol{\Gamma}$ has only finitely many elements of order two by Exercise C.4.11(3) $|G/G_s| < \infty$, so G_s is an open and closed subgroup. If $\gamma \in \overline{\boldsymbol{\Gamma}}$ is of order two, then $\gamma(g^2) = 1$. Hence, γ is constant on the cosets of G/G_s and thus is a continuous character. □

Proposition C.1.16. *Suppose* $\boldsymbol{\Gamma}$ *contains only finitely many elements of order two. For each non-empty, open set* $U \subseteq G$ *there is some* $\delta = \delta(U) > 0$ *such that*

$$\boldsymbol{\Gamma}_1 = \{\gamma \in \overline{\boldsymbol{\Gamma}} : \min\{|\gamma(u) - 1|, |\gamma(u) + 1|\} < \delta \text{ for all } u \in U\}$$

is finite. Moreover, if $\gamma \in \boldsymbol{\Gamma}_1$, *then* $\gamma(u) = \pm 1$ *for all* $u \in U$.

Proof. Let U be an open subset of G. By compactness, there exist finitely many $x_1, \ldots, x_J \in G$ such that $G = \bigcup_{j=1}^{J} x_j U$. Take $0 < \delta \leq \pi/4J$, consider $\gamma \in \overline{\boldsymbol{\Gamma}}$ and assume

$$\arg \gamma(u) \in (-\delta, \delta) \cup (\pi - \delta, \pi + \delta) \qquad\qquad (\text{C.1.1})$$

for every $u \in U$.

For every $g \in G$, there exist $u \in U$ and $1 \le j \le J$ such that $g = ux_j$. Thus, $\gamma(g) = \gamma(u)\gamma(x_j)$, and hence $\arg \gamma(g) \in T_0$, where if $t_j = arg\gamma(x_j)$, then

$$T_0 = \bigcup_{j=1}^{J} (t_j - \delta, t_j + \delta) \cup (t_j + \pi - \delta, t_j + \pi + \delta).$$

The range of a character is either a finite subgroup of the circle or is dense in $\mathbb{T}$. Since T_0 is not dense (indeed, there is an interval I of width $(\pi - 2J\delta)/J$ disjoint from T_0) it follows that range γ is a finite subgroup and therefore γ has finite order, say p. But then range γ must contain all pth roots of unity and since I will contain a pth root of unity for each p sufficiently large, say $p > p_0(\delta)$, we see that γ has order $\le p_0$. By reducing δ, if necessary, we can assume $(-\delta, \delta) \cup (\pi - \delta, \pi + \delta)$ contains no pth roots of unity for any $p \le p_0$, other than ± 1. With this choice of $\delta = \delta(U)$ we deduce that if (C.1.1) holds for all $u \in U$, then $\gamma|_U = \pm 1$. Thus, we have established that $\Gamma_1 = \{\gamma \in \overline{\Gamma} : \gamma|_U = \pm 1\}$ and, furthermore, that every element in Γ_1 is of order p for some $p \le p_0$.

Put $S = \{\gamma^2 : \gamma \in \Gamma_1\} \subseteq \overline{\Gamma}$. Then $U \subseteq S^\perp$ and thus $S^\perp$ is a subgroup of G with interior. Consequently, G is a finite union of translates of $S^\perp$ and this ensures that $\{\chi \in \overline{\Gamma} : \chi(x) = 1 \text{ for all } x \in S^\perp\}$ is finite. As this latter set contains S, S is a finite set of characters, $\{\chi_1, \ldots, \chi_K\}$. Hence, if $\gamma \in \Gamma_1$, then $\gamma^2 = \chi_k$ for some $1 \le k \le K$.

Let $G_s = \{g^2 : g \in G\}$. Because Γ contains only finitely many elements of order two, $G = \bigcup_{n=1}^{N} y_n G_s$ for suitable $y_1, \ldots, y_N \in G$. When $\gamma \in \Gamma_1$, $\gamma(y_n)$ is a pth root of unity for some $p \le p_0$, and thus the set of N-tuples,

$$\{(\gamma(y_1), \ldots, \gamma(y_N)) : \gamma \in \Gamma_1\} \subseteq \mathbb{T}^N,$$

is finite. Furthermore, if $\gamma, \beta \in \Gamma_1$ with $\gamma^2 = \beta^2 = \chi_k$ and $\gamma(y_n) = \beta(y_n)$ for $n = 1, \ldots, N$, then for each $g \in G$, say $g = y_n h^2$, we have

$$\gamma(g) = \gamma(y_n)\gamma^2(h) = \gamma(y_n)\chi_k(h) = \beta(y_n)\beta^2(h) = \beta(g).$$

Thus, $\gamma = \beta$ and this implies that Γ_1 is a finite set. $\qquad\qquad\square$

Corollary C.1.17. *Suppose Γ contains only finitely many elements of order two. Let $U \subset G$ be an e-neighbourhood and H be the subgroup of G generated by U. Then $\widehat{H}$ contains only finitely many characters of order two.*

Proof. Let $\delta > 0$ and $\Gamma_1 \subset \overline{\Gamma}$ be given by Proposition C.1.16 for U. If $\gamma \in \widehat{H}$ has order two, then $\gamma = \pm 1$ on U, so $\gamma \in \Gamma_1$. Since Γ_1 is finite, the corollary follows. $\qquad\qquad\square$

Corollary C.1.18. *Suppose* $\boldsymbol{\Gamma}$ *contains only finitely many elements of order two,* $U \subset G$ *is an e-neighbourhood and* $\gamma \in \overline{\boldsymbol{\Gamma}}$ *satisfies* $\gamma(u) = \pm 1$ *for all* $u \in U$. *Then* γ *is continuous.*

Proof. Let H be the open subgroup generated by U. Since $\gamma(u) = \pm 1$ for all $u \in U$, $\gamma|_H$ is of order two. Corollary C.1.17 implies $\widehat{H}$ contains only finitely many characters of order two and therefore Lemma C.1.15 (2) implies that each character on H of order two is a continuous character on H.   □

C.2 Basic Probability

Suppose Ω is a set and $\mathcal{M}$ (the *measurable sets*) is a σ-algebra of subsets of Ω. A *probability* is a positive measure, $\mathbb{P}$, defined on $\mathcal{M}$, with $\mathbb{P}(\Omega) = 1$. A probability space is a triple $(\Omega, \mathcal{M}, \mathbb{P})$. A measurable set is also called an *event*. An event E is said to occur almost surely (a.s.) if $\mathbb{P}(E) = 1$.

A *random variable* $X : \Omega \to \mathbb{C}$ is an $\mathcal{M}$-measurable function. By $\sigma(X)$ we mean the smallest σ-algebra that makes X measurable. The *expectation* (or *mean*) of a random variable X is denoted $\mathbb{E}(X)$ and is given by

$$\mathbb{E}(X) = \int_{\Omega} X(\omega) d\mathbb{P}(\omega).$$

An elementary fact is Markov's inequality, for which a proof is asked in Exercise C.4.1 (1).

Lemma C.2.1 (Markov's inequality). *For a random variable* X,

$$\mathbb{P}(|X| \geq d) \leq \frac{\mathbb{E}(|X|)}{d}. \tag{C.2.1}$$

Two important classes of random variables are the *Bernoulli random variables* and the *Poisson random variables*. Bernoulli random variables take on only the values 0 and 1, with probabilities p and $1 - p$, respectively. The expectation of such a random variable X is $\mathbb{E}(X) = p$. A Poisson random variable X with *parameter* $\lambda > 0$ takes on non-negative integer values and

$$\mathbb{P}(\omega : X(\omega) = k) = \frac{\lambda^k e^{-\lambda}}{k!} \text{ for } k = 0, 1, \ldots.$$

It is easy to see that $\mathbb{E}(X) = \lambda$ and that the sum of Poisson r.v.'s with parameters p_n is again Poisson, with parameter $\sum p_n$.

Sub-σ-algebras, $\sigma_1, \sigma_2, \ldots$ of $\mathcal{M}$, are said to be *independent* if whenever $Y_i \in \sigma_i$ and $i_1, \ldots, i_N$ are distinct, then

$$\mathbb{P}(Y_{i_1} \cap \cdots \cap Y_{i_N}) = \prod_{n=1}^{N} \mathbb{P}(Y_{i_n}).$$

Random variables $X_1, X_2, \ldots$ are said to be *independent* if their associated σ-algebras $\sigma(X_1), \sigma(X_2), \ldots$ are independent. If $X_1, X_2, \ldots$ are independent, then $\mathbb{E}(\prod_n X_{i_n}) = \prod_n \mathbb{E}(X_{i_n})$

Here is one application of Markov's inequality.

Lemma C.2.2. *Let $X_1, \ldots, X_K$ be real, mean 0, independent random variables, with $|X_k| \leq 1$ a.s. for all k. Suppose $c^2 \geq \sum_{k=1}^K \mathbb{E}(X_k^2)$. Then for all $a > 0, V$*

$$\mathbb{P}\left(\left|\sum_k X_k\right| \geq a\right) \leq 2\exp\left(-\frac{a^2}{2(a+c^2)}\right).$$

Proof. For $\tau > 0$,

$$\exp(\tau X_k) = 1 + \tau X_k + \sum_{j=2}^{\infty} \frac{\tau^j}{j!} X_k^j \leq 1 + \tau X_k + \sum_{j=2}^{\infty} \frac{\tau^j}{j!} X_k^2,$$

so

$$\mathbb{E}(\exp(\tau X_k)) \leq 1 + \mathbb{E}(X_k^2)(e^\tau - \tau - 1)$$
$$\leq \exp\left(\mathbb{E}(X_k^2)(e^\tau - \tau - 1)\right). \tag{C.2.2}$$

Markov's inequality (C.2.1) applied to $Y = \exp(\tau \sum_k X_k)$ and $d = e^{\tau a}$, and the independence of the X_k give

$$\mathbb{P}\left(\sum_k X_k \geq a\right) = \mathbb{P}\left(\exp\left(\tau \sum_k X_k\right) \geq \exp(\tau a)\right)$$
$$\leq e^{-\tau a} \prod_k \mathbb{E}(e^{\tau X_k}).$$

Applying (C.2.2) and the assumption that $c^2 \geq \sum \mathbb{E}(X_k^2)$ yields

$$\mathbb{P}\left(\sum_k X_k \geq a\right) \leq \exp(-\tau a + c^2(e^\tau - \tau - 1)).$$

The value $\tau = \log(1 + \frac{a}{c^2})$ minimizes the last term, and so

$$\mathbb{P}\left(\sum_k X_k \geq a\right) \leq \exp\left(a - (a + c^2)\log\left(1 + \frac{a}{c^2}\right)\right)$$
$$= \exp\left(a + (a + c^2)\log(1 - u)\right),$$

where $u = a/(a + c^2)$. Thus,

$$\mathbb{P}\left(\sum_k X_k \geq a\right) \leq \exp\left(a - (a + c^2)\left(u + \frac{u^2}{2}\right)\right) = \exp\left(-\frac{a^2}{2(a+c^2)}\right).$$

The same bound holds for $\mathbb{P}(\sum_k -X_k \geq a)$, and hence the result follows. $\square$

C.2.1 Zero-One Laws

Suppose $\{A_n\}_n$ is a sequence of events. By $\{A_n \text{ i.o.}\}$ (i.o. being short for *infinitely often*) we mean the measurable set

$$\{A_n \text{ i.o.}\} = \bigcap_m \bigcup_{n \geq m} A_n = \{\omega : \omega \in A_n \text{ for infinitely many } n\}.$$

A very useful result is the Borel–Cantelli lemma (Exercise C.4.1 (2)).

Lemma C.2.3 (Borel–Cantelli). *Suppose A_n are events.*

1. *If $\sum_{n=1}^{\infty} \mathbb{P}(A_n) < \infty$, then $\mathbb{P}(A_n \text{ i.o.}) = 0$.*
2. *If the A_n are independent and $\sum_{n=1}^{\infty} \mathbb{P}(A_n) = \infty$, then $\mathbb{P}(A_n \text{ i.o.}) = 1$.*

The Borel–Cantelli lemma is one example of a zero-one law. For another example, we first introduce the notion of the tail σ-algebra. Let $\{X_n\}_n$ be a sequence of random variables and let

$$\mathcal{T}_n = \sigma(X_n, X_{n+1}, \dots\} \text{ and } \mathcal{T} = \bigcap_n \mathcal{T}_n.$$

The σ-algebra $\mathcal{T}$ is called the *tail σ-algebra* of the sequence $\{X_n\}$ and contains many important events. Examples include:

1. $\{\omega : \lim_n X_n(\omega) \text{ exists}\}$
2. $\{\omega : \sum_n X_n(\omega) \text{ converges}\}$
3. $\{\omega : X_n(\omega) \in A_n \text{ i.o. }\}$ where $\{A_n\}$ is a sequence of measurable subsets of $\mathbb{C}$.

The Kolmogorov $0-1$ law is an important example of a zero-one law.

Theorem C.2.4 (Kolmogorov $0-1$ law). *Let $\{X_n\}_n$ be a sequence of independent random variables and let $\mathcal{T}$ be the associated tail σ-algebra. If $F \in \mathcal{T}$, then either $\mathbb{P}(F) = 0$ or $\mathbb{P}(F) = 1$.*

C.2.2 Martingales

Suppose X is a random variable with $\mathbb{E}(|X|) < \infty$ and $\mathcal{F}$ is a sub-σ-algebra of $\mathcal{M}$. The *conditional expectation of X with respect to $\mathcal{F}$*, denoted $E(X|\mathcal{F})$, is any $\mathcal{F}$-measurable, integrable, random variable Z such that

$$\int_A Z \, d\mathbb{P} = \int_A X \, d\mathbb{P} \text{ for every } A \in \mathcal{F}. \qquad \text{(C.2.3)}$$

The Radon–Nikodym theorem guarantees the existence of the conditional expectation and its uniqueness, up to sets of measure zero. In other words, if both Z_1 and Z_2 are $\mathcal{F}$-measurable, integrable random variables satisfying (C.2.3), then $Z_1 = Z_2$ a.s. In this sense we can speak of "the" conditional expectation $E(X|\mathcal{F})$. Listed below are some important properties of the conditional expectation, most of which follow easily:

1. $\mathbb{E}(E(X|\mathcal{F})) = \mathbb{E}(X)$
2. If X is $\mathcal{F}$-measurable, then $E(X|\mathcal{F}) = X$ a.s.
3. If X is independent of $\mathcal{F}$, then $E(X|\mathcal{F}) = \mathbb{E}(X)$ a.s.
4. If Z is $\mathcal{F}$-measurable and bounded, then $E(XZ|\mathcal{F}) = ZE(X|\mathcal{F})$ a.s.

Definition C.2.5. Suppose $\{X_n : n = 0, 1, \dots\}$ is a sequence of integrable random variables. Let $\mathcal{F}_n = \sigma\{X_0, X_1, \dots, X_n\}$. The sequence $\{X_n\}$ is called a *martingale* if

$$E(X_{n+1}|\mathcal{F}_n) = X_n \text{ a.s. for every } n \geq 0.$$

Example C.2.6. A typical martingale: Suppose $\{Y_n\}_{n\geq 1}$ are independent random variables with mean 1. Let $X_n = \prod_{k=1}^n Y_k$ and $X_0 = 1$. Let $\mathcal{F}_n = \sigma\{X_0, X_1, \dots, X_n\}$. Then $\{X_n\}$ is a martingale because the properties above yield the defining condition

$$E(X_{n+1}|\mathcal{F}_n) = E(X_n Y_{n+1}|\mathcal{F}_n) = X_n E(Y_{n+1}|\mathcal{F}_n) = X_n \mathbb{E}(Y_{n+1}) = X_n \text{ a.s.}$$

One reason for the importance of martingales is their good convergence properties.

Theorem C.2.7 (Martingale convergence). *Let $\{X_n\}$ be a martingale with $\sup \mathbb{E}(|X_n|) < \infty$. Then $\lim_{n\to\infty} X_n$ exists and is finite almost surely.*

Theorem C.2.8 (Martingale convergence for L^2 bounded martingales). *Let $\{X_n\}$ be a martingale with $\sup \mathbb{E}(|X_n|^2) < \infty$. Then X_n converges a.s. and in L^2.*

The first theorem says that L^1-bounded martingales converge. The second says that under the stronger assumption of L^2-boundedness of the martingale, the martingale even converges in L^2 (and hence in L^1) norm. These are deep, important results.

C.3 Remarks and Credits

Overview of Harmonic Analysis on Locally Compact Abelian Groups. Most of the results in Sect. C.1.1–C.1.4 can be found in standard references for harmonic analysis on locally compact abelian groups such as [87, 136, 167]. Theorem C.1.8 is due to Bochner [15] when $\Gamma = \mathbb{R}$ and Eberlein [34] in the general case. A proof is also given in [167, 1.9.1].

The Cauchy integral proof of the Wiener–Levy Theorem C.1.11 is Satz 20 in Gel'fand's fundamental—and still worth reading—paper [43]. That paper contains the proof of Gel'fand's Theorem C.1.12, as well. The statement and proof of Theorem C.1.12 can also be found in standard references on Banach algebras. See also [56, 9.1–2] and [167, D7] for further discussion on these results.

Spectral synthesis is discussed in many books, including [56, Chap. 3, Sect. 11.2], [88, Chap. 10], [106, Chap. 9] and [167, Chap. 7]. All those references contain a proof of the existence of sets that are not of spectral synthesis.

The results of Sect. C.1.5 are adapted from [62, 63].

Basic Probability. The probability definitions and results are standard and can be found in any graduate probability text, such as [25, 69].

Exercises. Many of the exercises can be found in the standard references, such as [56, 87, 88, 136, 167].

C.4 Exercises

In all exercises "G" will denote a locally compact abelian group (with possibly additional properties, as specified) and "Γ" its dual.

Exercise C.4.1. 1. Prove Markov's inequality, Lemma C.2.1.
  2. Prove the Borel–Cantelli Lemma C.2.3.

Exercise C.4.2. 1. Prove that every e-neighbourhood U of a topological group contains a symmetric e-neighbourhood V with $V^2 \subseteq U$.
  2. Prove that every open subgroup of a topological group is closed.
  3. Prove that every topological group is regular.
  4. Prove that every compact group is locally compact.

Exercise C.4.3. 1. Describe the dual group of $\mathbb{Z}_n$.
  2. Suppose $\gamma \in \Gamma$ has finite order. Show that $\{x : \gamma(x) = 1\}$ is open.

Exercise C.4.4. Construct a uniformly continuous function on $\mathbb{R}$ which is not almost periodic.

Exercise C.4.5. Let $\mathbf{S} \subseteq \Gamma$ be compact and $\varepsilon > 0$. Prove there is an e-neighbourhood U such that $|1 - \gamma(u)| < \varepsilon$ for all $\gamma \in \mathbf{S}$ and $u \in U$.

Exercise C.4.6. 1. Suppose $f \in L^p(G)$ and $g \in L^q(G)$ where $1 < p, q < \infty$ and $1/p + 1/q = 1$. Prove that $f * g \in C_0(G)$.
  2. Show that if $f \in L^1(G)$ and $g \in L^\infty(G)$, then $f * g$ is uniformly continuous and bounded.
  3. Show that if $f \in L^p(G)$ for $1 \le p \le \infty$ and $\mu \in M(G)$, then $\mu * f \in L^p$ and $\|\mu * f\|_p \le \|f\|_p \|\mu\|_{M(G)}$.

4. Prove that if A, B are Borel sets with $m_G(A) > 0, m_G(B) > 0$, then $A \cdot B$ has non-empty interior.

Exercise C.4.7. 1. Prove the local units Theorem C.1.6. Hint: Take functions $g, h \in L^2(G)$ whose Fourier transforms are the characteristic functions of V and $C \cdot V^{-1}$, respectively. (Why do these exist?) Put $f(x) = \frac{1}{m(V)} g(x) h(x)$.

2. Give an example in $\mathbb{R}$ to show that C compact and $m_{\mathbb{R}}(V) < \infty$ do not imply $m_{\mathbb{R}}(C + V) < \infty$.

Exercise C.4.8. 1. Prove that if S is any subset of G, then $S^{\perp} = \{\gamma \in \Gamma : \gamma(x) = 1 \, \forall x \in S\}$ is a closed subgroup of Γ.

2. Prove that every closed subgroup of Γ is equal to $H^{\perp}$ for some closed subgroup H of G.

Exercise C.4.9. Let $U \subseteq G$ be an e-neighbourhood and $\varepsilon > 0$. Find a trigonometric polynomial $p \geq 0$ such that $\widehat{p} \geq 0$, $\int p \, dx = 1$ and $|p(x)| \leq \varepsilon$ for $x \notin U$.

Exercise C.4.10. Assume G is compact and Γ_0 is the torsion subgroup of Γ.

1. Show that every element of Γ/Γ_0 has infinite order.
2. Show that G is connected if and only if Γ_0 is trivial if and only if G is divisible.
3. Show that Γ_0 is finite if and only if G has a finite number of connected components and that the annihilator, $\Gamma_0^{\perp}$, of Γ_0 is connected.
4. A topological group is said to be *locally connected* if it has a neighbourhood base of connected sets. Show that if G is locally connected, then Γ_0 is finite.

Exercise C.4.11. Assume G is compact.

1. Show that Γ has no elements of order two if and only if every element of G is a square. Show that "compact" is a necessary hypothesis.
2. Show that Γ has no elements of order two if and only if $\overline{\Gamma}$ has no elements of order two.
3. Show that Γ has only finitely many elements of order two if and only if the quotient of G by the subgroup of squares in G is finite.
4. Show that if G is not of bounded order, then G contains a dense set of elements of infinite order.

Exercise C.4.12. For a prime $p > 2$, formulate and prove p-versions of the results in Sect. C.1.5.

Exercise C.4.13. Give an example of G such that its dual group Γ has no elements of order 2, but G has a closed subgroup H whose dual has an infinite number of elements of order 2. Thus, "neighbourhood" (or non-empty interior) is essential in Corollary C.1.17.

Exercise C.4.14. Give an example of an open set in $\overline{\mathbb{Z}}$ that contains a non-trivial subgroup of $\mathbb{Z}$.

Exercise C.4.15. 1. Given any $\varepsilon > 0$ and $t_1, \ldots, t_N \in \mathbb{T}$, show there is some integer $n \neq 0$ such that $\left| e^{int_j} - 1 \right| < \varepsilon$ for all $j = 1, \ldots, N$.
  2. Show that every element of $\overline{\Gamma}$ is a cluster point of Γ. This generalizes the first part. Explain.

Exercise C.4.16. 1. Let $H \subseteq G$ be a compact sub-semigroup (i.e. closed under multiplication) of the compact group G. Show that H is a group.
  2. Show that 0 is a Bohr cluster point of $\{j, j+1, j+2, \ldots\}$ for all $j \geq 0$ in $\mathbb{Z}$.
  3. Show that $\{10^j\}_{j=1}^{\infty}$ and $\{10^j + j\}_{j=1}^{\infty}$ have a common cluster point.

Exercise C.4.17. 1. Show that a countable, locally compact group G is discrete. (This can be done in at least two quite different ways.) Hint: Consider $m_G(\{e\})$ or Baire category.
  2. Show that $\overline{\Gamma}$ is uncountable if Γ is infinite.

Exercise C.4.18. 1. Prove that G is metrizable if and only if Γ is σ-compact.
  2. Show that a compact group is metrizable if and only if the complement of $\{e\}$ is σ-compact.
  3. Show that every open subset of a compact metrizable group is σ-compact.
  4. Show that a compact metrizable group has a countable dense subgroup.

Exercise C.4.19. 1. Show that if Γ is not compact, then Γ contains a σ-compact, non-compact, open subgroup.
  2. Show that if every metrizable quotient of G is discrete, then G is discrete.
  3. Say that $G \in \mathcal{G}$ if $\Gamma = \overline{\Gamma}$ as discrete groups. Show that $G \in \mathcal{G}$ implies that every subgroup of G is closed and every quotient group of G belongs to $\mathcal{G}$.
  4. If Γ is not compact, prove $\overline{\Gamma} \neq \Gamma$ (as sets). Hint: G has a metrizable quotient.

Exercise C.4.20. 1. We say "$f \in \mathbf{E}$ locally at $\gamma \in \mathbf{E}$" if there exists a closed neighbourhood $\mathbf{U}$ of γ such that $f|_{\mathbf{U} \cap \mathbf{E}} \in A(\mathbf{U} \cap \mathbf{E})$. Show that if $\mathbf{E}$ is compact and $f \in \mathbf{E}$ locally at every point, then $f \in A(\mathbf{E})$.
  2. Formulate "$f \in \mathbf{E}$ locally at infinity". Then show that if $f \in \mathbf{E}$ locally at every $\gamma \in \mathbf{E}$ and at infinity, then $f \in A(\mathbf{E})$.
  3. Prove Theorem C.1.11.

References

[1] F. Albiac and N. J. Kalton. *Topics in Banach Space Theory*, volume 233 of *Graduate Texts in Mathematics*. Springer-Verlag, Berlin, Heidelberg, New York, 2006.

[2] M. E. Andersson. The Kaufman-Rickert inequality governs Rademacher sums. *Analysis (Munich)*, 23:65–79, 2003.

[3] J. Arias de Reyna. *Pointwise Convergence of Fourier Series*, volume 1785 of *Lecture Notes in Mathematics*. Springer-Verlag, Berlin, Heidelberg, New York, 2002.

[4] N. Asmar and S. Montgomery-Smith. On the distribution of Sidon series. *Arkiv för Math.*, 31(1):13–26, 1993.

[5] G. Bachelis and S. Ebenstein. On $\Lambda(p)$ sets. *Pacific J. Math.*, 54:35–38, 1974.

[6] S. Banach. Über einige Eigenschaften der lakunäre trigonometrischen Reihen, II. *Studia Math.*, 2:207–220, 1930.

[7] N. K. Bary. *A Treatise on Trigonometric Series*, volume I. MacMillan, New York, N. Y., 1964.

[8] F. Baur. Lacunarity on nonabelian groups and summing operators. *J. Aust. Math. Soc.*, 71(1):71–79, 2001.

[9] G. Benke. Arithmetic structure and lacunary Fourier series. *Proc. Amer. Math. Soc.*, 34:128–132, 1972.

[10] G. Benke. On the hypergroup structure of central $\Lambda(p)$ sets. *Pacific J. Math.*, 50:19–27, 1974.

[11] D. Berend. Parallelepipeds in sets of integers. *J. Combin. Theory*, 45 (2):163–170, 1987.

[12] K. G. Binmore. Analytic functions with Hadamard gaps. *Bull. London Math. Soc.*, 1:211–217, 1969.

[13] R. C. Blei. On trigonometric, series associated with separable, translation invariant subspaces of L^∞. *Trans. Amer. Math. Soc.*, 173:491–499, 1972.

[14] R. C. Blei. Fractional Cartesian products of sets. *Ann. Inst. Fourier (Grenoble)*, 29(2):79–105, 1979.

C.C. Graham and K.E. Hare, *Interpolation and Sidon Sets for Compact Groups*, 223
CMS Books in Mathematics, DOI 10.1007/978-1-4614-5392-5,
© Springer Science+Business Media New York 2013

[15] S. Bochner. Monotone Funktionen, Stieltjessche Integrale, und harmonische Analyse. *Math. Ann.*, 108:378–410, 1933.

[16] A. Bonami. Étude des coefficients de Fourier des fonctions de $L^p(G)$. *Ann. Inst. Fourier (Grenoble)*, 20(2):335–402, 1970.

[17] J. Bourgain. Propriétés de décomposition pour les ensembles de Sidon. *Bull. Soc. Math. France*, 111(4):421–428, 1983.

[18] J. Bourgain. Subspaces of ℓ_N^∞, arithmetical diameter and Sidon sets. In *Probability in Banach spaces, V (Medford, Mass., 1984)*, volume 1153 of *Lecture Notes in Math.*, pages 96–127, Berlin, Heidelberg, New York, 1985. Springer.

[19] J. Bourgain. Sidon sets and Riesz products. *Ann. Inst. Fourier (Grenoble)*, 35(1):137–148, 1985.

[20] J. Bourgain. A remark on entropy of abelian groups and the invariant uniform approximation property. *Studia Math.*, 86:79–84, 1987.

[21] J. Bourgan and V. Milman. Dichotomie du cotype pours les espace invariants. *C. R. Acad. Sci. Paris*, pages 435–438, 1985.

[22] L Carleson. On convergence and growth of partial sums of Fourier series. *Acta Math.*, 116:135–157, 1966.

[23] D. Cartwright and J. McMullen. A structural criterion for Sidon sets. *Pacific J. Math.*, 96:301–317, 1981.

[24] Mei-Chu Chang. On problems of Erdös and Rudin. *J. Functional Anal.*, 207(2):444–460, 2004.

[25] Y. S. Chow and H. Teicher. *Probability Theory. Independence, Interchangeability, Martingales.* Springer Texts in Statistics. Springer-Verlag, Berlin, Heidelberg, New York, 3rd edition, 1997.

[26] E. Crevier. Private communication. 2012.

[27] M. Déchamps(-Gondim). Ensembles de Sidon topologiques. *Ann. Inst. Fourier (Grenoble)*, 22(3):51–79, 1972.

[28] M. Déchamps(-Gondim). Densité harmonique et espaces de Banach ne contenant pas de sous-espace fermé isomorphe à c_0. *C. R. Acad. Sci. Paris*, 282(17):A963–A965, 1976.

[29] M. Déchamps(-Gondim). Densité harmonique et espaces de Banach invariants par translation ne contenant pas c_0. *Colloquium Math.*, 51: 67–84, 1987.

[30] M. Déchamps(-Gondim) and O. Selles. Compacts associés aus sommes de suites lacunaires. *Publ. Math. Orsay*, 1:27–40, 1996.

[31] R. Doss. Elementary proof of a theorem of Helson. *Proc. Amer. Math. Soc.*, 27(2):418–420, 1971.

[32] S. W. Drury. Sur les ensembles de Sidon. *C. R. Acad. Sci. Paris*, 271A: 162–163, 1970.

[33] S. W. Drury. The Fatou-Zygmund property for Sidon sets. *Bull. Amer. Math. Soc.*, 80:535–538, 1974.

[34] W. F. Eberlein. Characterizations of Fourier-Stieltjes transforms. *Duke Math. J.*, 22:465–468, 1955.

[35] R. E. Edwards and K. A. Ross. p-Sidon sets. *J. Functional Anal.*, 15: 404–427, 1974.

[36] R. E. Edwards, E. Hewitt, and K. A. Ross. Lacunarity for compact groups, III. *Studia Math.*, 44:429–476, 1972.

[37] W. Feller. *An Introduction to Probability Theory and its Applications.* John Wiley and Sons, New York, London, 2nd edition, 1957.

[38] A. Figá-Talamanca and D. G. Rider. A theorem of Littlewood and lacunary series for compact groups. *Pacific J. Math.*, 16:505–514, 1966.

[39] John J. F. Fournier and Louis Pigno. Analytic and arithmetic properties of thin sets. *Pacific J. Math.*, 105(1):115–141, 1983.

[40] W. H. J. Fuchs. On the zeros of a power series with Hadamard gaps. *Nagoya Math. J.*, 29:167–174, 1967.

[41] J. Galindo and S. Hernández. The concept of boundedness and the Bohr compactification of a MAP abelian group. *Fund. Math.*, 159(3): 195–218, 1999.

[42] J. Galindo and S. Hernández. Interpolation sets and the Bohr topology of locally compact groups. *Adv. in Math.*, 188:51–68, 2004.

[43] I. M. Gel'fand. Normierte Ringe. *Mat. Sb., N.S.*, 9:3–24, 1941.

[44] J. Gerver. The differentiability of the Riemann function at certain rational multiples of π. *Amer. J. Math.*, 92:33–55, 1970.

[45] J. Gerver. More on the differentiability of the Riemann function. *Amer. J. Math.*, 93:33–41, 1971.

[46] B. N. Givens and K. Kunen. Chromatic numbers and Bohr topologies. *Topology Appl.*, 131(2):189–202, 2003.

[47] D. Gnuschke and Ch. Pommerenke. On the radial limits of functions with Hadamard gaps. *Michigan Math. J.*, 32(1):21–31, 1985.

[48] E. Goursat. *A Course in Mathematical Analysis*, volume I. Ginn & Co., Boston, Chicago, London, New York, 1904. E. R. Hedrick, trans.

[49] W. T. Gowers. A new proof of Szemeredi's theorem. *Geom. Functional Anal.*, 11:465–588, 2001.

[50] C. C. Graham. Sur un théorème de Katznelson et McGehee. *C. R. Acad. Sci. Paris*, 276:A37–A40, 1973.

[51] C. C. Graham and K. E. Hare. ε-Kronecker and I_0 sets in abelian groups, III: interpolation by measures on small sets. *Studia Math.*, 171 (1):15–32, 2005.

[52] C. C. Graham and K. E. Hare. ε-Kronecker and I_0 sets in abelian groups, I: arithmetic properties of ε-Kronecker sets. *Math. Proc. Cambridge Philos. Soc.*, 140(3):475–489, 2006.

[53] C. C. Graham and K. E. Hare. ε-Konecker and I_0 sets in abelian groups, IV: interpolation by non-negative measures. *Studia Math.*, 177(1):9–24, 2006.

[54] C. C. Graham and K. E. Hare. I_0 sets for compact, connected groups: interpolation with measures that are nonnegative or of small support. *J. Austral. Math. Soc.*, 84(2):199–225, 2008.

[55] C. C. Graham and K. E. Hare. Characterizing Sidon sets by interpolation properties of subsets. *Colloquium Math.*, 112(2):175–199, 2008.

[56] C. C. Graham and O. C. McGehee. *Essays in Commutative Harmonic Analysis*. Number 228 in Grundleheren der Mat. Wissen. Springer-Verlag, Berlin, Heidelberg, New York, 1979.

[57] C. C. Graham and K. E. Hare. Characterizations of some classes of I_0 sets. *Rocky Mountain J. Math.*, 40(2):513–525, 2010.

[58] C. C. Graham and K. E. Hare. Sets of zero discrete harmonic density. *Math. Proc. Cambridge Philos. Soc.*, 148(2):253–266, 2010.

[59] C. C. Graham and K. E. Hare. Existence of large ε-Kronecker and $FZI_0(U)$ sets in discrete abelian groups. *Colloquium Math.*, 2012.

[60] C. C. Graham and A. T.-M. Lau. Relative weak compactness of orbits in Banach spaces associated with locally compact groups. *Trans. Amer. Math. Soc.*, 359:1129–1160, 2007.

[61] C. C. Graham, K. E. Hare, and T. W. Körner. ε-Kronecker and I_0 sets in abelian groups, II: sparseness of products of ε-Kronecker sets. *Math. Proc. Cambridge Philos. Soc.*, 140(3):491–508, 2006.

[62] C. C. Graham, K. E. Hare, and (L.) T. Ramsey. Union problems for I_0 sets. *Acta Sci. Math. (Szeged)*, 75(1–2):175–195, 2009.

[63] C. C. Graham, K. E. Hare, and (L.) T. Ramsey. Union problems for I_0 sets-corrigendum. *Acta Sci. Math. (Szeged)*, 76(3–4):487–8, 2009.

[64] D. Grow. A class of I_0-sets. *Colloquium Math.*, 53(1):111–124, 1987.

[65] D. Grow. Sidon sets and I_0-sets. *Colloquium Math.*, 53(2):269–270, 1987.

[66] D. Grow. A further note on a class of I_0-sets. *Colloquium Math.*, 53(1): 125–128, 1987.

[67] D. Grow and K. E. Hare. The independence of characters on non-abelian groups. *Proc. Amer. Math. Soc.*, 132(12):3641–3651, 2004.

[68] D. Grow and K. E. Hare. Central interpolation sets for compact groups and hypergroups. *Glasgow Math. J.*, 51(3):593–603, 2009.

[69] A. Gut. *An Intermediate Course in Probability*. Springer Texts in Statistics. Springer, Berlin, Heidelberg, New York, 2nd edition, 2009.

[70] J. Hadamard. Essai sur l'étude des fonctions données par leur développement de Taylor. *J. Math. Pures Appl.*, 8(4):101–186, 1892.

[71] A. Harcharras. Fourier analysis, Schur multipliers on S^p and non-commutative $\Lambda(p)$-sets. *Studia Math.*, 137(3):203–260, 1999.

[72] G. H. Hardy. Weierstrass's non-differentiable function. *Trans. Amer. Math. Soc.*, 17:301–325, 1916.

[73] K. E. Hare. Arithmetic properties of thin sets. *Pacific J. Math.*, 131: 143–155, 1988.

[74] K. E. Hare. An elementary proof of a result on $\Lambda(p)$ sets. *Proc. Amer. Math. Soc.*, 104:829–832, 1988.

[75] K. E. Hare and (L.) T. Ramsey. I_0 sets in non-abelian groups. *Math. Proc. Cambridge Philos. Soc.*, 135:81–98, 2003.

[76] K. E. Hare and (L.) T. Ramsey. Kronecker constants for finite subsets of integers. *J. Fourier Anal. Appl.*, 18(2):326–366, 2012.

[77] K. E. Hare and N. Tomczak-Jaegermann. Some Banach space properties of translation-invariant subspaces of L^p. In *Analysis at Urbana, I, 1986-1987*, volume 137 of *London Math. Soc. Lecture Notes*, pages 185–195. Cambridge Univ. Press, Cambridge, U. K., 1989.

[78] K. E. Hare and D. C. Wilson. A structural criterion for the existence of infinite central $\Lambda(p)$ sets. *Trans. Amer. Math. Soc.*, 337(2):907–925, 1993.

[79] S. Hartman. On interpolation by almost periodic functions. *Colloquium Math*, 8:99–101, 1961.

[80] S. Hartman. Interpolation par les mesures diffuses. *Colloquium Math.*, 26:339–343, 1972.

[81] S. Hartman and C. Ryll-Nardzewski. Almost periodic extensions of functions. *Colloquium Math.*, 12:23–39, 1964.

[82] S. Hartman and C. Ryll-Nardzewski. Almost periodic extensions of functions, II. *Colloquium Math.*, 15:79–86, 1966.

[83] S. Hartman and C. Ryll-Nardzewski. Almost periodic extensions of functions, III. *Colloquium Math.*, 16:223–224, 1967.

[84] H. Helson. Fourier transforms on perfect sets. *Studia. Math.*, 14:209–213, 1954.

[85] H. Helson and J.-P. Kahane. A Fourier method in diophantine problems. *J. d'Analyse Math.*, 15:245–262, 1965.

[86] C. Herz. Drury's lemma and Helson sets. *Studia Math.*, 42:205–219, 1972.

[87] E. Hewitt and K. A. Ross. *Abstract Harmonic Analysis*, volume I. Springer-Verlag, Berlin, Heidelberg, New York, 1963.

[88] E. Hewitt and K. A. Ross. *Abstract Harmonic Analysis*, volume II. Springer-Verlag, Berlin, Heidelberg, New York, 1970.

[89] E. Hewitt and H. S. Zuckerman. Some theorems on lacunary Fourier series, with extensions to compact groups. *Trans. Amer. Math Soc.*, 93:1–19, 1959.

[90] E. Hewitt and H. S. Zuckerman. Singular measures with absolutely continuous convolution squares. *Proc. Cambridge Phil. Soc.*, 62:399–420, 1966.

[91] E. Hewitt and H. S. Zuckerman. Singular measures with absolutely continuous convolution squares-corrigendum. *Proc. Cambridge Phil. Soc.*, 63:367–368, 1967.

[92] E. Hille. *Analytic Function Theory*, volume II. Chelsea, New York,, N. Y., 1973.

[93] B. Host and F. Parreau. Ensembles de Rajchman et ensembles de continuité. *C. R. Acad. Sci. Paris*, 288:A899–A902, 1979.

[94] B. Host, J.-F. Méla, and F. Parreau. *Analyse Harmonique des Mesures*, volume 135-136 of *Astérisque*. Soc. Math. France, Paris, 1986.

[95] R. A. Hunt. On the convergence of Fourier series. In *1968 Orthogonal Expansions and their Continuous Analogues (Proc. Conf., Edwardsville, Ill., 1967)*, pages 235–255. Southern Illinois Univ. Press, Carbondale, Ill., 1968.

[96] M. F. Hutchinson. Non-tall compact groups admit infinite Sidon sets. *J. Aust. Math. Soc.*, 23(4):467–475, 1977.

[97] J. Johnsen. Simple proofs of nowhere-differentiability for Weierstrass's function and cases of slow growth. *J. Fourier Anal. Appl.*, 16(1):17–33, 2010.

[98] G. W. Johnson and G. S. Woodward. On p-Sidon sets. *Indiana Univ. Math J.*, 24:161–167, 1974/75.

[99] J.-P. Kahane. Sur les fonctions moyennes-périodique bornées. *Ann. Inst. Fourier (Grenoble)*, 7:293–314, 1957.

[100] J.-P. Kahane. Ensembles de Ryll-Nardzewski et ensembles de Helson. *Colloquium Math.*, 15:87–92, 1966.

[101] J.-P. Kahane. *Séries de Fourier Absolument Convergentes*, volume 50 of *Ergebnisse der Math.* Springer, Berlin, Heidelberg, New York, 1970.

[102] J.-P. Kahane. Algèbres tensorielles et analyse harmonique. In *Séminaire Bourbaki, Années 1964/1965-1965/1966, Exposés 277-312*, pages 221–230. Société Math. France, Paris, 1995.

[103] J.-P. Kahane. Un théorème de Helson pour des séries de Walsh. In *Linear and Complex Analysis*, volume 226 of *Amer. Math. Soc. Transl. Ser. 2*, pages 67–73. Amer. Math. Soc., Providence, RI, 2009.

[104] J.-P. Kahane and Y. Katznelson. Entiers aléatoires et analyse harmonique. *J. Anal. Math.*, 105:363–378, 2008.

[105] J.-P. Kahane and Y. Katznelson. Distribution uniforme de certaines suites d'entiers aléatoires dans le groupe de Bohr. *J. Anal. Math.*, 105: 379–382, 2008.

[106] J.-P. Kahane and R. Salem. *Ensembles Parfaits et Séries Trigonométriques (Nouvelle Édition)*. Hermann, Paris, 1994.

[107] N. J. Kalton. On vector-valued inequalities for Sidon sets and sets of interpolation. *Colloq. Math.* , 64(2):233–244, 1993.

[108] Y. Katznelson. *An Introduction to Harmonic Analysis.* Cambridge Mathematical Library. Cambridge University Press, Cambridge, U. K., 3rd edition, 2004.

[109] Y. Katznelson and P. Malliavin. Vérification statistique de la conjecture de la dichotomie sur une classe d'algèbres de restriction. *C. R. Acad. Sci. Paris*, 262:A490–A492, 1966.

[110] R. Kaufman and N. Rickert. An inequality concerning measures. *Bull. Amer. Math. Soc.*, 72(4):672–676, 1966.

[111] A. Kechris and A. Louveau. *Descriptive Set Theory and the Structure of Sets of Uniqueness.* Number 128 in London Math. Soc. Lecture Notes. Cambridge Univ. Press, Cambridge, U. K., 1987.

[112] J. H. B. Kemperman. On products of sets in a locally compact group. *Fund. Math.*, 56:51–68, 1964.

[113] S. V. Kislyakov. Banach spaces and classical harmonic analysis. In *Handbook of the Geometry of Banach Spaces*, volume I, pages 871–898. Elsivier, Amsterdam, London, New York, 2001.

[114] M. Kneser. Summendmengen in lokalkompakten abelschen Gruppen. *Math. Zeitschrift*, 66:88–110, 1956.

[115] A. Kolmogorov. Une contribution à l'étude de la convergence des séries de Fourier. *Fund. Math.*, 5:96–97, 1924.

[116] K. Kunen and W. Rudin. Lacunarity and the Bohr topology. *Math. Proc. Cambridge Philos. Soc.*, 126:117–137, 1999.

[117] S. Kwapień and A. Pełczyński. Absolutely summing operators and translation-invariant spaces of functions on compact abelian groups. *Math. Nachr.*, 94:303–340, 1980.

[118] N. Levinson. *Gap and Density Theorems*, volume 26 of *A. M. S. Colloquium Publications*. American Math. Soc., Providence, R. I., 1940.

[119] D. Li and H. Queffélec. *Introduction à l'Étude des Espaces de Banach, Analyse et Probabilités*. Societé Mathématique de France, Paris, 2004.

[120] L.-Å. Lindahl and F. Poulsen. *Thin Sets in Harmonic Analysis*. Dekker, New York, N. Y., 1971.

[121] J. S. Lipiński. Sur un problème de E. Marczewski concernant des fonctions péroidiques. *Bull. Acad. Pol. Sci,. Sér. Math., Astr. Phys.*, 8:695–697, 1960.

[122] J. S. Lipiński. On periodic extensions of functions. *Colloquium Math.*, 13:65–71, 1964.

[123] J. López and K. A. Ross. *Sidon Sets*, volume 13 of *Lecture Notes in Pure and Applied Math*. Marcel Dekker, New York, N. Y., 1975.

[124] F. Lust(-Piquard). Sur la réunion de deux ensembles de Helson. *C. R. Acad. Sci. Paris*, 272:A720–A723, 1971.

[125] F. Lust(-Piquard). L'espace des fonctions presque-périodiques dont le spectre est contenu dans un ensemble compact dénombrable a la propriété de Schur. *Colloquium Math.*, 41:273–284, 1979.

[126] M. P. Malliavin-Brameret and P. Malliavin. Caractérisation arithmétique des ensembles de Helson. *C. R. Acad. Sci. Paris*, 264:192–193, 1967.

[127] M. B. Marcus and G. Pisier. *Random Fourier Series with Applications to Harmonic Analysis*, volume 101 of *Annals of Math. Studies*. Princeton University Press, Princeton, N. J., 1981.

[128] J.-F. Méla. Sur certains ensembles exceptionnels en analyse de Fourier. *Ann. Inst. Fourier (Grenoble)*, 18:31–71, 1968.

[129] J.-F. Méla. Sur les ensembles d'interpolation de C. Ryll-Nardzewski et de S. Hartman. *Studia Math.*, 29:168–193, 1968.

[130] J.-F. Méla. Approximation diophantienne et ensembles lacunaires. *Mémoires Soc. Math. France*, 19:26–54, 1969.

[131] J.-F. Méla. Private communication, 2010.

[132] Y. Meyer. Elargissement des ensembles de Sidon sur la droite. In *Seminaire d'Analyse Harmonique Orsay*, number 2 in Publ. Math. Univ. Paris VII, pages 1–14. Faculté des Sciences (Univ. Paris-Sud), Orsay, France, 1967/1968. http://www.math.u-psud.fr/ ~bib\discretionary-lio/numer\discretionary-isation/docs/Seminaire_d_analyse_harmoniquedOrsay_1967-1968/pdf/Seminaire_d_analyse_harmonique_d_Or\discretionary-say_1967-1968.pdf.

[133] I. M. Miheev. Series with gaps. *Mat. Sb.*, (N.S.) 98(140)(4(12)):538–563, 639, 1975. in Russian.

[134] I. M. Miheev. On lacunary series. *Mat. Sb.*, 27(4):481–502, 1975. translation of 'Series with gaps'.

[135] L. J. Mordell. On power series with circle of convergence as a line of essential singularities. *J. London Math. Soc.*, 2:146–148, 1927.

[136] S. A. Morris. *Pontryagin Duality and the Structure of Locally Compact Abelian Groups*. Number 29 in London Mathematical Society Lecture Notes. Cambridge University Press, Cambridge-New York-Melbourne, 1977.

[137] J. Mycielski. On a problem of interpolation by periodic functions. *Colloquium Math.*, 8:95–97, 1961.

[138] K. O'Bryant. A complete annotated bibliography of work related to Sidon sequences. *Electronic Journal of Combinatorics*, DS11, 2004.

[139] H. Okamoto. A remark on continuous, nowhere differentiable functions. *Proc. Japan Acad.*, 81(3):47–50, 2005.

[140] A. Pajor. Plongement de ℓ_1^n dans les espaces de Banach complexes. *C. R. Acad. Sci. Paris*, 296:741–743, 1983. http://perso-math.univ-mlv.fr/users/pajor.alain/recherche/pub.htm.

[141] A. Pajor. Plongement de ℓ_1^K complexe dans les espaces de Banach. In *Seminar on the geometry of Banach spaces, Vol. I, II (Paris, 1983)*, number 18 in Publ. Math. Univ. Paris VII, pages 139–148. Univ. Paris VII, Paris, 1984. http://perso-math.univ-mlv.fr/users/pajor.alain/recherche/pub.htm.

[142] A. Pajor. *Sous-espaces ℓ_1^n des Espaces de Banach*. Number 16 in Travaux en Cours. Hermann, Paris, 1985. ISBN 2-7056-6021-6. http://perso-math.univ-mlv.fr/users/pajor.alain/recherche/pub.htm.

[143] W. Parker. Central Sidon and central Λ_p sets. *J. Aust. Math. Soc*, 14: 62–74, 1972.

[144] M. Pavlović. Lacunary series in weighted spaces of analytic functions. *Archiv der Math.*, 97(5):467–473, 2011.

[145] L. Pigno. Fourier-Stieltjes transforms which vanish at infinity off certain sets. *Glasgow Math. J.*, 19:49–56, 1978.

[146] G. Pisier. Ensembles de Sidon et espace de cotype 2. In *Séminaire sur la Géométrie des Espaces de Banach*, volume 14, Palaiseau, France, 1977-1978. École Polytech. http://archive.numdam.org/ARCHIVE/SAF/SAF_1977-1978__/SAF_1977-1978___A11_0/SAF_1977-1978___A11_0.pdf.

[147] G. Pisier. Sur l'espace de Banach des séries de Fourier aléatoires presque sûrement continues. In *Séminaire sur la Géométrie des Espaces de Banach*, volume 17-18, Palaiseau, France, 1977-1978. École Polytech. http://archive.numdam.org/ARCHIVE/SAF/SAF_1977-1978___/ SAF_1977-1978___A13_0/SAF_1977-1978___A13_0.pdf.

[148] G. Pisier. Ensembles de Sidon et processus Gaussiens. *C. R. Acad. Sci. Paris*, 286(15):A671–A674, 1978.

[149] G. Pisier. De nouvelles caractérisations des ensembles de Sidon. In *Mathematical analysis and applications, Part B*, volume 7b of *Adv. in Math. Suppl. Studies*, pages 685–726. Academic Press, New York, London, 1981.

[150] G. Pisier. Conditions d'entropie et caractérisations arithmétique des ensembles de Sidon. In *Proc. Conf. on Modern Topics in Harmonic Analysis*, pages 911–941, Torino/Milano, 1982. Inst. de Alta Mathematica.

[151] G. Pisier. Arithmetic characterizations of Sidon sets. *Bull. Amer. Math. Soc.*, (N.S.) 8(1):87–89, 1983.

[152] Ch. Pommerenke. Lacunary power series and univalent functions. *Mich. Math. J.*, 11:219–223, 1964.

[153] J. Price. *Lie Groups and Compact Groups*. Number 25 in London Math. Soc. Lecture Notes. Cambridge Univ. Press, Cambridge, U. K., 1977.

[154] M. Queffélec. *Substitution Dynamical Systems–Spectral Analysis*, volume 1294 of *Lecture Notes in Mathematics*. Springer-Verlag, Berlin, Heidelberg, New York, 2nd edition, 2010.

[155] D. Ragozin. Central measures on compact simple Lie groups. *J. Func. Anal.*, 10:212–229, 1972.

[156] (L.) T. Ramsey. A theorem of C. Ryll-Nardzewski and metrizable l.c.a. groups. *Proc. Amer. Math. Soc.*, 78(2):221–224, 1980.

[157] (L.) T. Ramsey. Bohr cluster points of Sidon sets. *Colloquium Math.*, 68(2):285–290, 1995.

[158] (L.) T. Ramsey. Comparisons of Sidon and I_0 sets. *Colloquium Math.*, 70(1):103–132, 1996.

[159] (L.) T. Ramsey and B. B. Wells. Interpolation sets in bounded groups. *Houston J. Math.*, 10:117–125, 1984.

[160] D. G. Rider. Gap series on groups and spheres. *Canadian J. Math.*, 18: 389–398, 1966.

[161] D. G. Rider. Central lacunary sets. *Monatsh. Math.*, 76:328–338, 1972.

[162] D. G. Rider. Randomly continuous functions and Sidon sets. *Duke Math. J.*, 42:759–764, 1975.

[163] F. Riesz. Über die Fourierkoeffizienten einer stetigen Funktionen von beschränkter Schwankung. *Math. Zeitschr.*, 18:312–315, 1918.

[164] H. Rosenthal. On trigonometric series associated with weak* closed subspaces of continuous functions. *J. Math. Mech. (Indiana Univ. Math. J.)*, 17:485–490.

[165] J. J. Rotman. *An Introduction to the Theory of Groups*, volume 148 of *Graduate Texts in Mathematics*. Springer-Verlag, Berlin, Heidelberg, New York, 4th edition, 1995 (2nd printing, 1999).

[166] W. Rudin. Trigonometric series with gaps. *J. Math. Mech. (Indiana Univ. Math. J.)*, 9(2):203–227, 1960.

[167] W. Rudin. *Fourier Analysis on Groups*. Wiley Interscience, New York, N. Y., 1962.

[168] C. Ryll-Nardzewski. Remarks on interpolation by periodic functions. *Bull. Acad. Pol. Sci, Sér. Math., Astr., Phys.*, 11:363–366, 1963.

[169] C. Ryll-Nardzewski. Concerning almost periodic extensions of functions. *Colloquium Math.*, 12:235–237, 1964.

[170] S. Saeki. On the union of two Helson sets. *J. Math. Soc. Japan*, 23: 636–648, 1971.

[171] J. A. Seigner. Rademacher variables in connection with complex scalars. *Acta Math. Univ. Comenianae*, 66:329–336, 1997.

[172] A. Shields. Sur la mesure d'une somme vectorielle. *Fund. Math.*, 42: 57–60, 1955.

[173] S. Sidon. Ein Satz uber die absolute Konvergenz von Fourierreihen in dem sehr viele Glieder fehlen. *Math. Ann.*, 96:418–419, 1927.

[174] S. Sidon. Veralgemeinerung eines Satzes über die absolute Konvergen von Fourierreihen mit Lücken. *Math. Ann.*, 97:675–676, 1927.

[175] B. P. Smith. Helson sets not containing the identity are uniform Fatou–Zygmund sets. *Indiana Univ. Math. J.*, 27:331–347, 1978.

[176] S. B. Stečkin. On the absolute convergence of Fourier series. *Izv. Akad. Nauk. S.S.S.R.*, 20:385, 1956.

[177] J. D. Stegeman. On union of Helson sets. *Indag. Math.*, 32:456–462, 1970.

[178] A. Stöhr. Gelöste und ungelöste Fragen über Basen der natürlichen Zahlenreihe. I. *J. Reine Angew. Math.*, 194:40–65, 1955.

[179] A. Stöhr. Gelöste und ungelöste Fragen über Basen der natürlichen Zahlenreihe. II. *J. Reine Angew. Math.*, 194:111–140, 1955.

[180] E. Strzelecki. On a problem of interpolation by periodic and almost periodic functions. *Colloquium Math.*, 11:91–99, 1963.

[181] E. Strzelecki. Some theorems on interpolation by periodic functions. *Colloquium Math.*, 12:239–248, 1964.

[182] E. Szemeredi. On sets of integers containing no k elements in arithmetic progression. *Acta Arith.*, 27:199–245, 1975.

[183] A. Ülger. An abstract form of a theorem of Helson and applications to sets of synthesis and sets of uniqueness. *J. Functional Anal.*, 258(3): 956–977, 2010.

[184] E. K. van Douwen. The maximal totally bounded group topology on G and the biggest minimal G-space, for abelian groups G. *Topology Appl.*, 34(1):69–91, 1990.

[185] V. S. Varadarajan. *Lie groups and Lie algebras and their Representations*. Number 102 in Graduate Texts in Mathematics. Springer-Verlag, Berlin, Heidelberg, New York, 1984. ISBN 0-387-90969-9.

[186] N. Th. Varopoulos. Sur les ensembles parfaits et les séries trigonométriques. *C. R. Acad. Sci. Paris*, 260:A3831–A3834, 1965.

[187] N. Th. Varopoulos. Tensor algebras over discrete spaces. *J. Functional Anal.*, 3:321–335, 1969.

[188] N. Th. Varopoulos. Groups of continuous functions in harmonic analysis. *Acta Math*, 125:109–154, 1970.

[189] N. Th. Varopoulos. Sur la réunion de deux ensembles de Helson. *C. R. Acad. Sci. Paris*, 271:A251–A253, 1970.

[190] N. Th. Varopoulos. Une remarque sur les ensembles de Helson. *Duke Math. J.*, 43:387–390, 1976.

[191] R. Vrem. Independent sets and lacunarity for hypergroups. *J. Austral. Math. Soc.*, 50(2):171–188, 1991.

[192] K. Weierstrass. Über continuirliche Functionen eines reellen Arguments, die für keinen Werth des letzteren einen bestimmten Differentialquotienten besitzen. In *Mathematische Werke von Karl Weierstrass, Vol. 2*, pages 71–74. Meyer, Berlin, 1895.

[193] G. Weiss and M. Weiss. On the Picard property of lacunary power series. *Studia Math.*, 22:221–245, 1963.

[194] M. Weiss. Concerning a theorem of Paley on lacunary power series. *Acta Math.*, 102:225–238, 1959.

[195] B. Wells. Restrictions of Fourier transforms of continuous measures. *Proc. Amer. Math. Soc.*, 38:92–94, 1973.

[196] D. C. Wilson. On the structure of Sidon sets. *Monatsh. für Math.*, 101: 67–74, 1986.

[197] A. Zygmund. On the convergence of lacunary trigonometric series. *Fund. Math.*, 16:90–107, 1930.

[198] A. Zygmund. Sur les séries trigonométriques lacunaires. *J. London Math. Soc.*, 18:138–145, 1930.

[199] A. Zygmund. *Trigonometric Series*, volume I. Cambridge University Press, Cambridge, U. K., 2 edition, 1959.

[200] A. Zygmund. *Trigonometric Series*, volume II. Cambridge University Press, Cambridge, U. K., 2 edition, 1959.

Author Index

C.C. Graham and K.E. Hare, *Interpolation and Sidon Sets for Compact Groups*,   235
CMS Books in Mathematics, DOI 10.1007/978-1-4614-5392-5,
© Springer Science+Business Media New York 2013

Subject Index

C.C. Graham and K.E. Hare, *Interpolation and Sidon Sets for Compact Groups*, 241
CMS Books in Mathematics, DOI 10.1007/978-1-4614-5392-5,
© Springer Science+Business Media New York 2013

Notation Index

2-large, 28

$A(\mathbf{E})$ - restriction of Fourier transforms to $\mathbf{E}$, 50
$A(\mathbf{\Gamma})$ - Fourier algebra, xv
$\alpha(\mathbf{E})$ - weak angular Kronecker constant of $\mathbf{E}$, 21
$||A||_{op}$ - operator norm of matrix, 193
$AP(\mathbf{E})$, 50
$AP(\mathbf{E}, U, N, \varepsilon)$, 53
$AP_{+}(\mathbf{E}, U, N, \varepsilon)$, 53
$AP_{r}(\mathbf{E}, U, N, \varepsilon)$, 53
$\arg(t)$ - argument of complex number, 20

$\mathrm{Ball}(X)$ - the unit ball of X, xvi
$B(\mathbf{\Gamma})$ - Fourier–Stieltjes transforms, xv
$B_{d}(\mathbf{\Gamma})$ - Fourier–Stieltjes transforms of discrete measures, xv
$B(\mathbf{E})$ - restriction of FS transforms to $\mathbf{E}$, 50
$B_{d}(\mathbf{E})$ - restriction of FS transforms of discrete to $\mathbf{E}$, 50

$\mathbb{C}$ - complex numbers, xv
$\mathcal{C}(p^{\infty})$, 28
$C(X)$ - bounded, continuous $f : X \to \mathbb{C}$, xv
$C_{0}(X)$ - uniform closure of compactly supported continuous $f : X \to \mathbb{C}$, xv
$|X|$ - cardinality of X, 10
$\lceil \cdots \rceil$ - ceiling, 61

$d(w, z)$ - angular distance, 20
$\mathbb{D} = \mathbb{Z}_{2}^{\mathbb{N}}$, xv
$\widehat{\mathbb{D}}$ - dual of $\mathbb{D}$, xv
$\Delta = \{z \in \mathbb{C} : |z| \le 1\}$, 4
δ_{x} - unit point mass at x, 6

e - the identity of G, xv
$\varepsilon(\mathbf{E})$ - Kronecker constant of $\mathbf{E}$, 20

$\widehat{f}$ - Fourier transform of f, 209
$\lfloor \cdots \rfloor$ - floor, 119
FZI_{0}, 51
$FZI_{0}(U)$, 51
$FZI_{0}(U, N, \varepsilon)$, 57

G - lca group, usually compact with dual $\mathbf{\Gamma}$, xi
G_{2} - annihilator of characters of orders 2^{k}, 33
$\mathbf{\Gamma}$ - lca group, usually discrete with dual G, xi
$\mathbf{\Gamma}_{0}$ - torsion subgroup, 30
$\mathrm{gap}(I, J)$, 155
$\widehat{G}$ - dual object, 193

$\mathbb{H}^{\mathbf{E}}$, 53
$\mathbb{H}_{\gamma}$, 53

I_{0}, xi
$I_{0}(U)$, 51
$I_{0}(N, \varepsilon)$, 57
$I_{0}(U, N, \varepsilon)$, 57
$\mathfrak{Im}$ - imaginary part, 26

$\Lambda(p)$, 103
$\ll$ - $\mu \ll \nu$ if the measure μ is absolutely continuous wrt. ν, 211
$\ell^{\infty}(\mathbf{E})$ - bounded functions on $\mathbf{E}$, xv

$M(U)$ - measures concentrated on U, xi
$M^{+}(U)$ - non-negative measures on U, xi
$M_{0}(X)$, 41
$M_{d}(U)$ - discrete measures on U, xi
$M_{d}^{+}(U)$ - discrete non-negative measures on U, xi

C.C. Graham and K.E. Hare, *Interpolation and Sidon Sets for Compact Groups*, 247
CMS Books in Mathematics, DOI 10.1007/978-1-4614-5392-5,
© Springer Science+Business Media New York 2013

Open Problem Index

1. Is every Sidon set a finite union of: I_0 sets? ε-Kronecker sets? quasi-independent sets? v, xii, 102, 107, 114, 135, 153, 175

2. Can a Sidon set be dense in the Bohr group? v, xii, 114, 135, 175,

3. Is every Hadamard set an ε-Kronecker set? 44

4. Is every ε-Kronecker set $(\varepsilon < 2)$ Sidon? 20, 35

5. Determine the Kronecker constant of $\{1, \ldots, N\}$. 44

6. Is every dissociate set, quasi-independent set or independent union of $I_0(N, 1/2)$ sets an I_0 set? 101, 134

7. Is every I_0 set in $\mathbb{Z}$ a finite union of ε-Kronecker sets? 25

8. Does every infinite subset of Γ contain a FZI_0 subset of the same cardinality? 73

9. Suppose $\mathbf{E}$ is $I_0(U)$ with bounded constants. Is $\mathbf{E}$ $I_0(U)$ with bounded length? Is $\mathbf{E} \cup \mathbf{E}^{-1}$ $I_0(U)$? 79, 80

10. Is there always a co-finite subset of an $RI_0(U)$ set that is $FZI_0(U)$? 69, 87

11. Do any of the following classes of sets have zdhd? Sidon sets? quasi-independent sets? dissociate sets? ε-Kronecker sets? Sums (no repetitions) of Hadamard sets? finite unions of zdhd sets? 176, 179, 184

12. If $\mathbf{E}$ has zdhd, is $\overline{\mathbf{E}}$ a U_0 set? 178

13. When does a set of zero density have zhd? 186

14. Does Kahane's characterization of I_0 sets fail for non-metrizable Γ? 192

15. When $\mathbf{E}$ is I_0, is the Haar measure of $\overline{\mathbf{E}} \cdot \Gamma$ equal to zero? 188

16. Is every Sidon set a Sidon(U) set (non-abelian G)? 196

17. Does every local Sidon set have a Sidon (or FZI_0) subset of same cardinality? 197

18. Can the periodic extension results for $\mathbb{R}$ be extended to $\mathbb{R}^n$? 16

C.C. Graham and K.E. Hare, *Interpolation and Sidon Sets for Compact Groups*, 249
CMS Books in Mathematics, DOI 10.1007/978-1-4614-5392-5,
© Springer Science+Business Media New York 2013